Real-Time Environmental Monitoring

This lab manual is a companion to the second edition of the textbook *Real-Time Environmental Monitoring: Sensors and Systems*. Tested in pedagogical settings by the author for many years, it includes applications with the state-of-the-art sensor technology and programs in R, Python, Arduino, PHP, HTML, and SQL. It helps students and instructors in science and engineering better understand how to use and design a variety of sensors, and how to build systems and databases when monitoring a variety of environments such as soil, water, and air. Examples of low-cost and open-access systems are included and can serve as the basis of learning tools for the concepts and techniques described in the companion textbook. Furthermore, the manual provides links to websites and scripts in R that allow learning how to analyze a variety of datasets available from repositories and databases maintained by many agencies and institutions.

- The first hands-on environmental monitoring lab manual written in tutorial style and classroom tested.
- Includes 14 lab guides that parallel the theory developed in 14 chapters in the companion textbook.
- Provides clear step-by-step protocols to understand basic and advanced theory through applicable exercises and problems.
- Injects a practical implementation of the existing textbook. A valuable guide for students and practitioners worldwide engaged in efforts to develop, employ, and maintain environmental monitors.

Intended for upper-level undergraduate and graduate students taking courses in electrical engineering, civil and environmental engineering, mechanical engineering, geosciences, and environmental sciences, as well as instructors who teach these courses. Professionals working in fields such as environmental services, and researchers and academics in engineering will also benefit from the range of topics included in this lab manual.

Real-Time Environmental Monitoring

Sensors and Systems - Lab Manual

Second Edition

Miguel F. Acevedo

CRC Press
Taylor & Francis Group
Boca Raton London New York

CRC Press is an imprint of the
Taylor & Francis Group, an **informa** business

Second edition published 2024
by CRC Press
6000 Broken Sound Parkway NW, Suite 300, Boca Raton, FL 33487-2742

and by CRC Press
4 Park Square, Milton Park, Abingdon, Oxon, OX14 4RN

CRC Press is an imprint of Taylor & Francis Group, LLC

© 2024 Miguel F. Acevedo

First edition published by CRC Press 2015

Library of Congress Cataloging-in-Publication Data
Names: Acevedo, Miguel F., author.
Title: Real-time environmental monitoring : sensors and systems : lab
manual / Miguel F. Acevedo.
Description: Second edition. | Boca Raton : CRC Press, 2024. |
Includes bibliographical references and index.
Identifiers: LCCN 2023008155 (print) | LCCN 2023008156 (ebook) |
ISBN 9781032012681 (paperback) | ISBN 9781003184362 (ebook)
Subjects: LCSH: Environmental monitoring—Data processing—Laboratory
manuals. | Environmental monitoring—Equipment and supplies—Laboratory manuals.
Classification: LCC QH541.15.M64 A245 2024 (print) |
LCC QH541.15.M64 (ebook) | DDC 363.7/0630285—dc23/eng/20230705
LC record available at https://lccn.loc.gov/2023008155
LC ebook record available at https://lccn.loc.gov/2023008156

ISBN: 978-1-032-01268-1 (pbk)
ISBN: 978-1-003-18436-2 (ebk)

DOI: 10.1201/9781003184362

Typeset in Times
by codeMantra

Access the Instructor and Student Resources/Support Material: http://www.routledge.com/9781032545714

Contents

Preface to the Lab Manual

The first edition of the book *Real-Time Environmental Monitoring: Sensors and Systems,* written ten years ago, included a few hands-on and programming exercises, that inspired more complete exercises as I taught a regularly offered course on the subject for undergraduate and graduate students in the Electrical Engineering program at the University of North Texas (UNT). Motivated by making these experiences available in a second edition of the textbook, I have put together this lab manual following the chapter structure of the textbook.

Accordingly, this lab manual is organized into two parts, the first with a focus on methods (the first ten lab guides) but yet providing some practical examples to illustrate the applications of the methods learned. The second part (the last four lab guides) focuses on application domains namely, atmospheric, water, terrestrial, and wildlife monitoring, yet uses opportunities to introduce more methods. To support the new material included in the second edition of the textbook, the lab manual provides hands-on exercises on hardware, programming, data analysis and statistics, machine learning, remote sensing, and geographic information systems (GIS).

Pedagogically, the lab manual goes directly into hands-on and computer exercises that the students can conduct on their own, assuming they were exposed to the underlying concepts and theory available from the textbook. Nevertheless, the laboratory work would motivate the learner to review or refresh concepts that may have been missed on the first reading of the companion textbook.

A few years back, we developed a parts kit that the students could take home to conduct the hands-on experiments on their own and individually spending more quality time than a short session in the campus laboratory. This laboratory manual reflects this approach, and consequently each chapter is a lab guide that lists the materials needed for that session and illustrates graphically the step-by-step process to complete the exercises. More complete lists of parts, illustrations using full color pictures, and screenshot images are available online from the eResources offered by the publisher. Furthermore, increasing availability of repositories and databases maintained by many agencies and institutions, made it possible in this lab manual to develop computer exercises on analysis of a variety of environmental monitoring data. In this regard, the manual provides links to websites and scripts in R that allow learning how to analyze these datasets.

I would like to express my gratitude to many individuals that have made this lab manual possible. First and foremost, to the many students who have taken the environmental monitoring course following this material provided feedback that contributed to its continued improvement or have conducted projects on this subject. Indeed, some exercises are inspired by the work of students in senior capstone projects, master thesis, and doctoral dissertations. UNT's electrical engineering department supported experimenting with the idea of providing a lab kit for the students to develop the laboratory exercises individually. Sanjaya Gurung helped developing some of the hands-on exercises in wireless and wireless sensor networks. Special thanks are due to Breana Smithers who helped organize and maintain the laboratory kits over several semesters as well as preparing figures in the textbook and the lab manual. Irma Britton, editor for Environmental Science and Engineering, at CRC Press, was supportive of this project through the entire process and Chelsea Reeves, editorial assistant, provided help preparing the materials for production. Anonymous reviewers provided excellent feedback that helped to improve the approach followed in the textbook as well as in this lab manual.

Miguel F. Acevedo, Denton, Texas, January 2023

Preface to the First Edition

My aim in writing this book is to introduce the fundamentals of environmental monitoring based on electronic sensors, instruments, systems, and software that allow continuous and long-term ecological and environmental data collection. I have tried to accomplish two objectives, as reflected in the two major parts of this book. In the first part, I develop a story of how starting with sensors, we progressively build more complex instruments, leading to entire systems, and ending on database servers, web servers, and repositories. In the second part, once I lay out this foundation, I cover a variety of sensors and systems employed to measure environmental variables in air, water, soils, vegetation canopies, and wildlife observation and tracking.

I have attempted to present the state-of-the-art technology, while at the same time using a practical approach and being comprehensive including applications to many environmental and ecological systems. My preference has been to explain the fundamentals behind the many sensors and systems so that the reader can gain an understanding of the basics. As in any other endeavor, specialized references would supplement this basic material according to specific interests.

I have based this material on my experience developing systems for ecological and environmental studies, particularly those leading to ECOPLEX and the Texas Environmental Observatory (TEO). I have tried to provide a wide coverage and offer a broad perspective of environmental monitoring; naturally, I emphasize those topics with which I am more familiar. In the last few years, I have employed successive drafts of this book while developing a course in environmental modeling for undergraduate and graduate students in electrical engineering and environmental science.

Although my target is a textbook, I have also structured the material in such a way that could serve as a reference book for the monitoring practitioner. The material is organized into 14 chapters; therefore, when used as a textbook, it can be covered on a chapter-per-week basis in a typical 14-week semester. Part I includes problems that can be assigned as homework exercises.

I hope to reach out to students and practitioners worldwide interested and engaged in efforts to develop, employ, and maintain environmental monitors. The book includes examples of low-cost and open-access systems that can serve as the basis of learning tools for the concepts and techniques described in the book.

I would like to thank many individuals with whom I have shared experiences in this field, in a variety of projects, such as monitoring and assessment methods in lakes and estuaries, the startup of ECOPLEX, developing cyber-infrastructure approach to monitoring and TEO (NSF-funded projects CI-TEAM, CRI) and the TEO, and the NSF RET (Research Experiences for Teachers) on sensor networks. These individuals include faculty and students of several units of the UNT, such as the Institute of Applied Science, Electrical Engineering Department, Computer Science Department, School of Library and Information Sciences, University Information Technology (UIT), as well as colleagues of the City of Denton and the University of the Andes (Venezuela). Among many, I would like to mention Ken Dickson, Tom Waller, Sam Atkinson, Bruce Hunter, Rudy Thompson, David Hunter, Shengli Fu, Xinrong Li, Yan Huang, Bill Moen, Duane Gustavus, Phillip Baczewski, Ermanno Pietrosemoli, Michele Ataroff, Wilfredo Franco, Jue Yang, Carlos Jerez, Gilbert Nebgen, Chengyang Zhang, Mitchel Horton, Jennifer Williams, Andrew Fashingbauer, and Jarred Stumberg.

As an outcome of the CI-TEAM mini-courses, I developed a pilot of the environmental monitoring class that I currently teach using this book. I say thanks to several individuals who contributed guest lectures: Shengli Fu, Xinrong Li, David Hunter, Kuruvilla John, Rudy Thompson, Carlos Jerez, Sanjaya Gurung, and Jason Powell.

I would like to say special thanks to Breana Smithers for providing help in the field to maintain monitoring equipment and preparing the figures in this book.

I am very grateful to Irma Shagla-Britton, editor for Environmental Science and Engineering, at CRC Press for her enthusiasm for this project. Several reviewers provided excellent feedback that shaped the final version and approach of the manuscript.

Miguel F. Acevedo, Denton, Texas, May 2015

Author

Miguel F. Acevedo has over 40 years of academic experience, the last 27 of these at the University of North Texas (UNT) where he currently works as a Regents Professor. His career has been interdisciplinary and especially at the interface of science and engineering. He has served UNT as a faculty member in the Department of Geography, the Graduate Program in Environmental Sciences of the Biology Department, and the Electrical Engineering Department. He obtained his PhD in Biophysics from the University of California, Berkeley (1980) and master's degrees in Electrical Engineering and Computer Science from Berkeley (M.E., 1978) and the University of Texas at Austin (M.S., 1972). Before joining UNT, he was at the Universidad de Los Andes, Merida, Venezuela, where he served in the School of Systems Engineering, the graduate program in Tropical Ecology, and the Center for Simulation and Modeling. He has served on the Science Advisory Board of the U.S. Environmental Protection Agency and on many review panels of the U.S. National Science Foundation. He has received numerous research grants and written several textbooks, numerous journal articles, as well as many book chapters and proceeding articles. In addition to the Regents Professor rank, UNT has recognized him with the Citation for Distinguished Service to International Education, and the Regent's Faculty Lectureship. His research interests focus on environmental systems and sustainability. He has published four textbooks with CRC Press.

1 Introduction to R and Statistical Analysis

INTRODUCTION

The goal of this lab session is learning how to use R, from the RGui as well as from the RStudio Integrated Development Environment (IDE), as well as practicing descriptive and inferential statistics. This guide is a brief tutorial intended to help you get started with R using simple examples. We introduce R installation and setup, basic skills, simple statistics, data structures, reading data from files, plotting graphics, writing functions, and using packages. Using R, in addition to descriptive and inferential statistics, we learn simple linear regression analysis. For descriptive statistics, we emphasize the use of Exploratory Data Analysis (EDA), whereas for inferential statistics, we emphasize parametric tests. We use simple data to provide examples of EDA and to introduce parametric tests and simple examples for regression analysis.

MATERIALS

READINGS

For background, refer to Chapter 1 of the textbook *Real-Time Environmental Monitoring: Sensors and Systems - Textbook, Second Edition* (Acevedo 2024), which is a companion to this Lab Manual. There are many resources online to learn and use R, and you can also refer to other books where I have written detailed tutorials on using R (Acevedo 2012, 2013).

SOFTWARE (URL LINKS PROVIDED IN THE REFERENCES)

- R system: from the Comprehensive R Archive Network (CRAN) repository (R Project 2023)
- RStudio IDE, Integrated Development Environment (Posit 2023)

DATA FILES

In this guide, we use sample files in the zip archive `data.zip` available from the GitHub repository https://github.com/mfacevedol/rtem created for this book. Throughout the manual, we will refer to this as the RTEM or the RTEM GitHub repository, where RTEM stands for Real-Time Environmental Monitoring, the title of the book. We also use sample data in the R package `datasets`.

SUPPLEMENTARY SUPPORT MATERIAL

Supplementary support material including additional screenshots, images, and procedures are available from the publisher's eResources web page provided for this book.

DOI: 10.1201/9781003184362-1

INSTALLING R

Download R from the CRAN repository http://cran.us.r-project.org/ by looking for the *precompiled binary distribution* for your operating system (Linux, Mac, or Windows). This guide was developed using Windows, but all examples and exercises can be done using Mac or Linux. Thus, for Windows select *base* and then the executable download for the current release; for example, at the time this guide was last updated the release was R-4.1.1. Run the installer program; you will find it convenient to create a desktop shortcut and associate R with .RData files.

SETUP AND RUNNING R

Using Windows, the default R *working directory* is in the Documents folder within the Users directory. For example, c:\Users\yourusername\Documents. However, you may find it convenient to set up a working directory in a separate part of your file system. When working in a university lab, a good practice is to use a removable drive or a network drive that you can access from a different location to continue your work. Let us name the working directory as labs. For example, folder labs in a directory c:\ yourown\path of your file system; then the working directory would be c:\yourown\path\labs.

Double click the R shortcut to run R; this guide was developed using R x64. Once the R system loads, the R Graphical User Interface (RGui) opens up with the ">" prompt on the *R Console* (Figure 1.1). Go to menu item File and select Change Dir ... Browse to find your working directory and select it (Figure 1.2). The R *workspace* contains *objects* such as arrays and functions. A convenient option is to store the workspace in your working folder; to do this select menu File and Save workspace ... then save .RData in your working folder (Figure 1.3). Once you save this file, after exiting R, you can restart R by double clicking this file, which launches R from your working folder (Figure 1.3). Note that now when you select File|Change dir, the working folder will contain the file RData, and therefore there is no need to repeat a directory change.

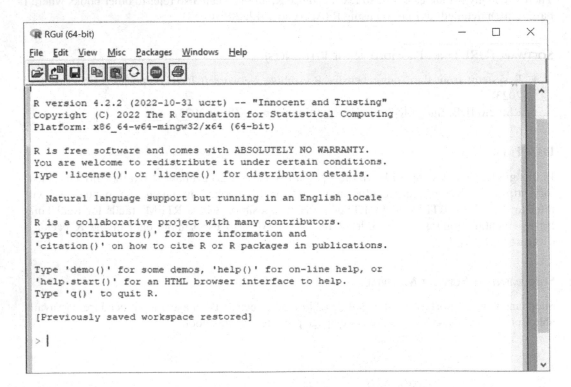

FIGURE 1.1   Running R: start of RGui and R console.

Browse For Folder ✕

Change working directory to:
C:\acevedo\courses\rtem\labs

> ☐ bookSecondEdition
> ☐ chicozip
> ☐ datasheets
> ☐ labs
> ☐ labs-dev
  ☐ manuals
  ☐ parts-labs

Folder: labs

Make New Folder   OK   Cancel

FIGURE 1.2 Change working directory by browsing to your folder.

Double click here to run R

Name | Dat
R .RData | 11/
.Rhistory | 11/

FIGURE 1.3 Workspace .RData file in your working folder and running R from the workspace file.

In the R console, you type R code upon the > prompt and receive text output (Figure 1.4), which is convenient for short sequences of code lines. However, to enter or edit longer segments of code you can use the Script facility of the Rgui or the editor of an IDE such as RStudio or Geany (Geany 2023).

To list objects in your workspace type ls() or using Misc|List objects of the RGui menu. Check your objects with ls() or with objects(); if this is your first run then you will not have existing objects and you will get character(0).

```
> ls()
character(0)
> objects()
character(0)
>
```

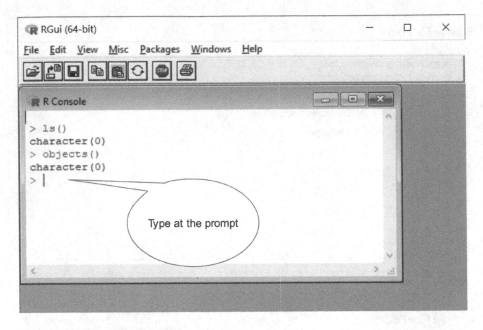

FIGURE 1.4   Typing code in the R Console.

R Manuals are available online in pdf and in html formats via the help menu item. PDFs (**P**ortable **D**ocument **F**iles) can be viewed and printed using the *Acrobat Reader*. HTML (**H**yper**T**ext **M**arkup **L**anguage) can be viewed using a web browser. To obtain help in HTML format use the `Help` menu item, select `Html help`. This will run a browser with the help files and manuals.

You can obtain help on specific Geany from the `Help` menu by selecting `R functions (text)` and then typing the function name on the dialog box. Alternatively, you can type `help(name of function)` at the prompt in the console; for example, `help(plot)` to get help on function `plot`. Also typing question mark followed by the function name would work, for example `? plot`. You can also launch the help with `help.start` function.

When editing through a long sequence of commands by using the arrow keys, one could use `History` to visualize all the previous commands at once. However, it is typically more convenient (especially if writing functions) to use a script.

Scripts: Using the RGui's R Editor

From the `File` menu, select `New Script` to open an editor window where you can type and edit a set of lines of code and then right click to run one line or a selection of lines by highlighting a section (Figure 1.5). To get started, create the script with just a couple of simple lines, say `ls()` and `help(plot)`, each preceded by a non-executable remark or comment line, which starts with the pound sign #; these comment lines are used to document your code.

```
# ls will list all objects in the workspace
ls()
# obtain help on a function
help("plot")
```

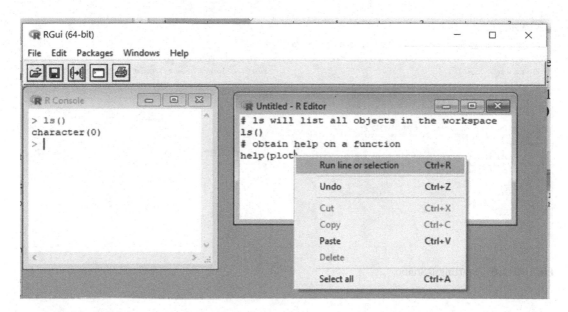

FIGURE 1.5 Script editor: right click to run a line or lines of the script.

We will keep building this script during the lab session and therefore we will save it for easy reference; it is convenient to keep program files organized in folders. For this purpose, in your working directory, you can create a new folder, e.g., R to store scripts. Let us now use `File|Save as …` to save the simple script we are developing as `lab1.R` in `labs/R`; scripts are saved as `*.R` files. Later you can open the script by `File|Open Script`, browsing to your folder, and selecting the file. Further edits can be saved using `File|Save`. We can double check our working folder by using `File| Display file(s)` and drill into R to verify that we have the `lab1.R` file.

It is a good idea to make sure, by using `File|Change Dir` or `File|Display file(s)`, that you are indeed in the working folder `labs` so that the path to the script file is relative to this folder. An alternate way of running your script is to use `File|Source R Code`, browse to folder, and select the script file, e.g., `lab1.R`. This is equivalent to using the function `source ("R/lab1.R")` to execute the code stored in file `lab1.R` within the R folder.

At this early stage, it is useful to say a word about the use of quotation marks in R programming. Single quotation mark ' and double quotation mark " can be used interchangeably, to delimit character constants or strings but double quotation marks are preferred and are used by R to print these constants. Please note that when you type the quotation mark inside the RGui editor and other text editors it corresponds to "and not to". The latter would correspond to using the quotation mark key in a word processor and will not produce the correct result if used in R code. A common typo that leads to errors in R code is pasting text from word processors that use ". For example, if you type `print("hello")`, the result will be [1] "hello" but if you paste or type `print("hello")`, the result would be `Error: unexpected input in "print("`.

Using an IDE: RStudio

The RStudio IDE is a set of integrated tools to facilitate using R and Python. It includes the R console, a source editor, workspace management, and other tools. There is an Open Source License free version that you can download from https://www.rstudio.com. Once you install it and run it, you would obtain a layout of panes or windows (Figure 1.6), where you would recognize the R Console in the left-hand side pane.

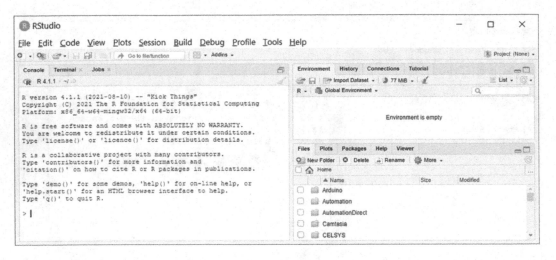

FIGURE 1.6　Starting RStudio.

In general, there are three other panes to consider and a variety of tabs for each. The upper right-hand side pane contains the `Environment tab`, empty for now. Underneath, the lower right-hand side pane contains the `Files tab` that would be at the "Home" directory (for example, `c:\Users\yourusername\Documents`) by default. The `Editor` pane would appear at the upper left-hand side once we open a file, and files will show as tabs in this pane. The panes are configurable from `View|Panes`.

First, set up your working directory by using the three dots "…" at the top right-hand side of the Files tab to navigate to your working directory. Use the `More` wheel icon and select `Set as Working Directory`. You will see that `setwd` was executed in the console. When set this way, the working directory will reset after closing the session you are currently working on. For a more permanent setup, change the default from Home to your own folder. Go to `Tools|Global Options` and type your working directory path in the `Default working directory` box. Whenever you restart RStudio, you would then see your working directory files on the `Files tab` (Figure 1.7).

Open the script file `lab1.R` you created before by using `File|Open File` and navigate to `R/lab1.R`. Now you will see the `Editor` pane in the upper left-hand side with the contents of your file and the console has moved to the lower left-hand side. You can run a line of code or selected lines using the `Run` icon located at the top of the editor or simply press keys `Ctrl-Enter`. The results are shown in the `Console` (Figure 1.7).

SAMPLE DATA FILES

In this Lab Manual, we will use sample files available from the RTEM GitHub repository https://github.com/mfacevedol/rtem; particularly, in this lab session, we will use files in the zip archive `data.zip`. Download this zip archive. Extract and save the folder `data` contained in data.zip in your working folder `labs`. Use the file manager or the R GUI's `File|Display file(s)` or the `Files` tab of `RStudio` to verify that you have a folder `labs/data` containing several files, such as `test100.txt`, `salinity.csv`, and several others (Figure 1.8). Now, we know all data files are in `labs/data` and all scripts in `labs/R`.

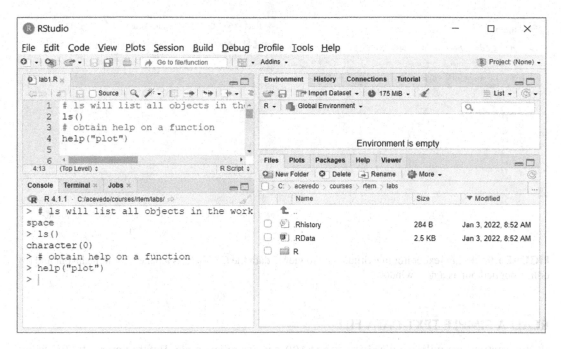

FIGURE 1.7 RStudio: lab1.R in the editor, output shown in the console and the Files tab shows the working directory.

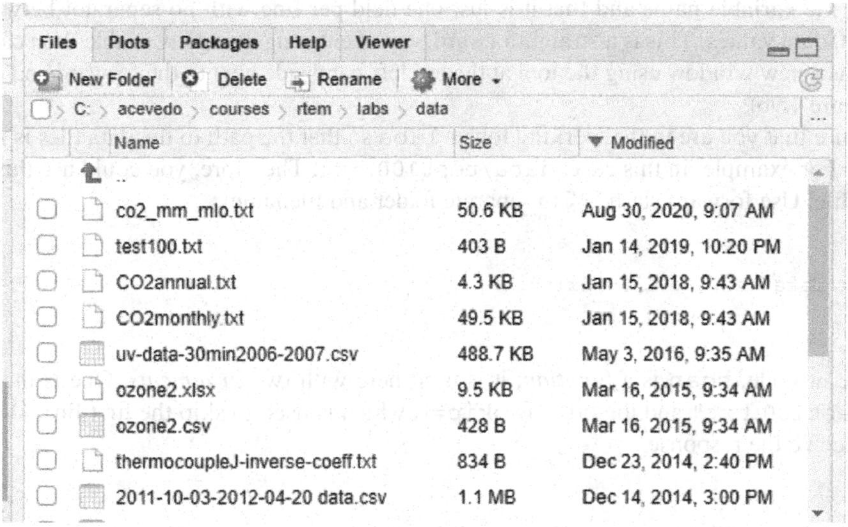

FIGURE 1.8 Archive data.zip extracted to working directory as data folder.

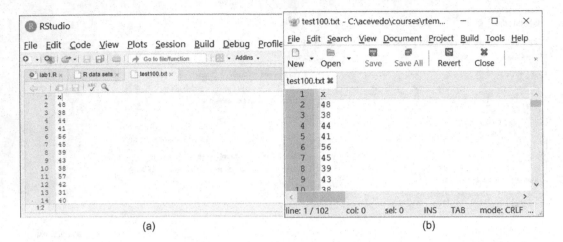

(a)   (b)

FIGURE 1.9 Using text editor in RStudio and to view a data file (a) editor as a pane within the layout and (b) editor popped out as a new window.

READ A SIMPLE TEXT DATA FILE

In this section, we will use file `data/test100.txt` as an example. Before you read a file using R (or any other programming language), it is good practice to examine the contents of the data file and understand it. RStudio's text editor helps examining text data files as well as typing and editing programs.

Open file `data/test100.txt` in RStudio (Figure 1.9a). You will see that it has a header x specifying the variable name and that it is just one field per line with no separator between fields and has 100 data values. This is a straightforward way of entering a single variable. You can pop out the editor as a new window using the tool at the top left-hand side next to the forward and backward arrow (Figure 1.9b).

Make sure that you are in the working folder `labs` so that the path to the data files is relative to this folder. For example, in this case, `data/test100.txt`. Therefore, you could use this name to `scan` the file. Use forward slash "/" to separate folder and filename).

```
> scan("data/test100.txt",skip=1)
```

In this line of code, `scan` is a *function*; it is used here with two *arguments*. One is the filename `"data/test100.txt"` and the other is `skip=1`, which is used to skip the first line. On the console, we receive the response

```
Read 100 items
  [1]  48 38 44 41 56 45 39 43 38 57 42 31 40 56 42 56 42 46 35 40 30 49 36 28 55
 [26]  29 40 53 49 45 32 35 38 38 26 38 26 49 45 30 40 38 38 36 45 41 42 35 35 25
 [51]  44 39 42 23 44 42 52 55 46 44 36 26 42 31 44 49 32 39 42 41 45 50 39 55 48
 [76]  49 26 50 46 56 31 54 26 29 32 34 40 53 37 27 45 37 34 32 33 35 50 37 74 44
```

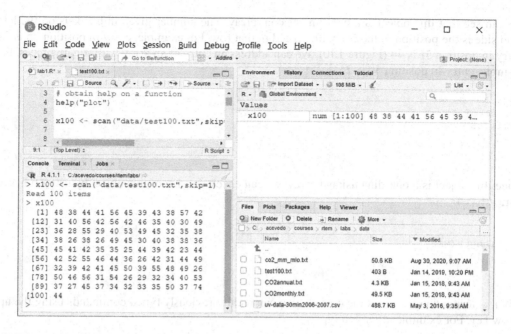

FIGURE 1.10 Reading a data file: showing the script in the Editor, the array contents in the Console, and the workspace containing the object in the Environment pane.

Add a line to the `lab1.R` script that creates object `x100`. Run this line and verify that it runs in the Console

```
x100 <- scan("data/test100.txt",skip=1)
```

object `x100` is assigned the results scanned from the file (Figure 1.10). The operator "`<-`" is used for `assignment`. Equivalently you can write the same using the equal sign "`=`". However, the equal sign is better used for other purposes, such as giving values to arguments of functions.

Double check that you have the newly created object by `Misc|List objects` (in The RGui) or looking at the Environment pane in RStudio, or using `ls( )`

```
> ls()
[1] "x100"
```

Object `x100` is stored in the workspace `labs\.Rdata` but file `test100.txt` resides in `labs\ data.` Double check the object contents by typing its name

```
> x100
  [1]  48 38 44 41 56 45 39 43 38 57 42 31 40 56 42 56 42 46 35 40 30 49 36 28 55
 [26]  29 40 53 49 45 32 35 38 38 26 38 26 49 45 30 40 38 38 36 45 41 42 35 35 25
 [51]  44 39 42 23 44 42 52 55 46 44 36 26 42 31 44 49 32 39 42 41 45 50 39 55 48
 [76]  49 26 50 46 56 31 54 26 29 32 34 40 53 37 27 45 37 34 32 33 35 50 37 74 44
```

We can see that this object is a one-dimensional array. The number given in brackets on the left-hand side is the position of the entry first listed in that row. For example, entry in position 26 is 29. Entry in position 51 is 44 (Figure 1.10). We can address an entry of an array with square braces. For example, entry in position 50 is 25.

```
> x100[50]
[1] 25
>
```

Since this object is a one-dimensional array, we can check the size of this object by using function length( )

```
> length(x100)
[1] 100
```

When entering commands at the console you can recall previously typed commands using the up-arrow key. For example, after you type

```
> x100
```

you can use up-arrow key and edit the line to add length

```
> length(x100)
```

Add another line to the Lab1.R script that stores the length in a variable n.x. Verify that n.x is in the workspace, and it has the correct value of 100.

SIMPLE STATISTICS

Now we can calculate sample mean, variance, standard deviation, and standard error of the data values in x100

```
> mean(x100)
[1] 40.86
> var(x100)
[1] 81.61657
> sd(x100)
[1] 9.034189
> sd(x100)/sqrt(length(x100))
[1] 0.9034189
>
```

It is good practice to round the results, for example, to three decimals

```
> round(mean(x100),3)
[1] 40.86
> round(var(x100),3)
[1] 81.617
> round(sd(x100),3)
[1] 9.034
> round(sd(x100)/sqrt(length(x100)),3)
[1] 0.903
>
```

We can concatenate functions in a single line by using the semicolon ";" character. Thus, for example, we can write one line to calculate the above statistics

```
> mean(x100); var(x100); sd(x100);sd(x100)/sqrt(length(x100))
[1] 40.86
[1] 81.61657
[1] 9.034189
[1] 0.9034189
>
```

Many times, it is important to be rigorous about how we express the significance of numbers that result from monitoring. For instance, when data are used for regulatory purposes, it is important to follow accepted standards. Take, for example, ozone at ground level, which is an important part of air quality. To decide on attainment of federal standards, ozone values are given in ppm using three decimal places, e.g., 0.064. One way of doing this is by *truncation*, i.e., to discard remaining decimals. For example, 0.0646 is truncated to three decimal places as 0.064. Note that this is different from *rounding* to three decimal places, which would make it 0.065. Ozone results are truncated to three decimal places by US federal regulatory standards, but it is rounded by some specific state standards.

Try the following and understand the difference

```
> trunc(0.00678*1000)/1000
[1] 0.006
> round(0.00678,3)
[1] 0.007
```

Create an array using c( ) which stands for column array and apply rounding and truncation to the average. Try the following

```
> x <- c(0.060,0.063,0.066,0.068,0.071,0.073,0.081)
> mean(x); round(mean(x),3); trunc(mean(x)*1000)/1000
[1] 0.06885714
[1] 0.069
[1] 0.068
>
```

Make sure you understand the difference between rounding and truncation. You can practice simple statistics by working on Exercise 1.1 at the end of the lab guide.

To obtain other descriptive stats such as median and quartiles, of the array x100 created above use the summary function

```
>summary(x100)
   Min. 1st Qu.  Median   Mean 3rd Qu.   Max.
  23.00   35.00   40.50  40.86   46.00  74.00
>
```

which yields the minimum and maximum, mean and median, first quartile and third quartile. In this case, the median is close to the mean indicating symmetry. We can obtain any one of these components by addressing it as an array. For example, the third quartile can be addressed with

```
> summary(x100)[5]
3rd Qu.
     46
>
```

The inter-quartile distance *iqd*, is the difference between the third and first quartile, iqd=3rd Qu. - 1st Qu. In this example, iqd=3rd Qu - 1st Qu=46−35=11. We can obtain the iqd by using

```
> iqd.x100 <- summary(x100)[5]-summary(x100)[2]
```

WORKING WITH ARRAYS

Besides addressing a single entry, we can refer to a set of entries of an array by a variety of combinations within the brackets or square braces. For example, the first ten positions of x100

```
> x100[1:10]
 [1] 48 38 44 41 56 45 39 43 38 57
```

Where the colon ":" is used to declare a sequence of entries. Or refer to entries at positions 1, 3, and 5 using c( ) and commas

```
> x100[c(1,3,5)]
[1] 48 44 56
>
```

Entries can be removed using the − sign. For example, x100[-1] removes the first entry

```
> x100[-1]
 [1] 38 44 41 56 45 39 43 38 57 42 31 40 56 42 56 42 46 35 40 30 49 36 28 55 29
> length(x100[-1])
[1] 99
```

Using a blank or no character within the brackets means that all entries are used

```
> x100[]
  [1] 48 38 44 41 56 45 39 43 38 57 42 31 40 56 42 56 42 46 35 40 30 49 36 28 55
 [26] 29 40 53 49 45 32 35 38 38 26 38 26 49 45 30 40 38 38 36 45 41 42 35 35 25
 [51] 44 39 42 23 44 42 52 55 46 44 36 26 42 31 44 49 32 39 42 41 45 50 39 55 48
 [76] 49 26 50 46 56 31 54 26 29 32 34 40 53 37 27 45 37 34 32 33 35 50 37 74 44
> length(x100[])
[1] 100
```

An array can be declared using `array()`. For example, as an empty array (containing no data NA) of length 10

```
> x <- array(,dim=10)
> x
 [1] NA NA NA NA NA NA NA NA NA NA
>
```

By default, dim is 1. Values can be updated in a variety of ways. For instance,

```
> x[5:7] <- c(1,2,3)
> x
 [1] NA NA NA NA  1  2  3 NA NA NA
>
```

We can also use other options, such as `seq` and `rep`. For example, a sequence from −1 to 1 in steps of 0.2

```
> x <- seq(-1,1,0.2)
> x
 [1] -1.0 -0.8 -0.6 -0.4 -0.2  0.0  0.2  0.4  0.6  0.8  1.0
>
```

Or using `rep` for a repeat of a value ten times

```
> x <- rep(1,10)
> x
 [1] 1 1 1 1 1 1 1 1 1 1
>
```

Function `array` can also be used to declare a matrix. For example, an empty matrix (containing no data NA) of two columns and three rows

```
> X <- array(,dim=c(3,2))
> X
     [,1] [,2]
[1,]   NA   NA
[2,]   NA   NA
[3,]   NA   NA
>
```

The same result can be achieved using matrix with arguments `ncol` and `nrow`

```
> X <- matrix(,ncol=2,nrow=3)
> X
     [,1] [,2]
[1,]   NA   NA
[2,]   NA   NA
[3,]   NA   NA
>
```

EXPLORATORY ANALYSIS: INDEX PLOT AND BOXPLOT

Graphical output is displayed in a "graphics device" or window that by default floats with respect to the R console in the RGui and that opens within the `Plots` tab in RStudio. When applying a graphics function, a new graph window will open by default if none is opened; otherwise, the graph is sent to the active graph window.

To start learning graphics, let us add more code to script `lab1.R`. Add a comment line (Recall use #) that says "# index plot of `x100`" and a line that applies function `plot` to the one-dimensional array object `x100` followed by `identify`.

```
#RStudio index plot of x100
plot(x100); identify(x100)
```

Note how pairs of parentheses are colored or highlighted when you focus on one of them. This helps you to ensure the syntax of opening and closing parenthesis is correct. Run and obtain the graph shown in Figure 1.11a. Because we have used identify, hover the cursor over the symbol for the very large observation that may be an outlier and press the Esc key. This will print the value 99 next to that symbol, indicating the observation number that corresponds to this point.

In the RGui, use `File|Save as` while the graphics window is selected to save in a variety of graphics formats, including metafile, postscript, pdf, png, bmp, TIFF, and jpeg. To capture on the clipboard and then paste on another application, you could simply use Copy and Paste from the graphics device. To do this, you can also use `File|Coopy to clipboard` and then paste the clipboard contents on the selected cursor position of the file being edited in the application. In RStudio, use `Export` and `Save to Image` to save in various formats including SVG.

Now, we will produce a boxplot and identify by repeating value 1 as the x coordinate for all values of the y coordinate.

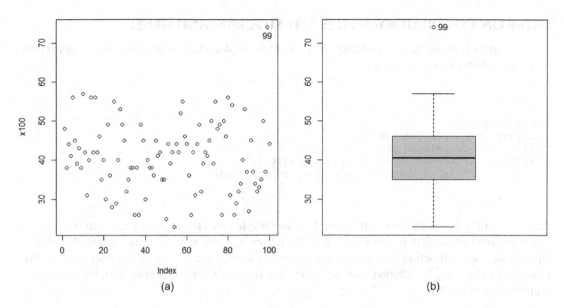

FIGURE 1.11   (a) Index plot and identification of potential outlier. (b) Boxplot and outlier.

```
# boxplot of x100
boxplot(x100);identify(rep(1,length(x100)),x100)
```

Run and obtain the graph shown in Figure 1.11b. Because we have used identify, hover the cursor over the outlier and press the Esc key. This will print the value 99 next to that symbol.

MULTIPLE PLOTS IN ONE GRAPHICS DEVICE AND MULTIPLE GRAPHICS DEVICES

Let us produce two figures on the same graphics device using function par with argument mfrow. For this purpose, we write par(mfrow=c(1,2)), which is one row of two figures. Add to your script using

```
par(mfrow=c(1,2))
plot(x100)
boxplot(x100)
```

to produce a figure like Figure 1.11 except for the outlier number because we did not include the identify function.

You can use arguments cex and cex.main to scale the size of the labels and main title.

```
par(mfrow=c(1,2),cex=0.8,cex.main=0.9)
plot(x100)
boxplot(x100)
```

We can open more graphics devices as needed for various plots using function win.graph(width, height), where width and height are arguments to control the size of the graphics windows. For example, win.graph(6,4) opens a new graphics window of width 6 and height 4 inches. The current graph window can be closed with dev.off(); all opened graphics windows can be closed with graphics.off().

MORE ON EXPLORATORY ANALYSIS: HISTOGRAM AND DENSITY

Now we apply function `hist` and `density` to `x100` to plot a histogram and density approximation together in a graphics window

```
win.graph(6,3)
par(mfrow=c(1,2),cex=0.8)
hist(x100,cex.main=0.8)
iqd.x100 <- summary(x100)[5]-summary(x100)[2]
plot(density(x100, width=2*iqd.x100),cex.main=0.8)
```

Run and obtain the graph shown in Figure 1.12. Bar heights are counts of how many measurements fall in the bin indicated in the horizontal axis. To represent the histogram as an estimate of the density, we have used function `density` using argument `width` equal to twice the inter-quartile distance (`width=2*iqd`). Alternatively, argument `bw` is specified by a character string denoting a method. For example,

```
>plot(density(x100,bw='SJ'))
```

Remember to close the graphics device.

STANDARD NORMAL

To obtain the probability density function (PDF) of the standard normal $N(0,1)$, generate a sequence of numbers from negative to positive. These will be in units of standard deviation σ, say from -5σ to $+5\sigma$ (every 0.1). Then apply functions `dnorm` and `pnorm` with this sequence and mean=0, $\sigma=1$, to graph the PDF in the left panel and the cumulative density function (CDF) in the right panel

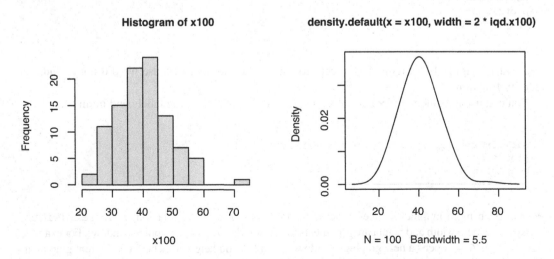

FIGURE 1.12   Histogram of `x100` and density approximation.

```
win.graph(6,4)
par(mfrow=c(1,2),cex=0.8)
Z <- seq(-5,+5,0.1)
plot(Z,dnorm(Z,0,1),type="l",ylab="p(Z) N(0,1)")
plot(Z,pnorm(Z,0,1),type="l",ylab="F(Z) N(0,1)")
```

as shown in Figure 1.13.

We can use the probabilities 0.1, 0.2, 0.3, ..., 0.9 to calculate the quantiles

```
Zp <- seq(0.1,0.9,0.1)
Zq <- qnorm(Zp,0,1)
```

and check these in the console

```
>Zq
[1] -1.2815516 -0.8416212 -0.5244005 -0.2533471  0.0000000  0.2533471
0.5244005
[8]  0.8416212  1.2815516
>
```

we get a cumulative probability of 0.1 at $z = -1.28$, a cumulative probability of 0.2 at -0.84, and so on. These values are the 10, 20, and 90 percentiles. For finer resolution, we can calculate all the 1, 2, 3, ..., 99 percentiles

```
Zp <- seq(0.01,0.99,0.01)
Zq <- qnorm(Zp,0,1)
```

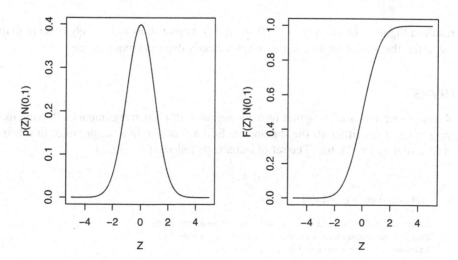

FIGURE 1.13 PDF and CDF of the standard normal.

yielding

```
> Zq
  [1] -2.32634787 -2.05374891 -1.88079361 -1.75068607 -1.64485363 -1.55477359
  [7] -1.47579103 -1.40507156 -1.34075503 -1.28155157 -1.22652812 -1.17498679

and so on ..

 [97]  1.88079361  2.05374891  2.32634787
```

We can plot the probabilities versus the quantiles using several arguments: `type="b"` to graph using both line and symbol, labeling the axis with `xlab` and `ylab`, and limiting the vertical axes to the range 0 to 1 using `ylim`.

```
plot(Zq,Zp,type="b",ylab="F(Z) N(0,1)", xlab="Z", ylim=c(0,1))
```

MORE ON EXPLORATORY ANALYSIS: ECDF AND Q–Q PLOTS

Next, we produce a plot of the empirical cumulative density function (ECDF) using `ecdf`, along with standardized variable (by subtracting the mean and divide by the standard deviation) and adding the theoretical $N(0,1)$ using lines. Then, we add quantile-quantile (Q-Q) plots to compare to normal using the functions `qqnorm` and `qqline`,

```
win.graph(6,4)
par(mfrow=c(1,2),cex=0.8)
# standardize observations before plot
z <- (x100-mean(x100))/sd(x100); plot(ecdf(z))
# add lines for plot theoretical standard normal N(0,1)
Z <- seq(-4,+4,0.1); lines(Z, pnorm(Z,0,1),lwd=2)
qqnorm(x100);qqline(x100)
```

See the result in Figure 1.14. Both the ECDF and the Q–Q plot show a relatively close fit to the normal; in the latter, the outlier we saw in the boxplot clearly departs from the line.

FUNCTIONS

We can develop an often-used program into a function with proper arguments, allowing us to run the program without rewriting all the statements. Edit a function in a script editor or a text editor window and save it as an `*.R` file. The set of statements follows this syntax

```
fnName <- function(x, other)
{
        Body of the function: a set of lines of code.
        This code operate on x and other arguments
        Arguments are datasets or other objects
}
```

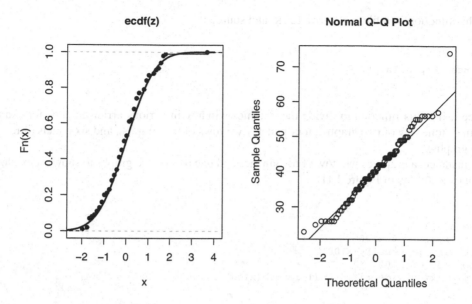

FIGURE 1.14 ECDF and Q–Q plots.

All we must do is declare object fnName as a function and then collect the statements within curly brackets using "{" to mark the start point and "}" to mark the end point. Indenting statements within the brackets makes the program more readable. Here, x and other are arguments to the function.

Once the file *.R is saved, we can use the RGui's File|Source R Code to source it or click on the source tool at the top of the RStudio editor to the right of Run. Alternatively, we can apply the source function

```
> source("fnName.R")
```

to this file name. For example, if we save the file as fnName.R, then

```
> fnName(data object for x, data object for other arguments)
```

Once your function is sourced, to use it we just type its name with its arguments. For example,

We illustrate how to take advantage of functions to produce multiple graphs in one window using layout to divide the graphics window in panels, using the space more efficiently by reducing the margins of each plot.

```
panels <- function (wd,ht,rows,cols,pty="m",int="r"){
       # default pty is maximal region and int is for axis regular
       np <- rows*cols # number of panels
       mat <- matrix(1:np,rows,cols,byrow=T) # matrix for layout
       layout(mat, widths=rep(wd/cols,cols), heights=rep(ht/rows,rows), TRUE)
       par(mar=c(4,4,1,.5),xaxs=int,yaxs=int,pty=pty)
}
```

Type this function in a file, say `panels.R`, and source it

```
source("R/panels.R")
```

Practice using this function to divide the graphics window in various arrangements, for example, two panels (one row of two graphs), four panels (two rows of two graphs), and six panels (three rows of two graphs).

For instance, a graphics window of 6×3 inches and one row of two graphs to show index plot and boxplot of `x100` (as in Figure 1.11)

```
panels(6,3,1,2)
# index plot of x100
plot(x100); identify(x100)
# boxplot of x100
boxplot(x100);identify(rep(1,length(x100)),x100)
```

Next, we will illustrate how to build a function using the code we have written above, to perform EDA, that is,

```
eda6 <- function (x){
        panels(6,6,3,2)
        # first row: index and boxplot
        plot(x); boxplot(x)
        # second row
        hist(x,cex.main=0.8); plot(density(x),cex.main=0.8)
        # third row: qqplot and ecdf vs. normal
        qqnorm(x);qqline(x)
        # standardize observations before plot
        z <- (x-mean(x))/sd(x); plot(ecdf(z))
        # add lines for plot theoretical standard normal N(0,1)
        Z <- seq(-4,+4,0.1); lines(Z, pnorm(Z,0,1),lwd=2)
}
```

You can type it in the RStudio text editor and save in `labs/R` as `eda6.R`. Here, we declared `eda6` as a function and then collected the statements within curly brackets. These braces are highlighted when you focus on one of them; this helps you to ensure the syntax is correct. The function span is marked on the left side by the arrowhead symbols.

We can source the function to store it in the workspace

```
> source("R/eda6.R")
```

Once your function is stored, then all you need to do to apply it is to type its name with its argument

```
>eda6(x100)
```

which will yield graphics as in Figure 1.15.

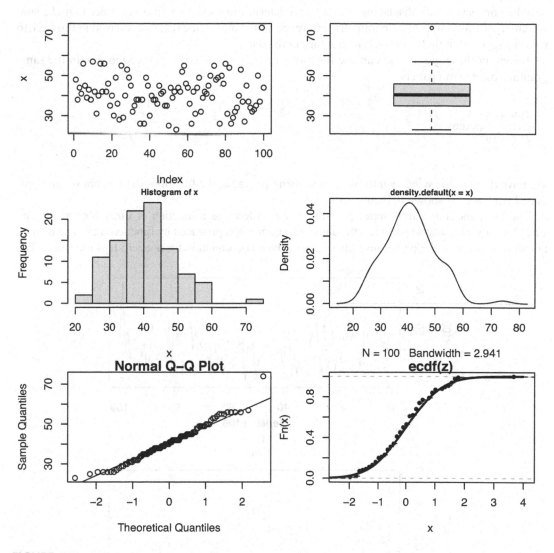

FIGURE 1.15   Results from eda6 function.

TIME SERIES AND AUTOCORRELATION PLOTS

A time series plot is a plot of the variable on the vertical axis and time along the horizontal axis (Figure 1.16a). A `correlogram` or `autocorrelation plot` is a plot of the sample autocorrelation on the vertical axis and lag on the horizontal axis (Figure 1.16b). The autocorrelation plot helps visualize the relation between successive data values. Note how the maximum value of autocorrelation is 1 and that it occurs for lag=0. Again, the lag does not have to be time. It could also be a distance. For example, values of soil moisture taken every 20 cm of depth, and values of vegetation cover taken every 10 m along a transect.

Later, we will be studying how to analyze serial data with more detail. In this lab session, we will only be concerned with displaying the data and determining serial structure in order to make sure we satisfy the assumptions of parametric inference. For example, the results displayed in Figure 1.16 would suggest that there is no serial structure in the data.

To obtain these graphs, we can use functions `ts.plot` and `acf`, arranged to plot in the same graphics page using panels.

```
panels(6,6, 2,1)
ts.plot(x100)
acf(x100)
```

We have divided the window into two panels using `panels`, the functions defined above, and produced two plots as shown in Figure 1.16.

The top panel has a time series plot (the values ordered as a function of time) generated using `ts.plot` and the second panel has the autocorrelation graph produced by function `acf`. The data do not show serial correlation because the spikes of the autocorrelation for nonzero lags are small. The

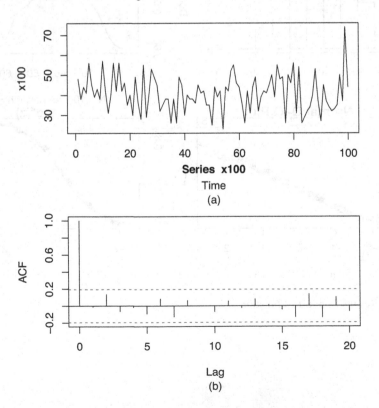

FIGURE 1.16   Time series plot (a) and autocorrelation plot (b).

results of this figure suggest that there is no major serial structure in the data and, therefore, it would be acceptable to apply parametric methods. You can practice these functions by working on Exercise 1.4.

INFERENCE

Assume a standard normal $N(0,1)$ and that we want to know whether the sample mean departs from the mean, that is, $H_0{:}\mu=0$. Select *Type I error* $\alpha=0.05\%$ or 5% and set up an unlikely event dividing 0.05 into two equal parts of 0.025. We find the values for a $N(0, 1)$ for these quantiles using qnorm,

```
> qnorm(c(0.025, 0.975), 0, 1)
[1] -1.959964  1.959964
>
```

Thus, the interval is all values less than −1.96 and >1.96. For example, suppose we get a value of −2.5. Then, we can reject H_0 at $\alpha=0.05\%$ or 5% error. We can give a specific value for the probability of getting this low value by using pnorm

```
> pnorm(c(-2.5),0,1)
[1] 0.006209665
>
```

Therefore, we would get a value up to −2.5 only 0.62% of the time.

PARAMETRIC TESTS: ONE SAMPLE T-TEST OR MEANS TEST

We will employ a t-test to check if the value of the mean of some observations corresponds to a given value. We need to check that the data are normally distributed, outlier free, and uncorrelated, i.e., assumptions for the validity of the t-test. Let us use the x100 object dataset for which we have used EDA already. From our previous work, we can see that the median is lower than the mean but not by much. From the boxplot and density, the data look symmetrical and normal, but not completely. There is only one outlier. The q–q plot and the ECDF suggest normality. The serial plots indicate that the data are uncorrelated. Therefore, we conclude that we can proceed with the t-test.

We will try a null hypothesis $H_0{:}\,\mu\leq39$. Apply t.test with H1: $\mu>39$ by selecting "greater" for the alternative argument (which has three options: two sided, greater, or less)

```
> t.test(x100, alternative="greater", mu=39)

        One Sample t-test

data:  x100
t = 2.0588, df = 99, p-value = 0.02107
alternative hypothesis: true mean is greater than 39
95 percent confidence interval:
 39.35997       Inf
sample estimates:
mean of x
    40.86


>
```

Thus, we have 99 degrees of freedom (df=99), because $n-1=100-1=99$ Note that we subtract one because one parameter (the mean) was estimated. The p-value indicates that we can reject the null hypothesis with a 0.021 probability of error. The 95% confidence interval of the estimate (40.86) of the mean does not include 39.00.

Now let us calculate the power of this t-test. We use function `power.t.test` in the following manner

```
> power.t.test(n=100, delta = 1, sd=sd(x100), type="one.sample", alternative =
"one.sided", sig.level=0.01)

    One-sample t test power calculation

              n = 100
          delta = 1
             sd = 9.034189
      sig.level = 0.01
          power = 0.1085085
    alternative = one.sided
>
```

where we have 100 observations and want to detect an effect size of delta=1, given the standard deviation, alternative is one-sided, and $\alpha=0.01$. The result is

```
    One-sample t test power calculation

              n = 100
          delta = 1
             sd = 9.034189
      sig.level = 0.01
          power = 0.1085085
    alternative = one.sided
```

We see that the power 0.11 is low. Note from the t-test results that the sample mean is 40.86. If we increase the effect size to the difference of 40.86−39.00, i.e., a delta 1.86, the power goes up to 0.38. We conclude that for this sample the t-test would have relatively low power given $\alpha=0.01$.

```
> power.t.test(n=100, delta = 1.86, sd=sd(x100), type="one.sample",
alternative = "one.sided", sig.level=0.01)

    One-sample t test power calculation

              n = 100
          delta = 1.86
             sd = 9.034189
      sig.level = 0.01
          power = 0.3837631
    alternative = one.sided


>
```

We can now determine how many more observations would be required to raise the power to 0.9 assuming the same variability, the same α, and the same effect size of delta$=1.86$.

```
> power.t.test(power=0.9, delta = 1.86, sd=sd(x), type="one.sample",
alternative = "one.sided", sig.level=0.01)

     One-sample t test power calculation

              n = 309.8002
          delta = 1.86
             sd = 9.034189
      sig.level = 0.01
          power = 0.9
    alternative = one.sided
```

Indicating that we require $n=310$ observations to achieve this power level given these conditions.

PACKAGE INSTALLATION AND LOADING

Packages are set up to use by the following two steps: (1) Installation from the Internet and (2) Loading it for use.

TO INSTALL AND LOAD FROM THE RGUI

First, go to `Packages|Install package(s)`, select a mirror site depending on your location, and select Install package. The download process starts and will give you messages about progress and success or not of the package installation. You do not need to repeat the installation as long as you do not re-install the R software or want to update the package.

The second step is to load the package. From the Rgui: go to `Packages|Load package`, and then select the package. Alternatively, you can run the function `library`. For example

```
library(nameofpackage)
```

Once a package is loaded, help on its functions becomes available. For example, go to HTML help, and when you click on packages, you will see the package you loaded.

R packages offer a multitude of useful sample datasets. We can obtain a list of datasets in a package applying function `data` to the package name

```
data(package=packagename)
```

which will generate a list of the package datasets in a pop-up text window. For example, for package `datasets`

```
data(package ='datasets')
```

To obtain a list for all packages installed

```
data(package = .packages(all.available = TRUE))
```

which will generate a list by package in a pop-up text window. There are arguments to data for
further control of searching for the packages.

To use a dataset simply apply function data and the name of the dataset. For example, there is
a dataset named CO2 in package datasets. We can use it by

```
> data(CO2)
> CO2
    Plant        Type  Treatment  conc  uptake
1    Qn1      Quebec  nonchilled   95   16.0
2    Qn1      Quebec  nonchilled  175   30.4
3    Qn1      Quebec  nonchilled  250   34.8
4    Qn1      Quebec  nonchilled  350   37.2
5    Qn1      Quebec  nonchilled  500   35.3
6    Qn1      Quebec  nonchilled  675   39.2

etc. up to 84 records
```

To Install and Load from RStudio

Go to Tools|Install Packages, use default for Repository (CRAN), and library directory.
Enter the name of the package in the Packages box. For example, datasets. Press Install. The
process starts and will give you messages about progress in the Console. The datasets package
would now be listed in the Packages tab of the Files/Packages, ... pane. To load the package, checkmark the package; for example, datasets. In the console, you will see the function
library() executed. For example, library(datasets). Help on functions provided by
the package is now available. Click on the package name in the Packages tab. You can make
this window larger to see more information at once. You will be at the documentation page
of the package; for example, datasets. You can list the datasets using the same function data
(package = "datasets").

Using Functions from Packages

R packages offer a multitude of useful functions. For example, install and load package car. A
convenient function in car is qqPlot, which allows one to obtain a Q–Q plot drawn vs. the standard normal (among other options), and includes confidence intervals. You can use it with data
x100 to obtain Figure 1.17.

```
qqPlot(x100)
[1] 99 54
```

The region bound by two lines around the straight line corresponds to the confidence intervals. We
can conclude that the data is likely normally distributed because it follows a straight line, except at
the high and low ends where some observations (99 and 54) depart and wander outside the confidence interval.

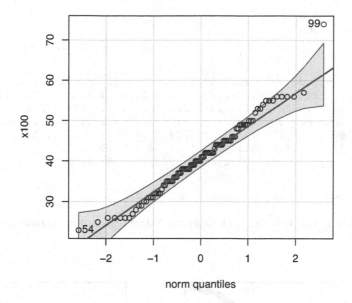

FIGURE 1.17 Q–Q plots using package `car`.

ENVIRONMENTAL DATA EXAMPLE: SALINITY, CSV FILES, AND NONPARAMETRIC TESTS

Next, we work with an example of data collected in the field. *Salinity* is an environmental variable of great ecological and engineering importance. In the context of water, it conditions the type of plants and animals that can live in a body of water and impacts the quality and potential use of water. At the interface between rivers and sea, such as estuaries, salinity experiences spatial and temporal gradients. It is traditional to express salinity in parts per thousand ‰ instead of percentage % because it is the same as approximately grams of salt per kilogram of solution. *Freshwater*'s salinity limit is 0.5‰, then water is considered *brackish* for the 0.5‰–30‰ range, above that we have *saline* water in the 30‰–50‰, and *brine* with more than 50‰.

A *comma separated values* (csv) file is a text file with fields separated by commas. In windows, a `csv` file will open in excel, but it is not an excel file. Let us open `salinity.csv` in the text editor of RStudio (Figure 1.18). As you can see, it is just one line with values separated by comma; the lines are wrapped; to wrap the lines in RStudio use Code| Soft wrap long lines. To read a CSV file into R, all you have to do is use scan with `sep=","` argument.

```
x<- scan("data/salinity.csv", sep=",")
```

Consider the salinity data read from the csv file and apply EDA

```
x<- scan("data/salinity.csv", sep=",")
eda6(x)
```

which results in Figure 1.19. From the index plot, we see a possible serial structure, and from the median line within the box in boxplot, we can see a lack of symmetry, but no outliers. From the histogram and density, we see that the data do not look normally distributed; moreover, it

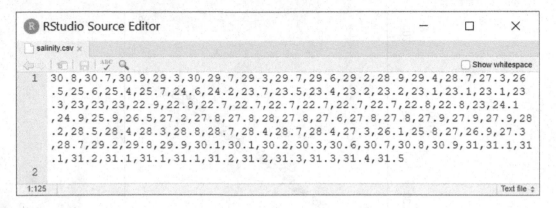

FIGURE 1.18   Salinity data in the `salinity.csv` file. Values separated by comma in one wrapped line.

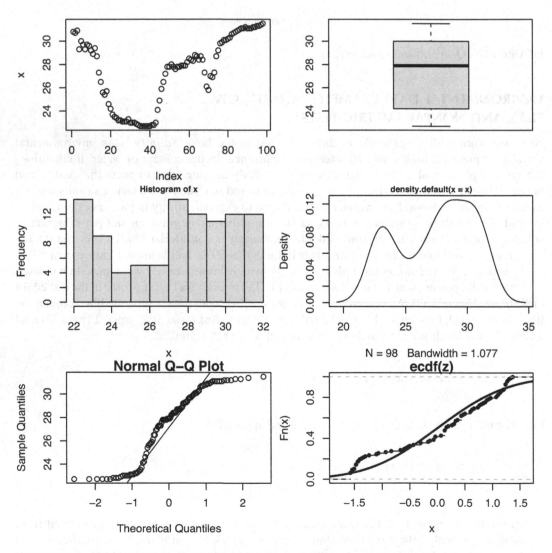

FIGURE 1.19   EDA results of salinity observations.

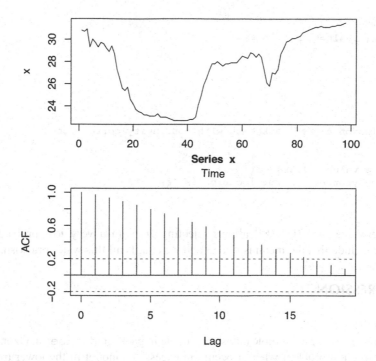

FIGURE 1.20   EDA serial check graphs.

seems to be bimodal. From the q–q plot and the ECDF, we see patterns in the departure from the theoretical normal distribution. We conclude that this salinity dataset does not seem to be normally distributed.

Now to explore the serial structure, using `ts.plot` and `acf`, we get Figure 1.20 that indicates serially structured data.

```
panels(6,6, 2,1)
ts.plot(x)
acf(x)
```

Therefore, we should not apply the parametric test to these data but a nonparametric one.

NONPARAMETRIC: WILCOXON TEST

First, determine the ranks of the values in the dataset and apply EDA

```
rx <- rank(x)
eda6(rx)
```

You would see that the ranks look more likely to be normally distributed than the observations (the figure is not shown here for the sake of space). The mean rank is about 50. Run the one-sample Wilcoxon signed rank test to see whether the mean is >30.

```
wilcox.test(x, alt="less", mu=30)
```

to obtain

```
        Wilcoxon signed rank test with continuity correction

data:  x
V = 520.5, p-value = 1.214e-11
alternative hypothesis: true location is less than 30
```

The very low p-value of 1.21×10^{-11} allows rejecting H_0 with a very low probability of error. Therefore, we conclude that the mean is not >3 and we could call this water brackish.

LINEAR REGRESSION

SCATTER PLOTS

We will use the `airquality` sample dataset available in package `datasets`. Ozone is an important urban air pollution problem when it occurs in excessive amount in the lower troposphere. Its concentration relates to meteorological variables.

Let us look at the contents of `airquality` dataset, It has 153 records and includes NA for "no data" for some records. Recall that you can address a component of the dataset by using the $ sign.

```
> airquality
    Ozone Solar.R Wind Temp Month Day
1      41     190  7.4   67     5   1
2      36     118  8.0   72     5   2
3      12     149 12.6   74     5   3
4      18     313 11.5   62     5   4
5      NA      NA 14.3   56     5   5
6      28      NA 14.9   66     5   6

and so on …

151    14     191 14.3   75     9  28
152    18     131  8.0   76     9  29
153    20     223 11.5   68     9  30
>
```

First, attach the dataset so that it becomes easier to use the component variables

```
>attach(airquality)
```

For example, instead of `airquality$Temp`, we can simply use `Temp`

```
> Temp
  [1]  67 72 74 62 56 66 65 59 61 69 74 69 66 68 58 64 66 57 68 62 59 73 61 61 57
 [26]  58 57 67 81 79 76 78 74 67 84 85 79 82 87 90 87 93 92 82 80 79 77 72 65 73
 [51]  76 77 76 76 76 75 78 73 80 77 83 84 85 81 84 83 83 88 92 92 89 82 73 81 91
 [76]  80 81 82 84 87 85 74 81 82 86 85 82 86 88 86 83 81 81 81 82 86 85 87 89 90
[101]  90 92 86 86 82 80 79 77 79 76 78 78 77 72 75 79 81 86 88 97 94 96 94 91 92
[126]  93 93 87 84 80 78 75 73 81 76 77 71 71 78 67 76 68 82 64 71 81 69 63 70 77
[151]  75 76 68
```

We will focus on ozone as a function of one of the meteorological variables; for example, air temperature. First do a scatter plot using the `plot()` function

```
> plot(Temp, Ozone)
```

To obtain the circles of Figure 1.21.

The observation numbers for the outliers were produced by applying function `identify`

```
> identify(Temp, Ozone)
```

Then moving the cursor (crosshair) to each outlier and clicking, the observation number will appear next to the point. Terminate the process by pressing the Escape key `Esc`. Upon termination, we have the observation number for each one of these points.

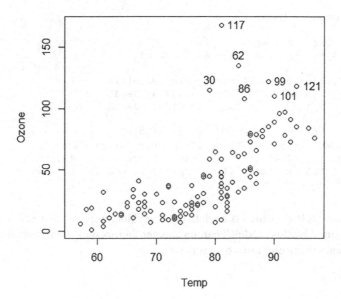

FIGURE 1.21 Ozone vs. air temperature. Scatter plot.

We plan to find the relationship between Temp and Ozone. An important thing to note is that a linear regression may not be the best choice. We will proceed with a linear model for now and later in Lab 12 see how to do nonlinear regression.

SIMPLE LINEAR REGRESSION

Simple linear regression is done using function lm(y~x) where y is the dependent variable and x is the independent variable. Symbol ~ denotes the functional relationship. A non-intercept regression, or regression through the origin, is done by adding character 0 in the expression. That is using lm(y~0+x).

Now we will use simple linear regression to explain ozone concentration from temperature lm(Ozone ~ Temp). Store the results in an object ozone.lm. We use extension lm to remind ourselves that this is an object of type lm. Then, we look at the object.

```
> ozone.lm <- lm(Ozone ~ Temp)
> ozone.lm
Call:
lm(formula = Ozone ~ Temp)
Coefficients:
(Intercept)          Temp
   -146.995          2.429
```

These are the regression coefficients b_0 (intercept) and b_1 (slope) yielding equation $y = -147 + 2.43x$, where y is the ozone concentration and x is the air temperature. This means that ozone increases with temperature and that when $x = 0$ we have a value of negative 147.

We get more information about the regression by looking at the summary

```
>summary(ozone.lm)
Call:
lm(formula = Ozone ~ Temp)
Residuals:
     Min        1Q   Median       3Q       Max
-40.7295 -17.4086  -0.5869  11.3062 118.2705
Coefficients:
            Estimate Std. Error t value Pr(>|t|)
(Intercept) -146.9955    18.2872   -8.038 9.37e-13 ***
Temp           2.4287     0.2331   10.418  < 2e-16 ***
---
Signif. codes:  0 '***' 0.001 '**' 0.01 '*' 0.05 '.' 0.1 ' ' 1
Residual standard error: 23.71 on 114 degrees of freedom
Multiple R-Squared: 0.4877,     Adjusted R-squared: 0.4832
F-statistic: 108.5 on 1 and 114 DF,  p-value: < 2.2e-16
```

For this model, the R^2 is 0.48, which is a relatively low value, it means that <50% of the variance in ozone can be explained by this model. Therefore, other factors must be important; hence, the common expression "something else must be going on".

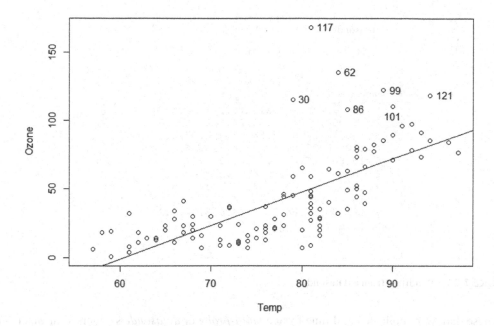

FIGURE 1.22   Ozone vs. air temperature. Scatter plot and regression line.

We can plot the regression line on top of the scatter plot using the function `abline`. We have already used plot to get the scatter plot and identified the outliers. Now add `abline` to obtain Figure 1.22.

```
abline(ozone.lm$coef)
```

Each coefficient has a standard error (standard deviation of its estimate) and a p-value of a t-test to evaluate its significance. For the example at hand we get $se_{b0}=18.28$, $t_0=8.04$ with p-value 9.4×10^{-13} and $se_{b1}=0.23$, $t_1=10.41$ with p-value 2×10^{-16}. These p-values are very low; therefore, we can reject the H_0, namely, that the coefficients are zero. Therefore, there is a nonzero slope and nonzero intercept. The residual standard error is 23.71. The degrees of freedom are df = 114 because two parameters were estimated ($116-2=114$). Note that $153-116=37$ values were no data (`NA`) and thus removed. The R^2 is 0.48. The F value is 108.5 with 1 and 114 degrees of freedom, which yields a p-value of 2.2×10^{-16}. The low p-value suggests that we reject H_0, and therefore the slope is different from zero.

MORE THAN ONE VARIABLE: READING FILES AND PLOT VARIABLES

Let us work on an example when we have several variables to read from file, typically organized in more than one column as in file `data/datasonde.csv`. When opening it using a text editor such as Geany, the few first lines look like Figure 1.23.

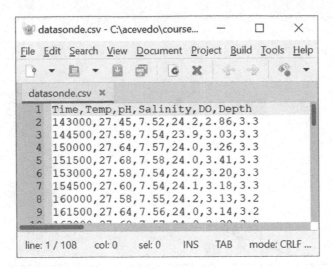

FIGURE 1.23   Readings from a datasonde.

These data were collected real time from a *multi-probe* or *datasonde* submersed in water. This instrument consists of multiple sensors read real-time by a datalogger. We will cover this with more detail in Lab 12.

We see that file `datasonde.csv` is a CSV file of six columns corresponding to `Time, Temp, pH, Salinity, DO, and Depth` as declared in the header (first row). When you scroll down you will see that we have 107 lines; therefore, discounting the header we have 106 lines of six data values each. Let us make an object named `x` from the data in the file named `datasonde.csv`. For this, use the `read.table` function with argument `header=TRUE` so that we use the file header to give names to the columns, and argument sep to declare that the separator is the comma character.

```
x <- read.table("data/datasonde.csv",header=TRUE,sep=",")
```

Here, `x` is a *data frame* configured from the data by rows. The components of `x` have names corresponding to the strings given in the header. When you print this data frame simply using `x`, the printout will quickly scroll on the console, and you will see just the end. There is an easy method to inspect the top, so that we can see the header, by printing a few rows such as `x[1:5,]`

```
> x[1:5,]
    Time   Temp   pH Salinity   DO Depth
1 143000 27.45 7.52     24.2 2.86   3.3
2 144500 27.58 7.54     23.9 3.03   3.3
3 150000 27.64 7.57     24.0 3.26   3.3
4 151500 27.68 7.58     24.0 3.41   3.3
5 153000 27.58 7.54     24.2 3.20   3.3
>
```

Or using `View` which will produce a spreadsheet-like view of the dataset in a new window. This is used directly in RStudio when querying a dataset.

```
View(x)
```

```
> x$Temp
  [1] 27.45 27.58 27.64 27.68 27.58 27.60 27.58 27.64 27.68 27.74 27.68 27.76
 [13] 27.74 27.74 27.72 27.77 27.81 27.85 27.89 27.93 27.93 27.91 27.99 27.99
 [25] 28.01 27.93 28.01 28.07 28.10 28.10 28.10 28.12 28.10 28.14 28.14 28.10
 [37] 28.14 28.14 28.12 28.08 28.16 28.14 28.14 28.20 28.18 28.18 28.16 28.20
 [49] 28.20 28.18 28.22 28.24 28.22 28.20 28.24 28.24 28.26 28.28 28.30 28.30
 [61] 28.26 28.24 28.26 28.24 28.24 28.26 28.26 28.28 28.26 28.28 28.24 28.30
 [73] 28.36 28.34 28.36 28.38 28.40 28.38 28.40 28.34 28.34 28.34 28.38 28.46
 [85] 28.44 28.49 28.51 28.57 28.55 28.55 28.55 28.53 28.53 28.53 28.57 28.55
 [97] 28.51 28.55 28.53 28.51 28.51 28.57 28.59 28.53 28.55 28.77
>
```

Another option is to make a `tibble` which will discuss in forthcoming labs.

We can refer to one of the variables by using the name of the dataset followed by $ and the name of the variable. For instance, `x$Time` is time and `x$Temp` is water temperature.

You can use the `dim` function to check the dimensions of the matrix

```
> dim(x)
[1] 106    6
```

We can use the names of the components directly if we `attach` the data frame.

```
> attach(x)
```

Once we attach `x` we can use the names of components of *x* without addressing them with the $ sign. Say, to plot variable `DO` as a function of `Time` (which is given in hhmmss)

```
panels(6,6,2,1)
plot(Time, DO)
plot(Time, DO, type="l")
```

The second call to `plot` uses a line graph by adding an argument `type="l"` to the plot function. Careful this is a letter "l" for `line` not number one "1" (Figure 1.24).

By default, the *x* and *y* axes are labeled with the variable name. These can be changed using `xlab=` and `ylab=` arguments. For example, we obtain the top panel of Figure 1.25 by applying the first call to plot function

```
panels(6,6,2,1)
plot(Time, DO, type="l",xlab="Time [hhmmss]", ylab="Dissolved Oxygen [mg/l]")
plot(Time, DO, type="l",xlab="Time [hhmmss]", ylab="Dissolved Oxygen [mg/l]",
     xlim=c(150000,240000),ylim=c(0,5))
```

The limits of the *x* and *y* axes can be changed using `xlim=c( )` and `ylim=c( )` as shown in the second call to plot above which produces Figure 1.25b. Recall `c( )` denotes a one-dimensional array. In this example, `xlim=c(150000, 240000)`, establishes the range from 150,000 to 240,000 and `ylim=c(0,5)` has two elements a minimum of 0 and a maximum of 5 (see Figure 1.25b).

One more detail to mention is that the axis can intersect at their minimum values by using argument `xaxs="i"` and `yaxs="i"`

```
plot(Time, DO, type="l",xlab="Time [hhmmss]", ylab="Dissolved Oxygen [mg/l]",
     xlim=c(150000,240000),ylim=c(0,5), xaxs="i",yaxs="i")
```

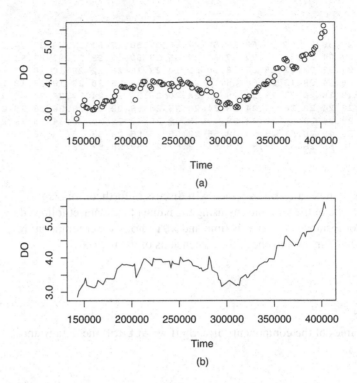

FIGURE 1.24 Plots of dissolved oxygen. Top panel (a) is produced by the default type (points). Bottom panel (b) is a line graph produced by `type="l"`.

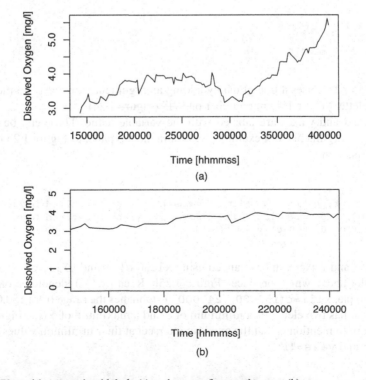

FIGURE 1.25 Plot with customized labels (a) and ranges for *x* and *y* axes (b).

The graph is not shown here for the sake of space, but you should try it and see the result.

To visualize several variables, use function `matplot` and follow it up with a function `legend` to identify the traces. The `col` argument is the line color and the `lty` argument is the line type. For instance, we can use `col=1` to obtain black color for all lines (Figure 1.26) and `lty= 1:` `length(names(x)[-1])` declaring the number of line types extracted from the length of the names of the dataset except the first name. Lastly, we use `lwd=2` to increase the width of the lines and obtain better visibility.

```
panels(6,3,1,1)
matplot(Time,x[,-1], type="l", col=1, lty=1:length(names(x)[-1]),
        xlab="Time [hhmmss]",ylab="Water Quality", lwd=2)
legend(250000,20,leg=names(x)[-1],col=1, lty=1:length(names(x)[-1]),lwd=2)
```

The first two arguments of `legend` are the *x*, *y* coordinates to place the legend, then `leg` argument is an array with the labels; here we have used the names of the dataset except the first which is `Time` (that is why we write −1). Arguments `col`, `lty`, and `lwd` match the selection of the matplot call. You can practice these concepts and those in the next section by working Exercise 1.8.

MORE ON GRAPHICS WINDOWS: PRODUCING A PDF FILE

Let us create a new folder, `output`, within the working folder `labs` to store graphics and analysis results, say `labs/output`, to keep the working folder organized. Now we have `data`, `output`, and R folders.

You could direct graphics to specific formats. For example, the water quality *x–y* plot can be directed to a pdf using function `pdf` with filename that includes the path to the output folder

```
pdf(file="output/datasonde.pdf")
  matplot(Time,x[,-1], type="l", col=1, lty=1:length(names(x)[-1]),
        xlab="Time [hhmmss]",ylab="Water Quality", lwd=2)
  legend(250000,20,leg=names(x)[-1],col=1, lty=1:length(names(x)[-1]),lwd=2)
dev.off()
```

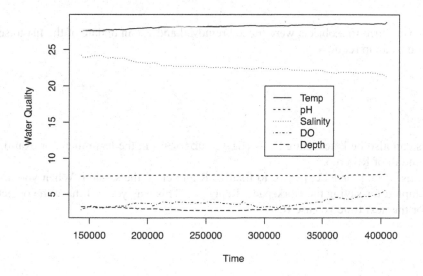

FIGURE 1.26 Plots with a family of lines.

In this case, it is very important to include `dev.off()` to close the pdf file. The pdf file will contain a page with the figure.

Several lines between the `pdf()` and the `dev.off()` functions are converted to multiple pages of the pdf file. For example,

```
pdf(file="output/test-2pages.pdf")
  plot(Time, DO, type="l",xlab="Time [hhmmss]", ylab="Dissolved Oxygen
  [mg/l]",xlim=c(150000,240000),ylim=c(0,5), xaxs="i",yaxs="i")
  matplot(Time,x[,-1], type="l", col=1, ylab="Water Quality")
  legend(350000,20,leg=names(x)[-1],lty=1:length(names(x)[-1]))
dev.off()
```

Once we are done with dataset `x`, we should use detach to remove the dataset from the first position of the workspace where it can mask other objects.

```
detach(x)
```

After detaching you will not be able to directly use the names of the dataset `x`.

CLEANUP AND CLOSE R SESSION

Many times, we generate objects that may not be needed after we use them. In this case, it is good practice to clean up after a session by removing objects using the `rm` function. A convenient way of doing this is to get a list of objects with `ls( )` and then see what we need to remove.

For example, suppose at this point we may want to keep object `x100` because they contain data that we may need later but remove object `x`.

```
> ls()
 [1] "x100"   .."x"
> rm(x)
```

You can also confirm that objects were indeed removed and get an update of the list to see if there is some more cleanup required.

```
> ls()
 [1]     "x100"
```

The objects can also be listed using `Misc|List Objects` in the Rgui menu or examined in the Environment tab of RStudio.

You can use `q( )` or `File|Exit` to finalize R and close the session. When you do this, you will be prompted for saving the workspace. Reply yes. This way you will have the objects created available for the next time you use R.

The appendix to Lab 1 available from the eResources contains additional information when using Windows, such as scripts using an external editor, specifically Notepad++, and how to modify the shortcut to launch R from working folder.

EXERCISES

Exercise 1.1 Simple statistics

Open your script `lab1.R`. Add lines to this script that calculate the mean, variance, standard deviation, and standard error of object `x100`, and store each one of these in a variable with names `avg.x`, `var.x`, `sd.x`, and `se.x`. Use rounding and truncation to one decimal. Add comment lines to document your code. Use function `summary` to describe `x100` and calculate inter-quartile distance storing it in object `iqd.x100`. Run the code and verify your results and that you have these variables in your workspace.

Exercise 1.2 Exploratory analysis

Compose all six EDA plots for array `x100`. Discuss each graph.

Exercise 1.3 Functions

Build functions panels and eda6. Source these, verify that they are in the workspace, and run eda6 on `x100`.

Exercise 1.4 Time series

Use the `x100` array to practice time series plot and autocorrelation.

Exercise 1.5 Using parametric test

Apply a t-test to object `x100` to test whether the mean is >42. Calculate how many observations are required to have a power of 0.9.

Exercise 1.6 Reading CSV files and processing data

Look at the `data/salinity.csv` file. Add lines to your `lab1.R` script to read the file, create an object, calculate descriptive statistics, obtain all EDA plots, and save the graph. Determine whether the values are normally distributed. Discuss whether this water is brackish or saline, taking into account that the limit between brackish and saline is considered 30‰. Use the Wilcoxon signed rank test to see whether the mean of the salinity dataset is >30.

Exercise 1.7 Linear regression

Perform a linear regression between ozone (`Ozone`) and temperature (`Temp`) of the air quality dataset, including a scatter plot with a superimposed regression line. Determine the value of the coefficients, evaluate this regression, and discuss the errors.

Exercise 1.8 Data frames, time series plots, sending graphical output to pdf

Use dataset x created from the file `data/datasonde.csv`. Produce histograms of all variables in dataset x (except `Time`) on the first page of a pdf file with the name `data-sonde-2pages.pdf` Produce plots of each variable in x vs. `Time` and one plot of all variables on the same graph vs. `Time`. Compose all six plots on the second page of the same pdf file.

REFERENCES

Acevedo, M.F. 2012. *Simulation of Ecological and Environmental Models.* Boca Raton, FL: CRC Press, Taylor & Francis Group, 464 pp.

Acevedo, M.F. 2013. *Data Analysis and Statistics for Geography, Environmental Science & Engineering.* Boca Raton, FL: CRC Press, Taylor & Francis Group, 535 pp.

Acevedo, M.F. 2024. *Real-Time Environmental Monitoring: Sensors and Systems - Textbook*, Second Edition. Boca Raton, FL: CRC Press, Taylor & Francis Group, 392 pp.

Geany. 2023. *Geany - The Flyweight IDE.* Accessed January 2023. https://www.geany.org/.

Posit. 2023. *RStudio IDE.* Accessed January 2023. https://posit.co/downloads/.

R Project. 2023. *The Comprehensive R Archive Network.* Accessed January 2023. http://cran.us.r-project.org/.

2 Programming and Single Board Computers

INTRODUCTION

In this lab session, we continue learning R, in particular control structures and functions `apply` to simplify loops, and start learning programming in Python, Arduino, HTML, and PHP. Then, we will study how to program and upload scripts to an Arduino, using the Integrated Development Environment (IDE) GUI, as well as the Command Language Interface (CLI). The latter implemented on a single board computer (SBC) by employing the Raspberry Pi Zero W, which is setup headless, and accessed using Secure Shell (SSH). To end the lab session, we practice Python programming from the Raspberry Pi (RPi). Basics of Linux and the nano editor are provided. To end the lab session, the Arduino is programmed to communicate serially to the RPi, which runs a Python script to read the serial stream.

MATERIALS

READINGS

For theoretical background, you can use Chapter 2 of the book Acevedo, M.F. 2024. *Real-Time Environmental Monitoring: Sensors and Systems - Textbook, Second Edition* which is a companion to these guides (Acevedo 2024). Other bibliographical references are cited throughout the guide.

COMPONENTS

- Protoboard and jumper wires
- Arduino Jumper wires
- Resistor 50 Ω, 1 W, to be used as heater
- Small cooling fan (5 V DC) to be used to cool the sensor

MAJOR COMPONENTS AND INSTRUMENTS

- Raspberry Pi. The guide is written based on a RPi Zero W. But you can use a RPi 3 or 4.
- Micro SD and adapter (e.g., 16 GB)
- Arduino UNO R3 with USB Cable
- Digital temperature probes, e.g., KETOTEK (Amazon 2023a) or KEYNICE Digital Thermometer

TOOLS (RECOMMENDED)

- Long nose pliers
- Wire strippers and clippers
- Alligator clips

DOI: 10.1201/9781003184362-2

EQUIPMENT AND NETWORK RESOURCES

- Laptop or PC with Wi-Fi and SD card reader/writer, as well administration rights to install software. The guides are written assuming Windows, but it can be adapted to Linux or Mac.
- Access to a small network (class C) provided by either a wireless router or Hotspot enabled from a smartphone.

SOFTWARE (URL LINKS PROVIDED IN THE REFERENCES)

- R, for data analysis (R Project 2023)
- RStudio, an IDE to use R (Posit 2023)
- Advanced IP scanner, to find IP addresses (Advanced IP Scanner 2023)
- PuTTY, for SSH (PuTTY 2023)
- WinSCP, to transfer files from the RPi to a windows PC (WinSCP 2023)
- Raspberry Pi imager, to install the Raspberry Pi OS (Raspberry Pi 2023)
- Python, Shell, and IDLE (Python 2022a)
- Arduino IDE, Interface to Arduino (Arduino 2016)
- Geany, IDE to edit programs as well as data (Geany 2023)
- Arduino-CLI, Arduino software to run using commands from the RPi (Arduino 2023b)
- PHP, for RPi.

SCRIPTS AND DATA FILES

- R, Python, and Arduino scripts contained in archive prgs.zip (available from the RTEM GitHub repository https://github.com/mfacevedol/rtem)
- Data files contained in archive data.zip (available via the RTEM GitHub repository already used for lab 1)

SUPPLEMENTARY SUPPORT MATERIAL

Supplementary support material including additional screenshots, images, and procedures are available from the publisher eResources web page provided for this book.

PROGRAMMING LOOPS IN R

PROGRAMMING LOOPS

We will use the simple example of evaluating a function of one variable for different values of a parameter. Say, evaluate the exponential function $x=\exp(rt)$ for several values of the coefficient r. First, declare the sequence of values of variable t

```
t<-seq(0,10,0.1)
```

Then, store values of r in an array

```
r <- c(-0.1,0,0.1)
```

now declare a two-dimensional array (matrix) to store the results; each column corresponds to the values of x for a given value of r

```
x <- matrix(nrow=length(t), ncol=length(r))
```

Here, nrow and ncol denote the number of rows and columns, respectively. Now use function for to perform a loop

```
for(i in 1: length(r)) x[,i]<-exp(r[i]*t)
```

For brevity we ask to see the results; the first few lines are

```
> x
          [,1] [,2]      [,3]
[1,] 1.0000000   1 1.000000
[2,] 0.9900498   1 1.010050
[3,] 0.9801987   1 1.020201
[4,] 0.9704455   1 1.030455
[5,] 0.9607894   1 1.040811
[6,] 0.9512294   1 1.051271
```

For brevity we round to two decimal places

```
> round(x,2)
       [,1] [,2] [,3]
[1,] 1.00    1 1.00
[2,] 0.99    1 1.01
[3,] 0.98    1 1.02
[4,] 0.97    1 1.03
[5,] 0.96    1 1.04
[6,] 0.95    1 1.05
```

FUNCTIONS APPLY, SAPPLY, VAPPLY, MAPPLY

A more effective and preferred way of applying loops in R is using functions sapply(), vapply(), and mapply(). For example, the calculation above using loops can be done using

```
x <- sapply(r, function(r) exp(r*t))
```

This simple line yields the same results in just one statement without having to predefine the matrix x.

Function apply works on a matrix. Function apply(X, margin, fun, ...) applies a function fun to entries of a matrix defined by margin row (1), column (2), or row and column (c(1,2)). For example, use apply over margin 2 (columns) invoking fun=mean.

```
y <- apply(x[,1:3],2,mean)
> y
[1] 0.6326388 1.0000000 1.7196907
>
```

EXAMPLE: AIR QUALITY

Ozone concentration is an important component of air quality. It is monitored frequently and at many locations; these data are used for compliance with State and US federal regulations. The standards are based on running eight-hour averages of ozone according to the US EPA procedure (US EPA 2018). In a day there are 24 hourly values of ozone. To be able to obtain a running eight-hour average for each hour that do not overlap with eight hours average of the previous or next day, we use only $24 - 8 + 1 = 17$ hours to start the average. For example, from hour 00:00 to 07:00, from 01:00 to 08:00, from 02:00 to hour 09:00, and so on until the period from 16:00 to hour 23:00. This may vary; it is also done starting at 7 am of one day and ending at 6 am of the next day (ASL Associates 2018). Ozone values are given by three decimal places in ppm using truncation to three decimal places by US federal regulatory standards, or rounded by some specific state standards. For this problem, we will use files data/ozone24hsample1.txt and data/ozone24hsample2.txt. These are not real data but made to simplify and illustrate the process. Inspect both files using a text editor. You can use the RStudio editor, or Geany (Geany 2023), or Notepad++ (Notepad++ 2023), which are convenient IDEs. As an example, Figure 2.1 shows data/ozone24hsample1.txt opened in Geany.

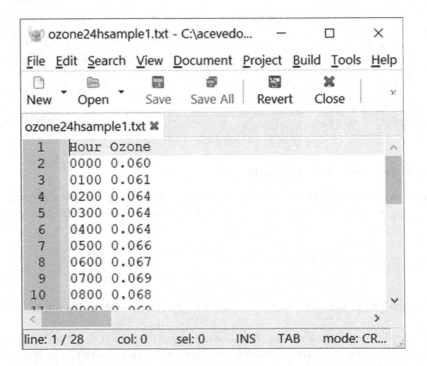

FIGURE 2.1   Data file open in the Geany IDE text editor.

There are 25 lines of data in each file. The time format is HHMM, ozone was sampled every hour, and the last record is the first hour of the next day. Read each file to a data frame using x1 and x2 for sample 1 and x2 for sample 2.

```
x1 <- read.table("data/ozone24hsample1.txt", header=TRUE)
x2 <- read.table("data/ozone24hsample2.txt",header=TRUE)
```

Start to write a function named oz8h with argument x corresponding to these objects. Use curly braces as placeholders for now.

```
oz8h <- function(x){

}
```

Create a matrix named x8 to store results in 17 columns and 8 rows.

```
# matrix to place all 8 values of each average
 x8 <- matrix(ncol=17,nrow=8)
```

Use mx8 <- array() to create a column vector to store the eight hour average

```
# vector to store the 8h average
 mx8 <- array()
```

Now add a loop with the following code

```
for(i in 8:24){
  j <- i-8+1 # starts at 1 ends at 24-8+1=17
  x8[,j] <- x[j:i,2]
  # truncating
  mx8[j] <- trunc(mean(x8[,j])*1000)/1000
}
```

This loop selects eight values at a time, calculates all the running eight hour averages. Add a line to calculate maximum maxx8 <- max(mx8) and add a line to return values from the function before closing return(list(avg.8h=mx8, max.avg.8h=maxx8)). Note: a list is used to return multiple results.

```
 maxx8 <- max(mx8)
 return(list(mean.8h=mx8, max.mean.8h=maxx8))
```

We now have the complete function

```
oz8h <- function(x){
# matrix to place all 8 values of each average
 x8 <- matrix(ncol=17,nrow=8)
# vector to store the 8h average
 mx8 <- array()
# loop
for(i in 8:24){
  j <- i-8+1 # starts at 1 ends at 24-8+1=17
  x8[,j] <- x[j:i,2]
  # truncating
  mx8[j] <- trunc(mean(x8[,j])*1000)/1000
 }
maxx8 <- max(mx8)
return(list(mean.8h=mx8, max.mean.8h=maxx8))
}
```

Save it as file oz8h.R in your R folder
    Source the function and run it for x1 and x2. That is, oz8h(x1) and oz8h(x2).

```
source("R/oz8h.R")

oz8h(x1)
oz8h(x2)
```

Show the result of your function for both x1 and x2 and explain how the results match the return code you used above.

```
> oz8h(x1)
$'mean.8h'
 [1] 0.064 0.065 0.066 0.066 0.067 0.068 0.068 0.069 0.069 0.069 0.069 0.069
[13] 0.069 0.069 0.068 0.067 0.067

$max.mean.8h
[1] 0.069

> oz8h(x2)
$'mean.8h'
 [1] 0.067 0.068 0.070 0.072 0.073 0.075 0.076 0.077 0.077 0.078 0.078 0.078
[13] 0.078 0.077 0.076 0.075 0.073

$max.mean.8h
[1] 0.078

>
```

We can see how the function returns a list with two components $mean.8h which has 17 values of
eight hour averages and $max.mean.8h which contains the max of the averages.

PYTHON

In this section, we learn how to write simple scripts in Python. A good tutorial is provided by the Python webpage (Python 2022b).

PYTHON PROGRAMMING USING THE SHELL AND IDLE

We demonstrate it using Windows, but you can use Mac or Linux. Download python from the web site (Python 2022a) and run the installer. Once installed, you can run the interpreter shell or the Integrated Development and Learning Environment (IDLE). Using the shell, at the >>> prompt (the primary prompt), and type a simple function

```
>>> print("Hello!")
Hello!
```

and receive the response displaying Hello!. In this case, we are working in interactive mode. We can use Python as a calculator

```
>>> 34+64
98
>>> 9/5
1.8
>>> 100*(50-5*8)
1000
```

More generically, you can assign values to a variable and use it for calculations

```
>>> width=1.0
>>> height=2
>>> area =width*height
>>> area
2.0
>>> square=width**2
>>> square
1.0
```

We can define *strings* and use function len() to obtain the length of a string.

```
>>> str1='environmental'
>>> str2=' monitoring'
>>> str1 + str2
'environmental monitoring'
>>> len(str1)
13
>>> len(str1+str2)
24
```

A versatile data structure is a *list* that can be created by comma-separated items between square brackets. It can be appended with new values by concatenation or using `.append()`.

```
>>> x =[1**2,2**2,3**2]
>>> x
[1, 4, 9]
>>> x = x+[4**2]
>>> x
[1, 4, 9, 16]
>>> x.append(5**2)
>>> x
[1, 4, 9, 16, 25]
```

Instead of interactive mode, we can create a script file and then execute it. Using the IDLE, we can create a file using `File>New File` and we can type code in the editor.

```
a,b=0,1
while a<10:
    print(a)
    a,b= b,a+b

for i in range(5):
    print(i)
```

It is convenient to show the line numbers using `Options`. This script shows how to write a loop using `while a<10:` Note that the lines after the : are indented, and there is no marking for the end of the loop because it is signaled by the last indented line. It is necessary to have the same number of space characters or tabs for all lines meaning to be indented. Another feature illustrated here is the *multiple assignment* `a,b=0,1` and `a,b=b,a+b`. Both variables are assigned values on the same line of code. You can create a new folder py in your working directory to store python scripts. Save this script as `simpleloop.py`. To run this script, use `Run` and receive response on the shell.

PYTHON PROGRAMMING USING RSTUDIO

We can use RStudio to create and run Python scripts. Install package `reticulate`, which is a Python interpreter in R and based on r-miniconda. In RStudio, we can use `File > New File > Python Script` to create a Python program. In the editor, type python code and run. You can also open an existing python script file; for example, open `py/simpleloop.py` and then Run to receive response on the Console.

ARDUINO UNO

BASICS

The Arduino boards are popular in a variety of fields since it has been designed to facilitate its use by non-experts in electronics, engineering, or programming (Arduino 2023a). It is easy to program using a cross-platform that runs on Windows and Mac and Linux. Both the hardware and software

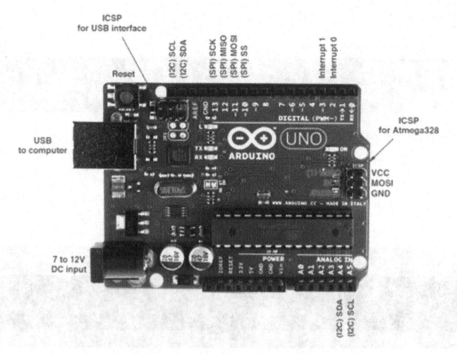

FIGURE 2.2 Arduino UNO R3. From Arduino webpage (Arduino 2023a).

are open source. The classic Arduino UNO (Figure 2.2) is a SBC based on the ATmega328 micro-controller; it includes 14 digital I/O pins, 6 analog inputs (10 bit ADC), a USB connection, a power jack (input 7–12 V), and a reset button. It will derive power when connected to a computer with a USB cable or from a DC power source, AC-to-DC adapter, connected to the power jack, or the VIN pin of the board (7–12V) (ARDUINO 2014).

The ATmega328 has 32 Kb of flash memory, 2 Kb of SRAM, and 1 Kb of EEPROM. The clock speed is 16 MHz. Besides the VIN pin and GND pins under the "power" label, you also find 5 V (regulated) and 3.3 V (regulated). Other features include pins for SPI, and I2C interface lines which we will employ in forthcoming lab sessions. There are also Arduino UNO clones, e.g., ELEGOO UNO R3 Board (Elegoo 2023).

IDE

The Arduino UNO is programmed using the Atmel IDE. Download the Arduino software from the web site Arduino (2016). Many examples in this book were conducted using version 1.8.13, which was the most recent available at the time. Currently, the IDE is at version 2.0.2. Once installed, you run the Arduino software and will see the script editor (scripts in Arduino are called sketches) (Figure 2.3). After variable declaration, an Arduino program has two functions: `setup()` and `loop()`. In the `setup()` function, we specify input and output pins, communication, and other configurations. In the `loop()` function, we perform instructions to be executed in a repeatedly manner.

Connect the Arduino to your PC using a USB cable. At the bottom right-hand side of the Arduino software window, you will see the COM port number used by the Arduino. Change this port as needed using `Tools|Port`.

```
sketch_feb17a | Arduino 1.8.13                          —    □    ×
File Edit Sketch Tools Help

  sketch_feb17a
void setup() {
  // put your setup code here, to run once:

}

void loop() {
  // put your main code here, to run repeatedly:

}

1                                              Arduino Uno on COM13
```

FIGURE 2.3   Opening the Arduino software.

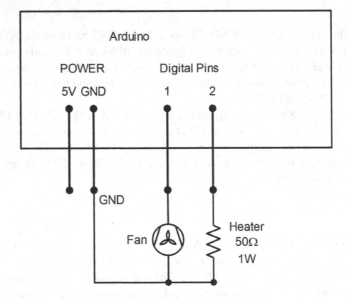

FIGURE 2.4   Warming or cooling driven by an Arduino.

EXAMPLE: DRIVING A FAN AND HEATER RESISTOR

We will systematically measure temperature changes in proximity of a 50 Ω (1 W) resistor, which by Joule heating acts as to increase the temperature, and a cooling fan to decrease the temperature. Both devices will be powered on and off by Arduino digital pins, which can deliver 5 V (V_S in Figure 2.4) when HIGH (limited to 40 mA). You will find the 5 V and GND pins under the "power" label and the digital pins in the DIGITAL (PWM~) area.

We will use this setup in Lab 3 to measure the resistance changes of a sensor. As an overview, Figure 2.5 shows how the Arduino interacts with the PC and the Raspberry Pi Zero W, which will be described in the next section.

Note that the power of the heater is $P_h = 0.04^2 \times 50 = 0.08$ W or approximately 8% of rated power of 1 W. The fan may be rated for 5 or 12 V, but the supplied current will be limited to 40 mA; therefore, it will run slowly. To measure temperature, you will use a digital thermometer such as the KETOTEK (Amazon 2023a) or KEYNICE (Amazon 2023b) and the degree Celsius (°C) scale. To measure temperature, use the probe close to the heater resistor, making thermal contact. Before we write an Arduino program to run the heater and fan, you can disconnect the pins and test the heater and fan using 5 V, verify raise and drop of temperature. Next, we will work on the program.

Declare variables for output pins as `fanPin` 9 and `heaterPin` 10,

```
// control fan and heater
// to measure temperature changes next to a resistor
int heaterPin = 10;
int fanPin =9;
```

define the timing

```
// edit the following values in minutes as needed
int fanTime=5;
int heatTime=5;
int repeatTime=5;
```

configure these pins for OUTPUT in `setup()` and start the serial port at 9600 baud.

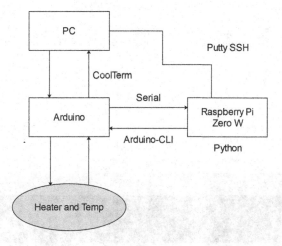

FIGURE 2.5 Overview of setup showing the Arduino and how it interacts with the PC and the Raspberry Pi Zero W.

```
void setup(){
        pinMode(heaterPin, OUTPUT);
        pinMode(fanPin, OUTPUT);
        Serial.begin(9600);
}
```

These pins can be used in the `loop()` function; for example, by using the `digitalWrite` function. A convenient feature of the IDE is to show line numbers in the editor window (Figure 2.6).

Then, we write a function `wait _ min()` that will message the user via the serial monitor and then perform a loop to wait several minutes. Here, the `delay()` function is in milliseconds (ms). Each minute is 60,000 ms.

```
void wait_min(int duration_min){
  // messages and allow time to measure
  Serial.println("You can read several values of temperature");
  Serial.println("Try to take readings every 0.1C or at least every 0.5C");
  for(int i=1; i<=duration_min;i++){
   delay(60000);
   Serial.print(i);
   Serial.print(" min ");
  }
  Serial.println("");
}
```

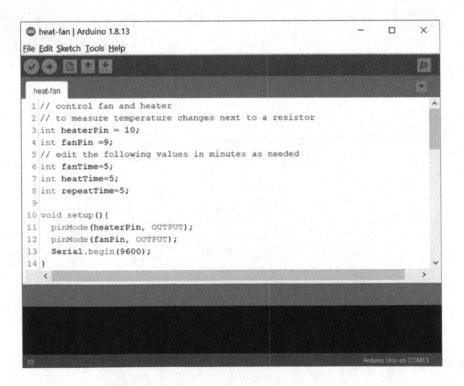

FIGURE 2.6   Declarations and setup.

See Figure 2.7. Then, we write the `loop()` function

```
void loop(){
  Serial.println("Started with heater off");
  digitalWrite(heaterPin,LOW);

      Serial.print("Now turning fan on for "); Serial.print(fanTime);
      Serial.println(" minutes");
  digitalWrite(fanPin,HIGH);
  wait_min(fanTime);

      Serial.println("Turned fan off");
      digitalWrite(fanPin,LOW);

  Serial.print("Now turning heat on for "); Serial.print(heatTime);
  Serial.println(" minutes");
  digitalWrite(heaterPin,HIGH);
  wait_min(heatTime);

      Serial.println("Turned the heater off");
  digitalWrite(heaterPin,LOW);
  Serial.print("The loop will repeat in "); Serial.print(repeatTime);
  Serial.println(" minutes");
      wait_min(repeatTime);
}
```

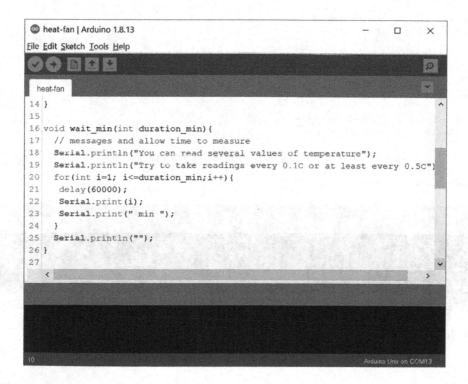

FIGURE 2.7 Function to message the user.

as shown in Figure 2.8. Once complete, you would have the script `heat-fan.ino`. You can also download from the RTEM GitHub repository. After saving, compile, upload, and click on the serial monitor using shortcuts (Figure 2.9) or menu items. At the serial monitor, we can see the results of running the script (Figure 2.10).

Measure each temperature using the temperature probe. When operating indoor at room temperature, you may observe temperature variations covering the range from 20°C to 30°C and include measurement of the reference ambient temperature T_0 of 25°C. You can type the values in Geany as records separated by commas. When you finish measuring, save the file as CSV (Figure 2.11). Once you have a data file, we can use the following R code to plot the data.

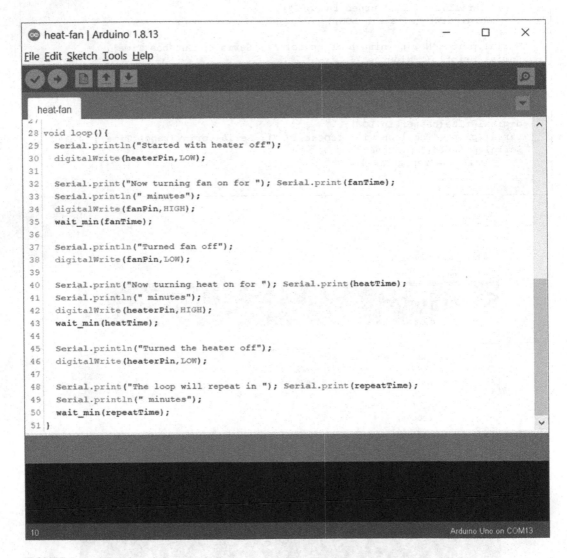

FIGURE 2.8 Loop function.

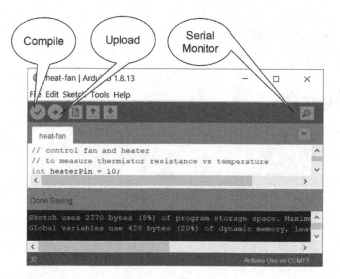

FIGURE 2.9 Shortcuts for menu items.

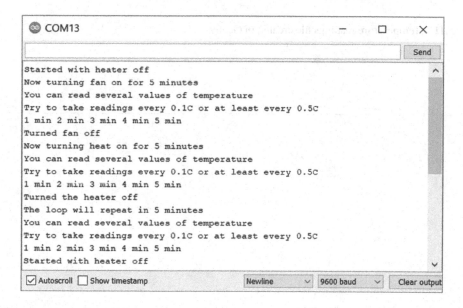

FIGURE 2.10 Output displayed in the serial monitor.

```
x <- scan("data/temp-readings.csv",sep=",")
plot(x)
```

See the results in Figure 2.12.

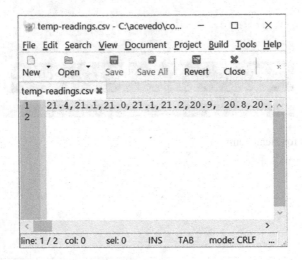

FIGURE 2.11   Temperature readings file created in Geany.

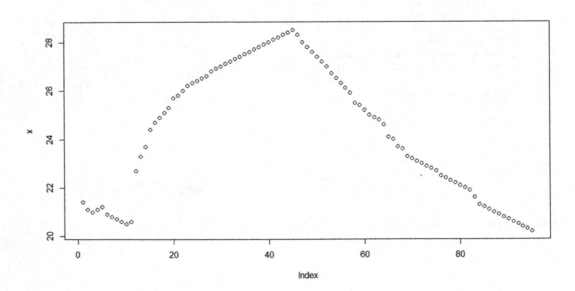

FIGURE 2.12   Plot of collected temperature data.

RASPBERRY PI ZERO

In this lab, we work with a Rpi Zero W. For briefness and whenever is clear from the context, we will also use RPi to refer to the Rpi Zero W.

RASPBERRY PI ZERO: SOFTWARE INSTALLATION IN THE PC

Before proceeding to set the RPi up, we need to install a Secure Shell (SSH) client program (PuTTY) on the PC. In the guide, we assume a windows installation, but it can be adapted to Linux or Mac. *PuTTY* facilitates running SSH to run commands on the RPi. Download from PuTTY (2023) and install it. In addition, before we proceed to set the RPi, you need to find information on the network that the RPi will connect wirelessly by Wi-Fi. The RPi Zero W will use 2.4 GHz, write down the network name (SSID) and password, and keep it handy for the next section.

RASPBERRY PI: HEADLESS SETUP

In this section, we set up the RPi Zero W in "headless" mode. This name is used for a configuration where the RPi is not connected to peripheral devices such as display, keyboard, and mouse. In this mode, we work with the RPi using SSH over a network in which the RPi is connected by Wi-Fi. There are web references useful to supplement this guide (Circuit Basics 2022; Desertbot 2022; Core Electronics 2022; Vorillaz 2022). The Raspberry Pi OS used to default to username `pi` with the password `raspberry`. This has changed and the default username and password are now set on first boot using a configuration wizard, when using peripherals. In headless mode, the default username and password can be set using the Raspberry Pi Imager.

Insert a micro-SD card in the PC SD reader/writer. For PCs having large format SD reader/ writer, insert the micro-SD in a SD adapter and then insert the adapter in the PC's SD reader/writer. You can download the Raspberry Pi Imager from the web site (Raspberry Pi 2023).

Install the Raspberry Pi Imager, open the imager, and press `Choose OS`, then select the recommended version `Raspberry Pi OS (32-bit)`. Now press `Choose Storage` and choose the SD card. Now press the settings wheel. Under advanced options, type a hostname (e.g., the default raspberrypi), click `Enable SSH`, click set username and password and type them (e.g., pi and a secure password), and of course make a note because you will need them later to login. Click `Configure wireless LAN` and type the SSID and password of your network (you wrote this down in the previous section). Select your country based on the ISO two-letter code, time zone, and keyboard. Now press `Save` and you will be back at the main screen of the Imager. Now press `Write` and proceed to format the SD to erase all data and get it ready to store the RPi OS image on the micro-SD.

Study the RPi connectors (Figure 2.13). There are two micro-USB connectors; one is for power and one for serial communication. Be careful not to connect power to the micro-USB port since this may damage the serial port. Assuming the micro-SD card is inserted into the RPi, go ahead and power the RPi up (i.e., connect the power supply to the RPi's power micro port).

On the windows PC, use the command window (search windows cmd) and then type `ipconfig` to obtain the IPv4 address, subnet mask, and gateway IP. For example,

```
Wireless LAN adapter Wi-Fi:
   Connection-specific DNS Suffix  . :
   Link-local IPv6 Address . . . . . : fe80::158e:3a7e:9709:57d4%16
   IPv4 Address. . . . . . . . . . . : 192.168.1.12
   Subnet Mask . . . . . . . . . . . : 255.255.255.0
   Default Gateway . . . . . . . . . : 192.168.1.1
```

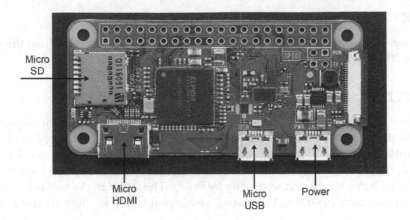

FIGURE 2.13   Raspberry Pi Zero W. (Original image from Adafruit web page.)

Now you can find other device on the same LAN using command `arp -a` of the Address Resolution Protocol (ARP). For example,

```
C:\>arp -a

Interface: 192.168.1.12 --- 0x10
  Internet Address      Physical Address      Type
  192.168.1.1           9c-c9-eb-5f-26-fc     dynamic
  192.168.1.3           28-16-ad-05-e0-bf     dynamic
  192.168.1.4           24-4c-e3-49-93-ef     dynamic
  192.168.1.5           04-c2-9b-e5-38-d6     dynamic
  192.168.1.6           b8-27-eb-10-0f-1e     dynamic
  192.168.1.255         ff-ff-ff-ff-ff-ff     static
```

Out of this list of IP and MAC addresses, we identify the RPi at IP address192.168.1.6. We would know this by inspecting the MAC address; in this case, the Organizational Unique Identifier (OUI) is B8:27:EB corresponding to RPi. The last three bytes 10:0F:1E is the Network Interface Controller (NIC) number assigned by the manufacturer (see Chapter 2 of the companion textbook). Take note of the IP and MAC addresses since you will need them later. In addition, note that the type of IP address for the RPi is "dynamic" because it was acquired by DHCP (Dynamic Host Configuration Protocol). At the end of this chapter, we provide guidance on how to establish a static IP address.

If the RPi does not show up in the list, you need to determine the problem and correct it. The most common issue is that you accidentally typed an incorrect name and password of your network when you configured the SD card, or the network settings were configured incorrectly for some other reason. To troubleshoot this type of issue, you can try a couple of options. First, you can power down the RPi, take the SD card out, and reinsert it in the PC to review (and redo if necessary) the network configuration performed with the Raspberry Pi imager program. There is no need to reinstall the complete image and write it to the SD; you only need to use the configuration tool provided by the settings (wheel icon) and look and edit the network configuration. Then re-insert the SD card in the RPi, power up, wait a few minutes and rerun `arp -ar` to verify that the RPi is on the network.

If that does not work, another option is to use a `wpa_supplicant.conf` text file. Power down the RPi, remove the SD card, and reinsert it in the PC. Using a text editor type the following contents and save it as `wpa_supplicant.conf` in the SD card.

```
country=US
  ctrl_interface=DIR=/var/run/wpa_supplicant GROUP=netdev
  update_config=1
  network={
      ssid="your network name"
      psk="password to your network"
}
```

Of course, replace the ssid and psk strings with your network information. Make sure you do not use file extension `.txt` and also make sure you use only a text editor. Do not use word processors because they will add hidden characters. Once you have copied this file to the SD card, insert it again in the RPi and power it up, wait a few minutes and rerun `arp -ar` to verify that the RPi is on the network.

Other reasons for the RPi not show up on the network is that you have two WiFi networks, one for 2.4 GHz and one for 5 GHz, with different names and passkeys, and that the PC is on one network and the RPi in another. The RPi Zero W has only 2.4 GHz, then the ssid for 2.4 GHz should be the one written to the RPi, and when you connect your PC to WiFi should be the ssid for 2.4 GHz.

If problems remain you can look at the eResources of this book for further troubleshooting tips.

While at the PC command window use `ping` to the RPi IP address to verify that you can find the RPi from the PC.

```
C:\>ping 192.168.1.6

Pinging 192.168.1.6 with 32 bytes of data:
Reply from 192.168.1.6: bytes=32 time=6ms TTL=64
Reply from 192.168.1.6: bytes=32 time=7ms TTL=64
Reply from 192.168.1.6: bytes=32 time=7ms TTL=64
Reply from 192.168.1.6: bytes=32 time=7ms TTL=64

Ping statistics for 192.168.1.6:
    Packets: Sent = 4, Received = 4, Lost = 0 (0% loss),
Approximate round trip times in milli-seconds:
    Minimum = 6ms, Maximum = 7ms, Average = 6ms
```

As an alternative to the ARP command, you can use the *Advanced IP Scanner* to scan your network. You can download and install from Advanced IP Scanner (2023). For example, you can set the first three octets of the range to 192.168.1 to match the PC's IP. Thus, scan the range to 192.168.1.1-254 to scan the subnet varying only by the last octet. Right click on the device, select `Tools`, and `Ping` to verify that you can find the RPi from the PC.

Once you verify that the RPi is reachable from the PC, you can use PuTTY to logon to it using SSH. Open PuTTY, select SSH, keep the port at 22, and type the IP address of the RPi (Figure 2.14).

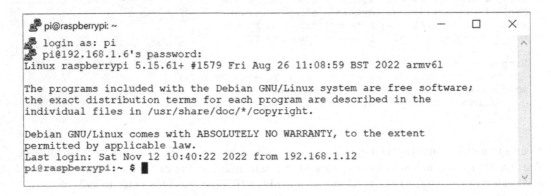

FIGURE 2.14 Running PuTTY.

Click Open and say yes to potential risks due to the certificate. Type the username and password that you created when creating the image.

Optionally, you can modify the colors and appearance of the terminal and save these as a session for consistency. It is also convenient to save "Default Settings" session with Port 22 to ensure using this port in subsequent use of the associate secure copy program PSCP. At this point, you are ready to run the RPi using SSH from your PC (Figure 2.15). The appendix to Lab2 available from the eResources contains information on how to add other networks to the RPi, using peripherals, and the VNC interface.

```
pi@raspberrypi: ~                                        —    □    ×
login as: pi
pi@192.168.1.6's password:
Linux raspberrypi 5.15.61+ #1579 Fri Aug 26 11:08:59 BST 2022 armv6l

The programs included with the Debian GNU/Linux system are free software;
the exact distribution terms for each program are described in the
individual files in /usr/share/doc/*/copyright.

Debian GNU/Linux comes with ABSOLUTELY NO WARRANTY, to the extent
permitted by applicable law.
Last login: Sat Nov 12 10:40:22 2022 from 192.168.1.12
pi@raspberrypi:~ $ ▮
```

FIGURE 2.15 Connecting to RPi by SSH.

LINUX COMMANDS

From this point on, you need to know some basic Linux commands. There are resources online to learn the most important commands (McKay 2022; Hostinger 2022). To get started, we will focus on a few basic commands

```
sudo
mkdir
ls
ls -la
cd
```

The sudo command stands for *SuperUser DO* and allows access to restricted parts of the system by temporarily elevating privileges of root user. For example, to further configure the RPi, at the prompt type

```
sudo raspi-config
```

will bring up a configuration menu. System Options allows changing the Hostname and password. Under Localization Options, you can select region and timezone. Under Interface Options, you can enable VNC and other interfaces SPI, I2C which we will use in forthcoming lab guides. Another example relates to updating the RPi. On the PuTTY terminal type

```
sudo apt-get update
```

We can use ls to get a short list of folders

```
ls
```

and with option -la to get all details

```
ls -la
```

We create directories or folder with mkdir. For example, create a directory named labs to store the files for the lab sessions. This will be our working directory in the RPi. Use

```
mkdir labs
```

and verify the result with ls

```
ls -la
```

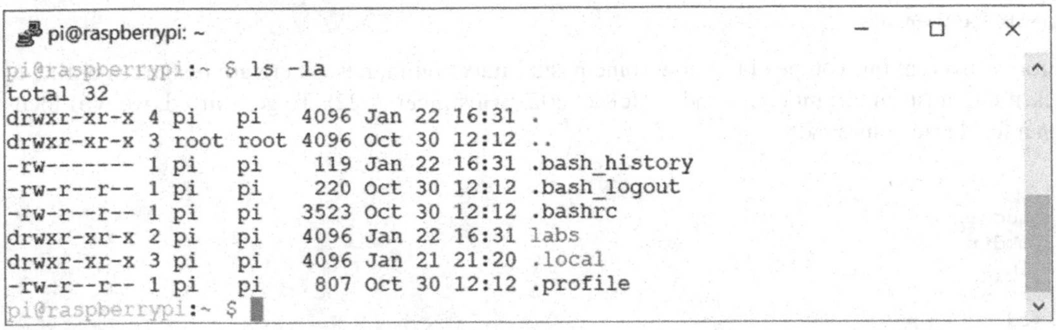

```
pi@raspberrypi: ~                                        —    □    ×
pi@raspberrypi:~ $ ls -la
total 32
drwxr-xr-x 4 pi    pi   4096 Jan 22 16:31 .
drwxr-xr-x 3 root root 4096 Oct 30 12:12 ..
-rw------- 1 pi    pi    119 Jan 22 16:31 .bash_history
-rw-r--r-- 1 pi    pi    220 Oct 30 12:12 .bash_logout
-rw-r--r-- 1 pi    pi   3523 Oct 30 12:12 .bashrc
drwxr-xr-x 2 pi    pi   4096 Jan 22 16:31 labs
drwxr-xr-x 3 pi    pi   4096 Jan 21 21:20 .local
-rw-r--r-- 1 pi    pi    807 Oct 30 12:12 .profile
pi@raspberrypi:~ $ ▮
```

FIGURE 2.16 List of directories.

as shown in Figure 2.16. We can change directory using cd. For example,

```
cd labs
```

which for now is empty. To return to the directory above use two dots cd ..

```
cd ..
```

To organize our working directory, create a directory data within labs. This will be labs/data to store data files.

```
mkdir labs/data
```

We will learn more Linux commands as we progress through the lab guides.

PYTHON SCRIPTS

We will use Python to run programs on the RPi. Use the text editor nano to create a python text file (extension .py). There are online tutorials on nano (How-To Geek 2022).

```
nano labs/test.py
```

Once you are in the nano editor type

```
print("Hello!")
```

See Figure 2.17. Use Ctrl X to exit and answer Y to save. Check that the file is in directory labs using

```
ls labs
```

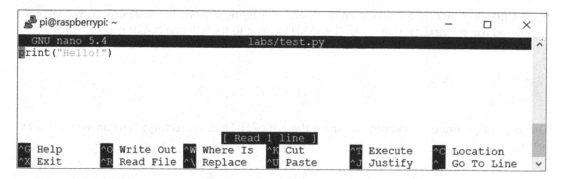

FIGURE 2.17 Text editor nano.

Now run the script using `python3`

```
python3 labs/test.py
```

Verify that it works. Now using nano compose the Python script `singleloop.py` that we used above when learning python from IDLE and run it.

TRANSFERRING FILES BETWEEN WINDOWS PC AND RPI

We can use secure copy PSCP of PuTTY that is available from the command tool in Windows. Let us transfer the file `data\datasonde.csv` from the windows working directory labs to the RPi directory labs/data

```
C:\ labs>pscp data\datasonde.csv pi@192.168.1.132:labs/data
pi@192.168.1.132's password:
datasonde.csv              | 3 kB |   3.4 kB/s | ETA: 00:00:00 | 100%
```

It is possible that you may get an error "ssh_init: Network error: Cannot assign requested address" when PuTTY has not saved the port number on the machine. To avoid this, save "Defaults Settings" in PuTTY with Port 22 as mentioned before.

Alternatively, if using Windows, you can use WinSCP to transfer the file `datasonde.csv` to the RPi. To do this, download and open WinSCP, select File protocol as SCP. Then for host-name enter the RPi's IP address (or the Hostname you selected for the RPi), enter username, and password. You will see two panes. The left side is your local directory (the PC in this case). The right side is the remote directory (the RPi in this case). WinSCP facilitates navigation within each side and copying file from one side to the other. To facilitate repeated use of WinSCP, you can go to `Session|Save Workspace` and save the server information. For example, save as `rtem-labs`. The next time you open WinSCP, it will go directly to a login window. Press login and enter password.

RUNNING A WEB SERVER FROM THE RPI

One of the goals of these lab sessions is to learn to display environmental monitoring output from a variety of sensors on a web page. The RPi could be used to run a *web server* when it collects data from dataloggers or microcontrollers reading and processing sensors. For this purpose, we will now install and run the `nginx` web server. Use SSH to login to the RPi and install `nginx`

```
sudo apt install nginx
```

The `nginx` program should now be under the `/etc/init.d/` directory; you can verify this by listing

```
ls /etc/init.d
```

We will now start `nginx` as a service

```
sudo /etc/init.d/nginx start
```

To test the service, use your browser on the PC and go to `http://192.168.1.132` (or whatever the RPi IP address is); you should see the `nginx` default welcome page. There should also be a directory `/var/www/html` verify its contents

```
ls /var/www/html
```

The html file `index.nginx-debian.html` is the page displayed on the browser. In addition, there should be a directory `/etc/nginx`

```
ls /etc/nginx
```

You should see several directories; one of these is `sites-available`, let us check the content

```
ls /etc/nginx/sites-avai*
```

We will edit the `default` file to setup our web page.

```
sudo nano /etc/nginx/sites-avai*/default
```

Make a simple change for now; edit the root line from `root /var/www/html` to `/home/pi/labs`, which is your working directory under user pi. If you have changed the username, then use that name instead. Once the revised default is saved, we need to restart nginx

```
sudo service nginx restart
```

In the next section, we will create a simple `index.html` page in the working directory `home/pi/labs`.

HTML AND PHP

In this section, we practice the elements of web pages and php discussed in Chapter 2 of the companion textbook using the HTML and PHP examples given there. Create a directory www within labs. We will create a simple `index.html` page in the working directory `home/pi/labs`.

```
nano labs/index.html
```

with this html content

```
<!DOCTYPE html>
<html>
<head>
 <title>Real-Time Environmental Monitoring (RTEM)</title>
</head>
<body>
 <h1>RTEM Test Page</h1>
 <h2>Learning HTML</h2>
 <p>Examples of URL links: external web page and a local file</p>
 <p><a href=https://github.com/mfacevedol/rtem>RTEM GitHub Repository</a></p>
 <p><a href="data/datasonde.csv">Datasonde file</a></p>
</body>
</html>
```

Refresh the web page you had with URL `http://192.168.1.132` and you will see the display of this HTML document (Figure 2.18). Click on the links to verify that they work correctly.

To continue, we will install `php` 7.4, which was the latest version of `php` at the time of writing this guide. We install four packages

FIGURE 2.18   Web page displayed by the HTML example.

```
sudo apt-get install php7.4-fpm php7.4-cgi php7.4-cli php7.4-common
```

The configuration files will be in /etc/php/7.4; list contents with

```
ls /etc/php/7.4
```

and verify that you have four directories, cgi, cli, fpm, and mods-available. Each one of these directories contains configuration files. Next, we will enable php by editing the nginx settings in the /etc/nginx/sites-available/default file

```
sudo nano /etc/nginx/sites-avai*/default
```

Find the area related to php

```
#location ~ \.php$ {
#include snippets/fastcgi-php.conf;
#
#        # With php-fpm (or other unix sockets):
#fastcgi_pass unix:/run/php/php7.4-fpm.sock;
#        # With php-cgi (or other tcp sockets):
#        fastcgi_pass 127.0.0.1:9000;
#}
```

and uncomment four lines: location ..., include ..., fastcgi _ pass unix ..., and }. The edited segment would be

```
location ~ \.php$ {
        include snippets/fastcgi-php.conf;
#
#        # With php-fpm (or other unix sockets):
        fastcgi_pass unix:/run/php/php7.4-fpm.sock;
#        # With php-cgi (or other tcp sockets):
#        fastcgi_pass 127.0.0.1:9000;
}
```

We also need to insert index.php to the index line as follows

```
index index.php index.html index.htm index.nginx-debian.html;
```

After saving the edited default file, restart the `nginx` service

```
sudo service nginx restart
```

In the working directory, change file name `/labs/index.html` to `/labs/index.php`

```
mv labs/index.html labs/index.php
```

and edit its contents by inserting a piece of php code that will echo the datalog file to the web page. Write a <p> tag to announce the file contents

```
<p>File contents using php</p>
<?php
$lines = file("data/datasonde.csv");
foreach ($lines as $line_num=> $line) {
    echo $line . "<br />\n";
}
?>
```

Insert this piece of code within the body tags of the html document, after the <a href line, as shown below

```
<!DOCTYPE html>
<html>
<head>
 <title>Real-Time Environmental Monitoring (RTEM)</title>
</head>
<body>
 <h1>RTEM Test Page</h1>
 <h2>Learning HTML</h2>
 <p>Examples of URL links: external web page and a local file</p>
 <p><a href=https://github.com/mfacevedol/rtem>RTEM GitHub Repository</a></p>
 <p><a href="data/datasonde.csv">Datasonde File</a></p>
<p>File contents using php</p>
<?php
$lines = file("data/datasonde.csv");
foreach ($lines as $line_num=> $line) {
    echo $line . "<br />\n";
}
?>
</body>
</html>
```

At the browser, refresh the URL `http://192.168.1.132` (Figure 2.19). You may have to clear the browser's cache if the web page remains showing the previous `index.html`. Scrolling down the page you would see the contents of the file.

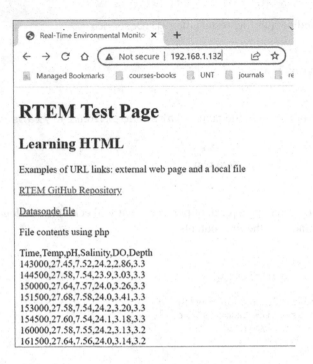

FIGURE 2.19 Web page displaying the contents as given in the php example.

INSTALLING AND USING ARDUINO COMMAND LINE (CLI)

Now we are going to learn how to program the Arduino from the RPi using the Command Line Interface or CLI. For this purpose, we need the Arduino-cli library, which is available from the web site (Arduino 2023b). This is the preferred way to program an Arduino from a Raspberry Pi Zero. The appendix to Lab2 available from the eResources contains information on how to install and use the Arduino IDE on a Raspberry Pi 3 and 4.

First, make a directory to store binary executables

```
mkdir bin
```

Double check that you have a bin directory and include this directory in your path. If you need help with path, you can check this reference (Linuxize 2022).

```
export PATH="$HOME/bin:$PATH"
```

Now verify the path

```
echo $PATH
```

you should see `/home/pi/bin`. Now install the `arduino-cli` library using `curl`

```
curl -fsSL https://raw.githubusercontent.com/arduino/arduino-cli/master/
install.sh | BINDIR=~/bin sh
```

Next, configure the `arduino-cli`. Type

```
arduino-cli config init
```

After this, you will see the `.arduino15` directory by using `ls -la`

```
ls -la
```

Check the configuration file named `arduino-cli.yaml` using `cat`.

```
cat .arduino15/*yaml
```

Create an Arduino sketch or script named `test` in your labs directory.

```
arduino-cli sketch new labs/test
```

Verify using `ls labs/test` that you have a file named `test.ino`. Check contents of `test.ino` using the nano text editor

```
nano labs/test/test.ino
```

you will see the empty sketch. Fill contents using this test sketch (see Figure 2.20)

```
void setup() {
  pinMode(LED_BUILTIN, OUTPUT);
}
```

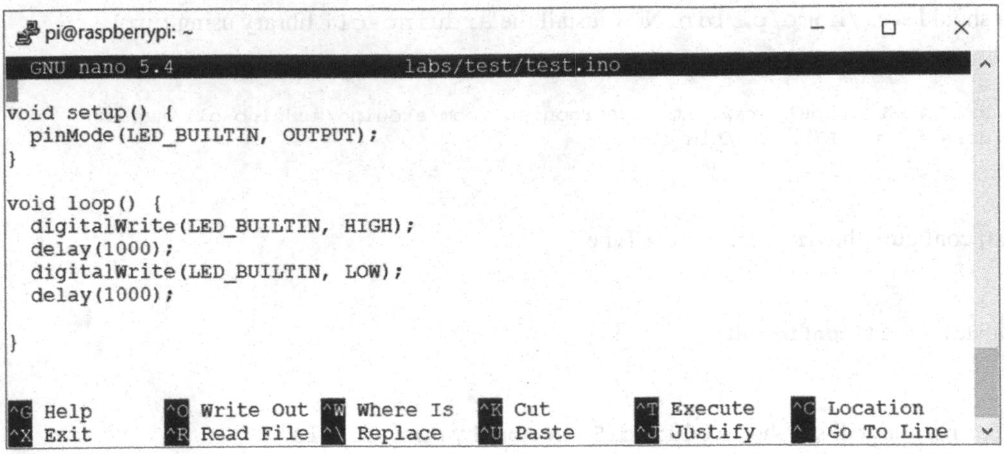

```
pi@raspberrypi: ~                                              —    □    ×
GNU nano 5.4                        labs/test/test.ino                      ^
void setup() {
  pinMode(LED_BUILTIN, OUTPUT);
}

void loop() {
  digitalWrite(LED_BUILTIN, HIGH);
  delay(1000);
  digitalWrite(LED_BUILTIN, LOW);
  delay(1000);

}

^G Help       ^O Write Out  ^W Where Is  ^K Cut      ^T Execute  ^C Location
^X Exit       ^R Read File  ^\ Replace   ^U Paste    ^J Justify  ^  Go To Line  ˅
```

FIGURE 2.20   Test sketch.

```
void loop() {
  digitalWrite(LED_BUILTIN, HIGH);
  delay(1000);
  digitalWrite(LED_BUILTIN, LOW);
  delay(1000);
}
```

Ctrl X to exit and answer y to save. Finally, update the `arduino-cli` core

```
arduino-cli core update-index
```

USING THE ARDUINO FROM THE RPI

First, connect the Arduino to the RPi using the micro-USB adapter or adapter cable. Now at the raspberry ssh terminal, we will type a command to identify the Arduino board connected to the RPi

```
arduino-cli board list
```

We have identified the board FQBN `arduino:avr:uno` and the core for this board `arduino:avr` (Figure 2.21). Now, we install the core for this board

```
arduino-cli core install arduino:avr
```

Several packages will be downloaded and installed; this may take longer to install. Verify installation (Figure 2.22).

```
arduino-cli core list
```

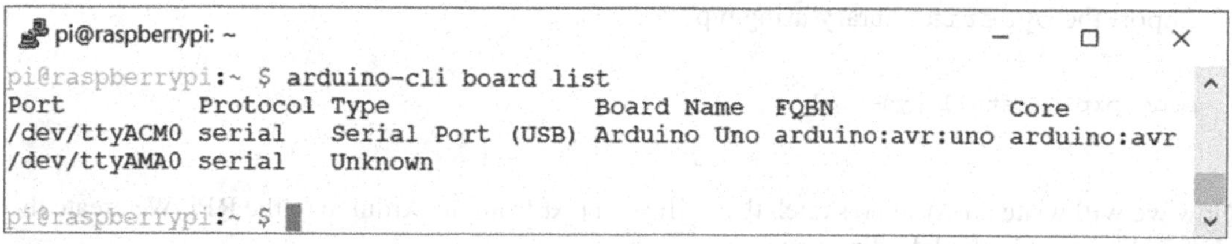

```
pi@raspberrypi: ~                                          —    □    ×
pi@raspberrypi:~ $ arduino-cli board list
Port           Protocol Type            Board Name FQBN         Core
/dev/ttyACM0 serial     Serial Port (USB) Arduino Uno arduino:avr:uno arduino:avr
/dev/ttyAMA0 serial     Unknown

pi@raspberrypi:~ $ █
```

FIGURE 2.21 Listing the devices.

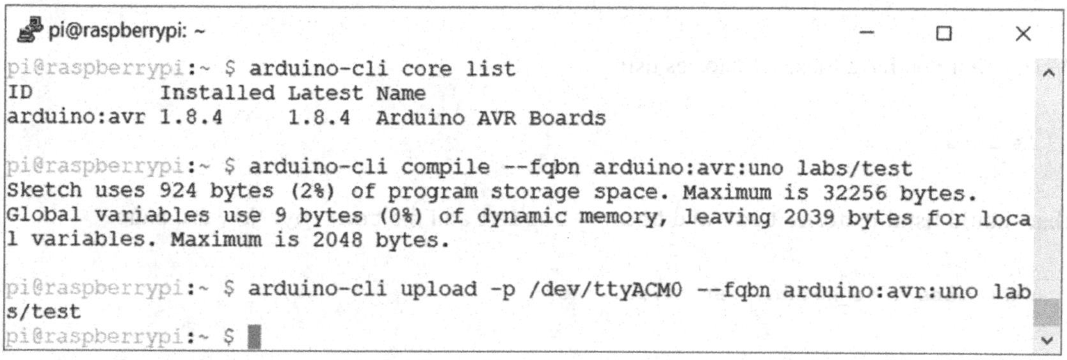

```
pi@raspberrypi: ~                                          —    □    ×
pi@raspberrypi:~ $ arduino-cli core list
ID            Installed Latest Name
arduino:avr 1.8.4       1.8.4  Arduino AVR Boards

pi@raspberrypi:~ $ arduino-cli compile --fqbn arduino:avr:uno labs/test
Sketch uses 924 bytes (2%) of program storage space. Maximum is 32256 bytes.
Global variables use 9 bytes (0%) of dynamic memory, leaving 2039 bytes for loca
l variables. Maximum is 2048 bytes.

pi@raspberrypi:~ $ arduino-cli upload -p /dev/ttyACM0 --fqbn arduino:avr:uno lab
s/test
pi@raspberrypi:~ $ █
```

FIGURE 2.22 Core list, compile, and upload.

Now we compile the sketch using Arduino-cli (Figure 2.22).

```
arduino-cli compile --fqbn arduino:avr:uno labs/test
```

Once the sketch has been compiled, we upload it to the Arduino board (Figure 2.22).

```
arduino-cli upload -p /dev/ttyACM0 --fqbn arduino:avr:uno labs/test
```

Verify that the sketch works as expected; the LED should blink every second.

SERIAL COMMUNICATION ARDUINO AND RPI

Now we will learn how to establish serial communication between the RPi and the Arduino (WUSL 2022; Instructables 2022). For this purpose, we need to install additional Python libraries. Installing additional Python libraries is facilitated by pip, which is the package manager for Python. It allows you to install and manage additional packages that are not part of the Python standard library

```
sudo apt install python3-pip
```

This install may take several minutes. Be patient. Your terminal informs you of progress, and later a percent of progress, followed by successive pound signs. Once it reaches 100% and shows finalized, you have installed the pip facility to install python libraries.

Import the pyserial library using pip

```
sudo pip3 install pyserial
```

Now we will write an Arduino sketch that will send text from the Arduino to the RPi. We create the sketch directory in the labs directory

```
arduino-cli sketch new labs/test_serial
```

Verify that you have these directories using

```
ls labs
```

that should list directories test and test _ serial and file test.py. Now use nano

```
nano labs/test_serial/test_serial.ino
```

the nano editor would show an empty sketch and you can add this content to setup() and loop()

```
void setup() {
Serial.begin(9600);                    //Starting serial communication
}

void loop() {
   Serial.println("Hello World from Arduino");    // print text
   delay(2000);                        // wait 2 sec
}
```

which would show as in Figure 2.23. Press Ctrl-X and save. Compile and upload the sketch

```
arduino-cli compile --fqbn arduino:avr:uno labs/test_serial
arduino-cli upload -p /dev/ttyACM0 --fqbn arduino:avr:uno labs/test_serial
```

producing results as in Figure 2.24.

Create a Python script to read the incoming serial data from the Arduino. Use the nano editor,

```
nano labs/test_serial.py
```

and type the following Python code in nano (Figure 2.25).

```
import serial
ser = serial.Serial('/dev/ttyACM0',9600)
while 1:
    if(ser.in_waiting >0):
        line = ser.readline()
        print(line.decode('utf-8'),end= "")
```

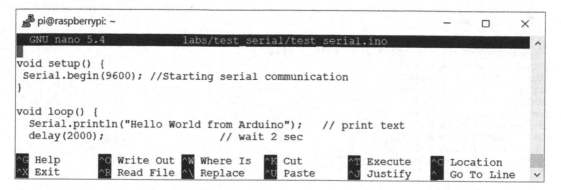

FIGURE 2.23 Simple Arduino test serial code.

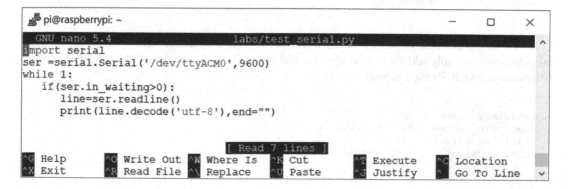

FIGURE 2.24 Compile and upload.

Here, we import module `serial` to handle serial communication, from the Arduino at port `tty-ACM0`, with 9600 baud. Then, we start a constantly running loop using `while`, which monitors the serial port and every time there are incoming data, it reads one line and prints it after decoding text as utf-8. Note that the print function uses an argument end="" to end with a blank, because by default print will add an extra line to the output.

Now, use `Ctrl-X`, save, and run it with `python3`

```
python3 labs/test_serial.py
```

FIGURE 2.25 Python script to obtain serial output.

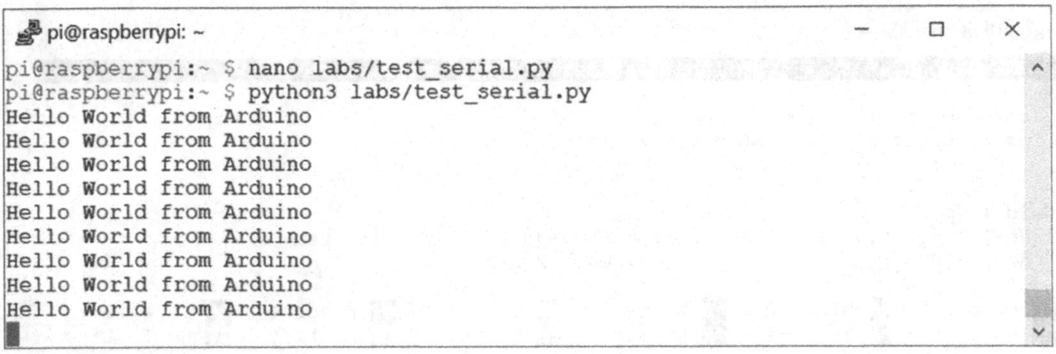

```
pi@raspberrypi: ~                                    —    □    ✕
pi@raspberrypi:~ $ nano labs/test_serial.py
pi@raspberrypi:~ $ python3 labs/test_serial.py
Hello World from Arduino
Hello World from Arduino
Hello World from Arduino
Hello World from Arduino
Hello World from Arduino
Hello World from Arduino
Hello World from Arduino
Hello World from Arduino
Hello World from Arduino
```

FIGURE 2.26   Serial output streamed at the RPi using Python script `test _ serial.py`.

The test message will appear repeated every two seconds on the RPi terminal (Figure 2.26). You can interrupt this using `Ctrl-C`.

ASSIGNING A STATIC IP ADDRESS

This is a more advanced exercise, and it may be skipped without impacting the rest of the lab guides. As mentioned above, the RPi connected to the LAN using DHCP (Dynamic Host Configuration Protocol) which assigns a *dynamic* IP to the RPi. This means the IP may differ every time the RPi connects to the network. Since we will be requiring connecting to the RPi often, it is convenient to assign a *static* IP, which will not vary as we disconnect and connect to the LAN.

A network router has an area reserved for DHCP, say, for example, 192.168.1.2 to 192.168.1.99. Thus, we can pick an address outside the range when assigning a static address, for example, 192.168.1.132. To find out the DHCP range of IP addresses for the router, you can use a browser and in the URL box type the IP address of your router, e.g., http://192.168.1.1. Sign in with administration username and password credentials of your router, which are provided by the manufacturer and typically listed online; for example, NETGEAR, D-Link, Linksys, and Cisco routers links for credentials can be obtained from the web page by Livewire (2023a). For instance, the default IP address, username, and password for NETGEAR router by model are available from Livewire (2023b). Once you login to the router as administrator, you will go to the DHCP setup, which is typically under a tab for advanced setup and LAN or network. You will see the starting and ending IP address for the DHCP server. Make a note of this to select a static address for the RPi outside this range.

To assign a static address to the RPi, we edit the DHCP configuration file

```
sudo nano /etc/dhcpcd.conf
```

You will see an example of static address with lines preceded by a comment symbol, we can edit this segment or simply add the following to the end of the file (here instead of 192.168.1.132 type an IP outside the DHCP range of your router)

```
interface wlan0
static ip_address=192.168.1.132/24
static routers=192.168.1.1
static domain_name_servers=192.168.1.1
```

and now reboot the RPi.

```
sudo reboot
```

Verify that the RPi is now connected to the LAN with that address using `arp-a` from the command window and reconnect to the RPi using PuTTY.

EXERCISES

Exercise 2.1 Eight-hour ozone.
   Calculate eight-hour ozone following the procedure explained in the guide and perform the calculation for the data files `data/ozone24hsample1.txt` and `data/ozone-24hsample2.txt`.

Exercise 2.2 Simple Python script.
   Using the IDLE, write the Python script `singleloop.py` and execute. Show the results and explain the code.

Exercise 2.3 Arduino script.
   Run Arduino script `heat-fan` for at least one cycle of heating and cooling. Record your temperature readings in a CSV text file using Geany. Plot your data. Show your script, data file, and graph.

Exercise 2.4 RPi in headless mode adn SSH.
   Configure the RPi in headless mode. Run a terminal using ssh. Create the Python script `singleloop.py` and run it.

Exercise 2.5 HTML, PHP and data files.
   Transfer datasonde file to the RPi using pscp. Install the `nginx` web server. Run the HTML and PHP examples.

Exercise 2.6 Arduino CLI.
   Install the Arduino-CLI software in the RPi Zero W.

Exercise 2.7 Using Arduino from RPi.
   Connect and use the Arduino from RPi Zero W.

Exercise 2.8 Serial communication Arduino and RPi.
   Use Arduino-cli to establish serial communication between Arduino and RPi Zero W.

REFERENCES

Acevedo, M.F. 2024. *Real-Time Environmental Monitoring: Sensors and Systems - Textbook, Second Edition.* Boca Raton, FL: CRC Press, Taylor & Francis Group, 392 pp.

Advanced IP Scanner. 2023. *Scan a Network in Seconds.* Accessed January 2023. https://www.advanced-ip-scanner.com/.

Amazon. 2023a. KETOTEK Digital Thermometer Temperature Meter Gauge. Accessed January 2023. https://www.amazon.com/KETOTEK-Thermometer-Temperature-Waterproof-Fahrenheit/dp/B01DVSIYN2/ref=asc_df_B01DVSIYN2?tag=bngsmtphsnus-20&linkCode=df0&hvadid=79989588513714&hvnetw=s&hvqmt=e&hvbmt=be&hvdev=c&hvlocint=&hvlocphy=&hvtargid=pla-4583589114951828&th=1.

Amazon. 2023b. *KEYNICE Digital Thermometer, Temperature Sensor USB Power Supply, Fahrenheit Degree and Degrees Celsius Color LCD Display, High Accurate-Black*. Accessed January 2023. https://www. amazon.com/Keynice-Thermometer-Temperature-Fahrenheit-Accurate-Black/dp/B01H1RDJOI/ref=sr_ 1_5?keywords=keynice+digital+thermometer&qid=1675003178&sr=8-5.

ARDUINO. 2014. *Arduino Uno*. Accessed October. http://arduino.cc/en/Main/ArduinoBoardUno.

Arduino. 2016. *Download the Arduino Software*. Accessed August 2016. https://www.arduino.cc/en/Main/ Software.

Arduino. 2023a. *Arduino*. Accessed January 2023. https://www.arduino.cc/.

Arduino. 2023b. *Arduino-CLI*. Accessed January 2023. https://github.com/arduino/arduino-cli.

ASL Associates. 2018. *Calculating the 8-Hour Ozone Standard*. http://www.asl-associates.com/cal8hr.htm.

Circuit Basics. 2022. *How to Setup a Raspberry Pi without a Monitor or Keyboard*. Accessed January 2022. https://www.circuitbasics.com/raspberry-pi-basics-setup-without-monitor-keyboard-headless-mode/.

Core Electronics. 2022. *How to Setup Raspberry Pi Zero W Headless WiFi*. Accessed January 2022. https:// core-electronics.com.au/tutorials/raspberry-pi-zerow-headless-wifi-setup.html.

Desertbot. 2022. *Headless Pi Zero W Wifi Setup (Windows)*. Accessed January 2022. https://desertbot.io/blog/ headless-pi-zero-w-wifi-setup-windows.

Elegoo. 2023. *Arduino Kits*. Accessed January 2023. https://www.elegoo.com/collections/arduino-learning-sets.

Geany. 2023. *Geany - The Flyweight IDE*. Accessed January 2023. https://www.geany.org/.

Hostinger. 2022. *34 Linux Basic Commands Every User Should Know*. Accessed January 2022. https://www. hostinger.com/tutorials/linux-commands.

How-To Geek. 2022. *The Beginner's Guide to Nano, the Linux Command-Line Text Editor*. Accessed January 2022. https://www.howtogeek.com/howto/42980/the-beginners-guide-to-nano-the-linux-command-line-text-editor/.

Instructables. 2022. *Raspberry Pi - Arduino Serial Communication*. Accessed February 2022. https://www. instructables.com/Raspberry-Pi-Arduino-Serial-Communication/.

Linuxize. 2022. *How to Add a Directory to PATH in Linux*. Accessed February 2022. https://linuxize.com/post/ how-to-add-directory-to-path-in-linux/.

Livewire. 2023a. *How to Connect to Your Home Router as an Administrator*. Accessed February 2023. https:// www.lifewire.com/accessing-your-router-at-home-818205.

Livewire. 2023b. *NETGEAR Default Password List*. Accessed February 2023. https://www.lifewire.com/ netgear-default-password-list-2619154.

McKay, D. 2022. *37 Important Linux Commands You Should Know*. Accessed January 2022. https://www. howtogeek.com/412055/37-important-linux-commands-you-should-know/.

Notepad++. 2023. *What is Notepad++*. Accessed January 2023. https://notepad-plus-plus.org/.

Posit. 2023. *RStudio IDE*. Accessed January 2023. https://posit.co/downloads/.

PuTTY. 2023. *Download PuTTY*. https://www.putty.org/.

Python. 2022a. *Python*. Accessed January 2022. https://www.python.org/.

Python. 2022b. *The Python Tutorial*. Accessed January 2022. https://docs.python.org/3/tutorial/.

R Project. 2023. *The Comprehensive R Archive Network*. Accessed January 2023. http://cran.us.r-project.org/.

Raspberry Pi. 2023. *Raspberry Pi OS*. Accessed January 2023. https://www.raspberrypi.com/software/.

US EPA. 2018. *Air Data: Air Quality Data Collected at Outdoor Monitors Across the US*. Accessed January 2018. https://www.epa.gov/outdoor-air-quality-data.

Vorillaz. 2022. *Raspberry Pi Zero W Headless Setup*. Accessed January 2022. https://dev.to/vorillaz/ headless-raspberry-pi-zero-w-setup-3llj.

WinSCP. 2023. *Free Award-Winning File Manager*. Accessed January 2023. https://winscp.net/eng/index.php.

WUSL. 2022. *Serial Communication between Raspberry Pi & Arduino*. Accessed February 2022. https://classes.engineering.wustl.edu/ese205/core/index.php?title=Serial_Communication_ between_Raspberry_Pi_%26_Arduino.

3 Sensors and Transducers: Basic Circuits

INTRODUCTION

In this lab session, we will use R to analyze and design transducers based on active and passive sensors. Specifically, we will study (1) an example of an active sensor: temperature transducers based on a thermistor and a voltage divider circuit and (2) an example of a passive sensor: a thermocouple. We introduce static specifications, i.e., when the sensor is at steady state conditions, which means when the measurand is not changing. We will discuss sensitivity and linearity as we work on the examples. We will also work hands-on implementing transducers using thermistors and voltage divider circuits employing an Arduino for the experimental setup. We learn how to read the transducer input using Arduino, how to capture Arduino output from a serial port, and how to read and use the captured text files to produce time series.

MATERIALS

READINGS

For theoretical background, you can use Chapter 3 of the book Acevedo, M.F. 2024. *Real-Time Environmental Monitoring: Sensors and Systems - Textbook, Second Edition* (Acevedo 2024), which is a companion to this lab manual. Other bibliographical references are cited throughout the guide.

COMPONENTS

- Protoboard and jumper wires
- Resistors (two 10 kΩ, 1/8 W, 1%) to be used in the voltage divider circuit
- Thermistor (10 kΩ nominal) to be used as a sensor
- Higher power 50 Ω resistor (1 W) to be used as a heater
- Small cooling fan (5 V DC) to be used to cool the sensor

MAJOR COMPONENTS AND INSTRUMENTS

- Raspberry Pi (RPi). The guide is written based on a Raspberry Pi Zero W. But you can use a Raspberry Pi 3 or 4.
- Micro SD and adapter (e.g., 16 GB)
- Arduino UNO R3 with USB Cable
- Digital temperature probes, e.g., KETOTEK (Amazon 2023) or KEYNICE Digital Thermometer
- Digital multimeter, e.g., ADM02 DMM or CAMWAY DMM 6000 (with thermocouple temperature probe)

DOI: 10.1201/9781003184362-3

TOOLS (RECOMMENDED)

- Long nose pliers
- Wire strippers and clippers
- Alligator clips

SOFTWARE

- R, for data analysis (R Project 2023)
- RStudio, an IDE to use R (Posit 2023)
- Advanced IP scanner, to find IP addresses (Advanced IP Scanner 2023)
- PuTTY, for Secure Shell (ssh) (PuTTY 2023)
- WinSCP, to transfer files from the RPi to a windows PC (WinSCP 2023)
- Raspberry Pi imager, to install the Raspberry Pi OS (Raspberry Pi 2023)
- Python, Shell, and IDLE (Python 2022)
- Arduino IDE, Interface to Arduino (Arduino 2016)
- Geany, IDE to edit programs as well as data (Geany 2023)
- Arduino-CLI, Arduino software to run using commands from the RPi (Arduino 2023)
- PHP, for RPi.
- CoolTerm for serial stream capture (Meier 2023)

SCRIPTS AND DATA FILES

All available from the GitHub RTEM repository https://github.com/mfacevedol/rtem

- R scripts contained in archive `R-temp-transducers.zip`
- Arduino scripts contained in archive `arduino-temp-transducers.zip`
- Python scripts `python-rpi.zip`
- JavaScript using D3.js in `plotd3.zip`
- Data file `thermocoupleJ-inverse-coeff.txt` contained in archive `data.zip` (already used for labs 1 and 2)

SUPPLEMENTARY SUPPORT MATERIAL

Supplementary support material including additional screenshots, images, and procedures are available from the publisher eResources web page provided for this book.

THERMISTOR RESISTANCE AS A FUNCTION OF TEMPERATURE

We will examine resistance as a nonlinear function of temperature following the B parameter model

$$R = R_0 \exp\left(B\left(T^{-1} - T_0^{-1}\right)\right) \tag{3.1}$$

where T is the temperature in K, R is the thermistor resistance in Ω, T_0 is the nominal value of T (25°C = 298 K), R_0 is the nominal value of R at T_0, and B is a parameter in units of K (Acevedo 2024).

First, build the following R function `B.param` and store in your R folder as `thermistor-functions.R`. We will be adding more functions to this file as we continue the lab session. You can also download this file from the GitHub RTEM repository. The argument `Rt.nom` is a list with the nominal temperature T_0 and resistance at nominal R_0. The argument `T.C` is temperature T in °C, and B is the B coefficient. The function converts T from °C to K, applies equation (3.1), and returns a list of thermistor resistance in kΩ and temperature in °C.

```
B.param <- function(Rt.nom,T.C,B){
 # B parameter model
 # Rt.nom is list(T0.C, R0.k)
 T.K <- T.C + 273; T0.K <- Rt.nom$T0.C+273
 Rt.k <- Rt.nom$R0.k*exp(B*(1/T.K - 1/T0.K))
 return(list(Rt.k=round(Rt.k,2),T.C=T.C))
}
```

Now open a new script lab3.R to type R code for this lab session. Source the thermistor-functions.R so that you can use the function.

```
# edit according to the path of your working folder
source("R/lab3/thermistor-functions.R")
```

We can obtain a graph of resistance vs. temperature, for nominal 10 kΩ at 25°C, in the 0°C–50°C, with $B=4100$, for the B model using the following code. The abline function allows drawing vertical and horizontal lines at desired locations on the axis.

```
# temp range moderate 0 to 50
Rt.nom <- list(T0.C = 25, R0.k=10)
T.C=seq(0,50,0.1);B=4100
RT <- B.param(Rt.nom,T.C,B)
plot(RT$T.C, RT$Rt.k, type="l", xlab="Temp(°C)", ylab="Rt(kΩ)")
abline(v=Rt.nom$T0.C,lty=2,col="grey")
abline(h=Rt.nom$R0.k,lty=2,col="grey")
```

As we narrow down the temperature range, say 20°–30°, the thermistor response would appear nearly linear because of the smaller variation of resistance with temperature. See Figure 3.1. We can produce this figure with the following code.

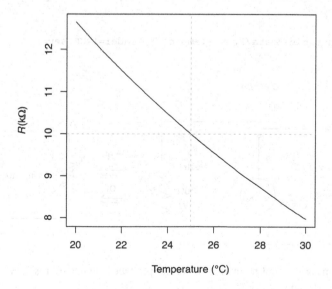

FIGURE 3.1 B model thermistor resistance vs. temperature in the 20°C–30°C range.

```
# narrower range
T.C=seq(20,30,0.1);B=4100
RT <- B.param(Rt.nom,T.C,B)
plot(RT$T.C, RT$Rt.k, type="l", xlab="Temp(°C)", ylab="Rt(kΩ)")
abline(v=Rt.nom$T0.C,lty=2,col="grey")
abline(h=Rt.nom$R0.k,lty=2,col="grey")
```

MEASURING THERMISTOR RESISTANCE AS A FUNCTION OF TEMPERATURE

We will systematically measure temperature and thermistor resistance to verify thermistor response. For this purpose, we re-use the heat and fan setup developed in Lab 2. Recall that we produce heat from a 50 Ω (1 W) resistor and cool down using a fan, and both devices are powered on and off by Arduino digital pins which can deliver 5 V (V_S in Figure 3.2) when HIGH (limited to 40 mA). Recall that you will find the 5 V and GND pins under the "power" label and the digital pins in the DIGITAL (PWM~) area.

As we learned in Lab 2, to measure temperature, you will use the KETOTEK or KEYNICE Digital Thermometer (which itself is based on a thermistor) in the degree Celsius (°C) scale and powered using a USB adapter. To measure temperature, use the probe close to the thermistor to be tested and the heater resistor, making thermal contact. Connect the DMM in ohmmeter mode to measure the resistance of the test thermistor. Pictures are available in the supplementary material.

We will also re-use the Arduino program to run the heater and fan developed in Lab 2. You can write the code or download from the RTEM GitHub repository. When running the code, we will read the temperature as in Lab 2; however, in addition, we will read the thermistor resistance. We can make comments in the code to alert the user to read the thermistor resistance as well. After saving, compile, upload, and click on the serial monitor using shortcuts or menu items.

Measure each temperature using the temperature probe and resistance using the ohmmeter setting of the digital multimeter. Vary the temperature to cover the range from 20°C to 30°C and include measurement of the reference ambient temperature T_0 of 25°C. You can type the values in Geany as two columns separated by commas (Figure 3.3). When you finish measuring, save the file as `Rt-vs-Temp.csv` in your data folder of your working directory. Once you have this data file, we can use the following line code to read the data.

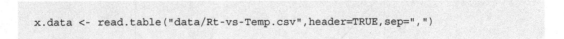

```
x.data <- read.table("data/Rt-vs-Temp.csv",header=TRUE,sep=",")
```

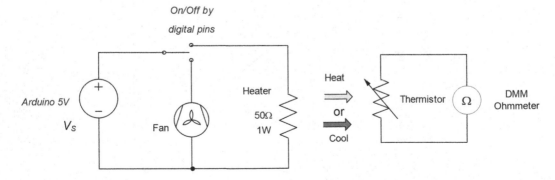

FIGURE 3.2 Warming or cooling a thermistor to produce change of resistance measured by the ohmmeter.

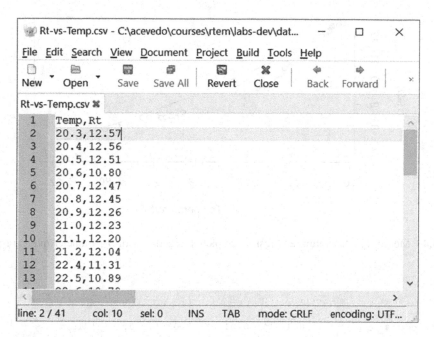

FIGURE 3.3   Data recorded as text using Geany.

Now, we use linear regression based on logarithmic transformation to obtain a statistical estimate B for this thermistor. As discussed in Chapter 3 of the companion textbook (Acevedo 2024), the parameter B can be estimated from data of R and T using linear regression. For this purpose, denote $x = \dfrac{1}{T} - \dfrac{1}{T_0}$ as an independent variable and $y = \ln\left(\dfrac{R}{R_0}\right)$ as a dependent variable, then $y = Bx$, and therefore B can be estimated by linear regression through the origin. Now, we write R code to perform this task

```
y <- log(x.data$Rt/Rt.nom$R0.k)
x <- 1/(x.data$Temp+273) -1/(Rt.nom$T0.C+273)
plot(x,y)
summary(lm(y ~ 0+x))
```

which yields the following results.

```
> summary(lm(y ~ 0+x))

Call:
lm(formula = y ~ 0 + x)

Residuals:
      Min        1Q     Median        3Q       Max
-0.130018 -0.084865 -0.051045 -0.004125  0.021700

Coefficients:
  Estimate Std. Error t value Pr(>|t|)
x     4116        278   14.81   <2e-16 ***
```

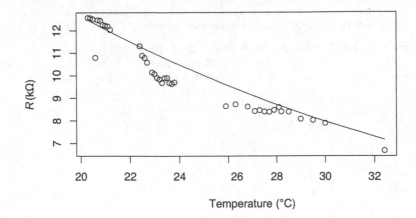

FIGURE 3.4 Measured temperature and resistance plotted together with *B*-parameter model response for *B* = 4116.

```
---
Signif. codes:  0 '***' 0.001 '**' 0.01 '*' 0.05 '.' 0.1 ' ' 1

Residual standard error: 0.0641 on 38 degrees of freedom
Multiple R-squared:  0.8523,  Adjusted R-squared:  0.8484
F-statistic: 219.2 on 1 and 38 DF,  p-value: < 2.2e-16
```

We have determined that the *B* parameter value for this thermistor is 4116, with an excellent p-value of 2×10^{-16} and a relatively good R^2 of ~0.85. We will use this estimated value $B = 4116$ for modeling the thermistor. We can use the following code to plot the data together with the response of the *B* model for $B = 4116$ (Figure 3.4).

```
plot(x.data$Temp,x.data$Rt, xlab="Temp(°C)", ylab="Rt(kΩ)")
x.model <- B.param(Rt.nom,T.C=x.data$Temp,B=4116)
lines(x.model$T.C,x.model$Rt.k)
```

SIMPLE TRANSDUCER: VOLTAGE DIVIDER

Consider a simple transducer build from a thermistor sensor (Figure 3.5): a voltage divider circuit where the sensor is connected to a fixed resistor *Rf* and a resistor R_1 which compensates for large differences in resistance for a wide temperature range. The transducer output voltage V_f is related to the input voltage source V_s by

$$V_{\text{out}} = V_f = V_s \frac{R_f}{R_t + R_1 + R_f}$$

Next, we add a function to calculate the output voltage, given the circuit parameter values, to the file `thermistor-functions.R` (also available from the RTEM repository).

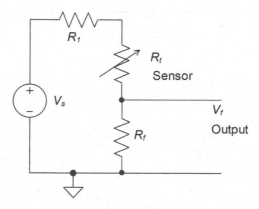

FIGURE 3.5 Voltage divider circuit with sensor and additional resistor R_1.

```
ckt.div <- function(T.C,Rt.nom,B,Rf.k,R1.k,Vs){
 Rt.k <- B.param(Rt.nom,T.C,B)$Rt.k
 R <- R1.k + Rt.k + Rf.k
 Vout <- Vs*Rf.k/R
 Pow <- (Vs/R)^2*Rt.k
 return(list(T.C=T.C,Vout=round(Vout,3),Pow.mW=round(Pow,3)))
}
```

We also calculate power (in mW) dissipated by the thermistor because it will be relevant to calculate self-heating later. We now call this function to calculate V_{out} for several values of design R_f and plot vs. temperature.

```
T.C=seq(0,50,0.1)
Rf.k=c(5,10,15); R1.k <- 10; Vs=5; B=4116;kt=1.5
y <- sapply(Rf.k, function(Rf.k) ckt.div(T.C,Rt.nom,B,Rf.k,R1.k,Vs)$Vout)
matplot(T.C,y,type="l",ylim=c(0,Vs),lty=1:3,col=1,lwd=2,xlab="Temp
(°C)",ylab="Vout (V)")
legend('topleft',leg=paste("Rf=",Rf.k,"kΩ"),lty=1:3,col=1)
abline(v=Rt.nom$T0.C,lty=2,col="grey")
abline(h=Vs/2,lty=2,col="grey")
```

The results are shown in Figure 3.6. Then, we use the result obtained from the ckt.div function to two other functions specs.transd to calculate the transducer specifications and self.heat to calculate the temperature offset due to the self-heating effect. Recall from Chapter 3 of the companion textbook the calculations of full scale (FS), sensitivity and linearity error. These specifications are calculated in the specs.transd function below. Add this function to your file thermistor-functions.R (also available from the RTEM repository).

```
specs.transd <- function(x){
 #T.C is seq(min,max,step)
 #Vout is output from ckt div
```

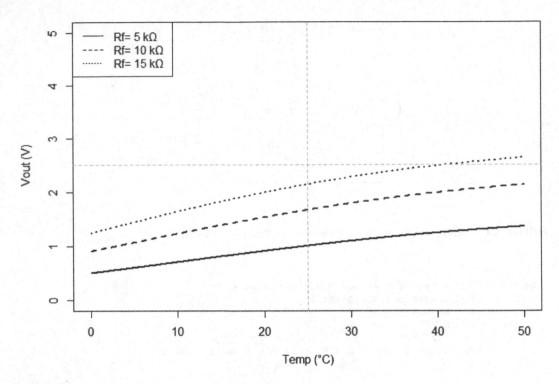

FIGURE 3.6 V_{out} vs. temperature for $V_s = 5$ V, $R_1 = 10$ kΩ, and for several values of R_f.

```
T.C <- x$T.C;  Vout <- x$Vout
xlm <- lm(Vout~T.C)
Vout.lsq <- xlm$coefficients[1]+xlm$coefficients[2]*T.C

nT <- length(T.C); range.C <- T.C[nT]-T.C[1]
FS <- Vout[nT]-Vout[1]
sens <- FS/range.C
# linearity end-points
Vout.end <- sens*(T.C-T.C[1]) + Vout[1]

# max deviance of response with respect to linear
dev.lsq <- abs(Vout-Vout.lsq); max.dev.lsq <- max(dev.lsq)
dev.end <- abs(Vout-Vout.end); max.dev.end <- max(dev.end)
lin.err.lsq <- 100*max.dev.lsq/FS
lin.err.end <- 100*max.dev.end/FS

rms.lsq <- sqrt(sum((Vout-Vout.lsq)^2)/nT)
rms.end <- sqrt(sum((Vout-Vout.end)^2)/nT)
rms.err.lsq <- 100*rms.lsq/FS
rms.err.end <- 100*rms.end/FS

result <- list(FS.mV=round(FS*1000,2),range.C=round(range.C,2),
               sens.mV=round(sens*1000,2),
               lin.err.lsq=round(lin.err.lsq,2),
               lin.err.end=round(lin.err.end,2),
               rms.err.lsq=round(rms.err.lsq,2),
               rms.err.end=round(rms.err.end,2),
```

```
                    Vout=round(Vout,5),T.C=T.C,
                    Vout.end=round(Vout.end,5),Vout.lsq=round(Vout.lsq,5),
                    dev.lsq=round(dev.lsq,5),dev.end=round(dev.end,5))
  return(result)
}
```

The resulting object obtained after executing `specs.transd` will be used by a function `specs.plot` to plot the results.

```
specs.plot <- function(X){

 matplot(X$T.C, cbind(X$Vout,X$Vout.end,X$Vout.lsq),type="l", col=1,
xlab="Temp(°C)",ylab="Vout(V)")

 pos <- which(X$dev.end==max(X$dev.end))
 lines(c(X$T.C[pos],X$T.C[pos]),c(X$Vout.end[pos],X$Vout[pos]),lty=2,
lwd=2,col="grey")

 pos <- which(X$dev.lsq==max(X$dev.lsq))
 lines(c(X$T.C[pos],X$T.C[pos]),c(X$Vout.lsq[pos],X$Vout[pos]),lty=3,
lwd=2,col="grey")

 mtext(text=paste("FS(mV)=",X$FS.mV," Sens(mV/°C)=",X$sens.mV,
                  " Lin Err Ends(%FS)=", X$lin.err.end,
                  " Lin Err LSQ(%FS)=", X$lin.err.lsq),
                  side=3,line=-1,cex=0.6)
 mtext(text=paste(" Lin Err RMS Ends(%FS)=", X$rms.err.end,
                  " Lin Err RMS LSQ(%FS)=", X$rms.err.lsq),
                  side=3,line=-2,cex=0.6)
 legend("bottomright",leg=c("Actual","Linear Ends", "Linear
LSQ"),lty=1:3,col=1,cex=0.8)

}
```

Now in the `lab3.R` script, we source the functions and write code to execute them, which calculates sensitivity and linearity.

```
source("R/lab3/thermistor-functions.R")
# edit according to the path of your working folder
Rf.k=10; R1.k=10; Vs=5; B=4116;kt=1.5
x<- ckt.div(T.C,Rt.nom,B,Rf.k,R1.k,Vs)
X<- specs.transd(x)
specs.plot(X)
```

The result of `specs.plot` is shown in Figure 3.7, illustrating that the response is nonlinear over this range of temperature. We illustrate the departure of the response with respect to the linear one. We see that the sensitivity is 24.64 mV/°C and end-points maximum linearity is poor (12.21% of FS).

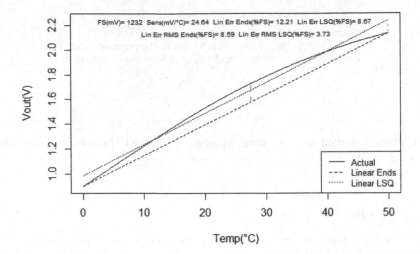

FIGURE 3.7   Response of voltage divider circuit annotated for FS, sensitivity, and linearity error.

To calculate self-heating, we use another function

```
self.heat <- function(x,kt){
 Vout <- x$Vout
 Ta.C <- x$T.C - x$Pow.mW/kt
 panels(6,6,2,1,pty="m")
 plot(x$T.C,x$T.C,type="l",xlab="Temp(°C)",ylab="Temp(°C)")
 lines(x$T.C,Ta.C,lty=2)
 legend("bottomright",leg=c("1:1 line","Actual"),lty=1:2,col=1,cex=0.8)

 matplot(X$Vout, cbind(x$T.C,Ta.C),type="l", col=1,
xlab="Vout(V)",ylab="Temp(°C)")
 legend("bottomright",leg=c("Tt","Ta"),lty=1:2,col=1,cex=0.8)
 }
```

Which we can apply to the results of the `ckt.div` function

```
self.heat(x,kt)
```

When you examine the resulting plot, the effect of self-heating in this case is not noticeable due to the low V_s voltage. To illustrate this concept, let us calculate the effect of self-heating using a voltage source of $V_s = 12\,V$ (Figure 3.8). We can see how T_a will be consistently under-estimated because the temperature of the thermistor is higher. To further illustrate the need for correction where we see the difference between T_a and T_t as a function of V_f once we select a voltage source V_s and a R_f resistor value. In this case, we can see the gap between T_a and T_t showing again the underestimation of T_a.

The appendix to this lab guide, available from the supplementary material, contains information on how all these concepts apply to a commercial probe.

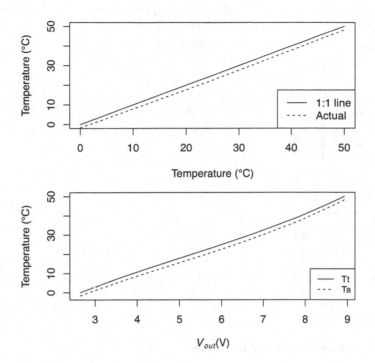

FIGURE 3.8   Self-heating effect for $V_s = 12$ V.

IMPLEMENTING THE TRANSDUCER

Implement the circuit of Figure 3.5 with the thermistor, and the design values $R_f = 10$ kΩ and $R_1 = 10$ kΩ. Test the response using $V_s = 5$ V. Increase and decrease temperature gradually as in the previous exercise by using a 50 Ω (1 W) resistor and a fan connected and disconnected to the 5-V power supply by means of the Arduino pins.

As a first test of your transducer, measure the transducer output with the DMM together with temperature measured by the temperature probe. Cover the range from 20°C to 30°C. One of the levels measured should be your reference ambient temperature T_0, the other should be below and above T_0. Prepare a csv file like the one shown in Figure 3.9. Save this file as `vout-vs-temp.csv`. Now you can read and plot measured response using the following code in your lab3.R.

```
x <- read.table("data/vout-vs-temp.csv",header=TRUE,sep=",")
plot(x$Temp,x$Vout)
```

Next, we will see how to perform these readings using the Arduino analog pin A0. Connect the output of the transducer to pin A0. We will modify our previous script as `heat-fan-read-sensor`. This script is also available in the RTEM repository. In the declarations, we now include sampling time, `voutPin` as pin A0, the voltage $V_{AD} = V_s = 5.00$ V, voltage resolution for an ADC of 10 bits $V_{res} = \dfrac{V_s}{2^{10} - 1} = \dfrac{V_s}{1024 - 1} = \dfrac{V_s}{1023}$ (see Chapter 3), the value of resistors in the divider, and the thermistor parameters R_{t0}, T_0, and B.

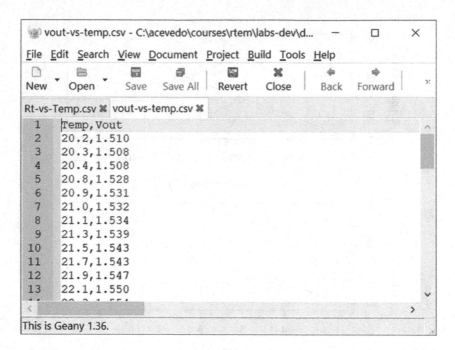

FIGURE 3.9 V_{out} and temperature recorded manually.

```
// control fan and heater
// to measure thermistor resistance vs temperature
int heaterPin = 10;
int fanPin =9;
// edit the following times in minutes as needed
int fanTime=10;
int heatTime=10;
int repeatTime=10;
// edit the following time in seconds as needed
int samplingTime = 10;

// analog pin to read the voltage
int voutPin = A0;
// voltage measure and edit if not 5.00 V
float Vs=5.0; // Vad for Arduino Uno
//volt resolution use ADC conv 2^10-1, 1024-1
float vRes = Vs/1023;
// voltage divider
float R1_k = 10.0;
float Rf_k=10.0;
//thermistor
float T0_K = 25.0+273.0;
float Rt0_k =10;
float B= 4116.0;
```

In the setup, we add `pinMode (voutPin, INPUT)` to configure the pin.

```
void setup(){
        pinMode(heaterPin, OUTPUT);
        pinMode(fanPin, OUTPUT);
  //analog set to input
  pinMode(voutPin,INPUT);
  Serial.begin(9600);
}
```

We add a new function to read the sensor instead of the wait _ min we used before.

```
void readSensor(int period_sec, int duration_min){
  Serial.print("will read the sensor every ");
  Serial.print(period_sec); Serial.println(" sec");
  Serial.println("vout, Rt_k, T_C");
  int nreadings = (duration_min/period_sec)*60;
  for(int i=1; i<=nreadings;i++){
  //read digital number
  int din = analogRead(voutPin);
  //convert to voltage
  float vout= vRes*din;
  float Rt_k = Vs*Rf_k/vout - R1_k - Rf_k;
  float T_K =1/(1/T0_K+(1/B)*log(Rt_k/Rt0_k));
  float T_C = T_K - 273.0;
  Serial.print(vout); Serial.print(",");
  Serial.print(Rt_k); Serial.print(",");
  Serial.println(T_C);

  //repeat every period_sec
  delay(period_sec*1000);
  }
}
```

In this function readSensor, we read the analog pin as a digital number din (a result of ADC conversion), convert to voltage using the voltage resolution $V_{out} = d_{in} \times V_{res}$, apply a calculation of thermistor resistance from the voltage divider circuit $R_t = V_s \dfrac{R_f}{V_{out}} - R_f - R_1$, apply the B parameter equation to calculate temperature from thermistor resistance $T = \left(T_0^{-1} + B^{-1} \ln\left(\dfrac{R_t}{R_0} \right) \right)^{-1}$, and convert from K to Celsius degrees. After calculations, we send the output voltage V_{out}, thermistor resistance, and temperature to the serial monitor, separated by commas. Using this separator allows for writing a CSV file later. We modify the loop to execute the read sensor function instead of the wait _ min.

```
void loop(){
  Serial.println("making sure heater is off");
  digitalWrite(heaterPin,LOW);
  Serial.print("turning fan on for "); Serial.print(fanTime);
```

```
    Serial.println(" minutes");
    digitalWrite(fanPin,HIGH);
    readSensor(samplingTime,fanTime);

        Serial.println("turning fan off");
        digitalWrite(fanPin,LOW);

    Serial.print("turning heat on for "); Serial.print(heatTime);
    Serial.println(" minutes");
    digitalWrite(heaterPin,HIGH);
    readSensor(samplingTime,heatTime);

    Serial.print("it will repeat the loop in "); Serial.print
(repeatTime);
    Serial.println(" minutes");
    digitalWrite(heaterPin,LOW);
    Serial.println("it will repeat the loop in duration_minutes");
        readSensor(samplingTime,repeatTime);
}
```

Compile and upload. The Arduino will control the fan and heater, plus read the sensor and print on the serial monitor. Compare the values printed to the readings of the thermometer and multimeter (Figure 3.10).

FIGURE 3.10   Serial monitor output.

DISPLAYING GRAPHICS ON THE ARDUINO IDE SERIAL PLOTTER

Use the same circuit setup and Arduino script `heat-fan-read-sensor` and add comment symbols // to all lines sending commentary text to the serial print. Save as `heat-fan-read-sensor-plot` (Figure 3.11). Alternatively, you can use the file with the same name provided in the archive downloaded from the repository. Compile, upload, and select Tools and Serial Plotter. You should see graphs of voltage, thermistor resistance, and temperature. See Figure 3.12 for an example, where we have added text labels to identify the lines.

CAPTURING ARDUINO OUTPUT TO A TEXT FILE

Use the same circuit setup and Arduino script `heat-fan-read-sensor-plot` as in the previous exercise, but we will comment out the line printing the header and add a `Serial.print(",")` to the line printing the data, so that a comma will precede the data (we do this because we will add a timestamp

```
heat-fan-read-sensor-plot

  Serial.begin(9600);
}


void readSensor(int period_sec, int duration_min){
  // Serial.print("will read the sensor every ");
  // Serial.print(period_sec); Serial.println(" sec");
  Serial.println("vout, Rt_k, T_C");
  int nreadings = (duration_min/period_sec)*60;
  for(int i=1; i<=nreadings;i++){
    //read digital number
```

FIGURE 3.11 Arduino IDE: editing the sketch to comment out printing text other than data.

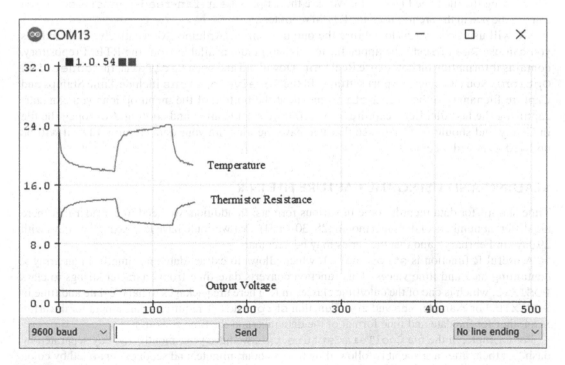

FIGURE 3.12 Arduino IDE: Serial Plotter graphs.

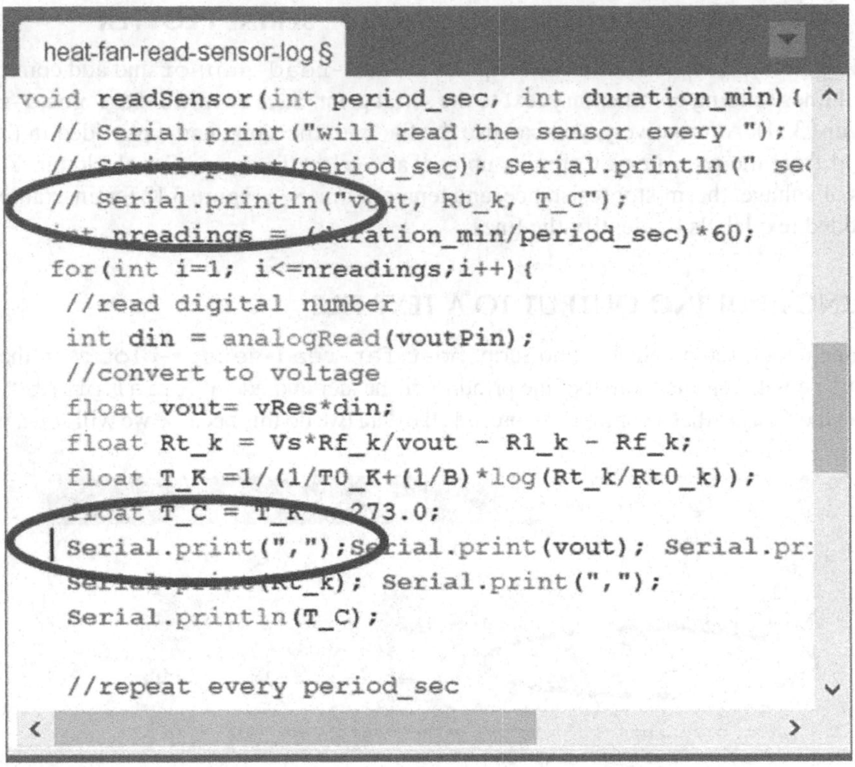

```
heat-fan-read-sensor-log §                              ▼

void readSensor(int period_sec, int duration_min){  ^
  // Serial.print("will read the sensor every ");
  // Serial.print(period_sec); Serial.println(" sec
  // Serial.println("vout, Rt_k, T_C");
  int nreadings = (duration_min/period_sec)*60;
  for(int i=1; i<=nreadings;i++){
    //read digital number
    int din = analogRead(voutPin);
    //convert to voltage
    float vout= vRes*din;
    float Rt_k = Vs*Rf_k/vout - R1_k - Rf_k;
    float T_K =1/(1/T0_K+(1/B)*log(Rt_k/Rt0_k));
    float T_C = T_K - 273.0;
    Serial.print(",");Serial.print(vout); Serial.pr:
    Serial.print(Rt_k); Serial.print(",");
    Serial.println(T_C);

    //repeat every period_sec                          v
<                                                       >
```

FIGURE 3.13 Removing the header and adding a preceding comma to the record.

when saving the file). See Figure 3.13. We save the script as `heat-fan-read-sensor-log`, or you can use the one in the archive from the RTEM repository.

We will use `CoolTerm` to capture the output from the Arduino. Alternatively, for Windows you can use `RealTerm`. The appendix to this lab guide, available from the RTEM repository, contains information on how to use RealTerm. Download and open `CoolTerm`. In `Connection Options`, you can enter the port settings. In the `Receive options` include Time Stamps and Capture filename. In the `Connection` menu, at the bottom of the menu options you can start capturing the text file. Let it capture for several cycles of heating and cooling. Now open the file in Geany and should be able to see that you have the file with your data (Figure 3.14). It will be updated as new data comes in.

READING AND USING THE CAPTURE FILE IN R

Time stamps for data records come in various formats. In addition, we need to consider that there are different numbers of days in a month (28, 30, or 31), that we could have leap years (i.e., days with 29 days in February), and that the series may be irregular.

A useful R function is `strptime( )`, which allows to extract date and time from an array `x` containing date and time values. This function converts date-time from character strings to class `POSIXlt`, which is one of the date/time classes in R. There are packages to handle date and time in `POSIXlt`. For example, `xts` and `zoo`. Function `strptime( )` requires an argument for the array `x`, another for the date and time format of the data in x, and the time zone (tz).

For example, in the file `CoolTermCapture.txt`, we have year, month, and day separated by dash "-"; then, after a space, it is followed by time as hour, minute, and seconds separated by colon ":". We can read the file and then use strptime as follows to generate the time stamps tt as a string array, which we can use as input to time series packages.

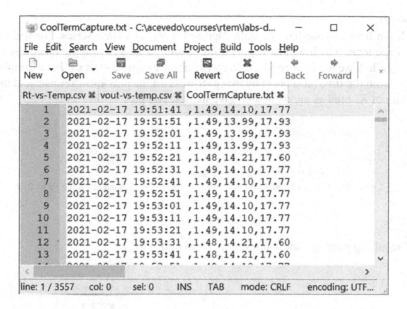

FIGURE 3.14 CoolTerm capture txt file.

```
X <- read.table("data/CoolTermCapture.txt", header=FALSE,sep=",")
# read time sequence from file
tt <- strptime(X[,1], format="%Y-%m-%d %H:%M:%S",tz="America/Chicago")
Y <- X[,2:4]
```

The resulting string array `tt` contains the time stamps

```
> tt
  [1] "2021-02-17 19:51:41 CST" "2021-02-17 19:51:51 CST" "2021-02-17
19:52:01 CST" "2021-02-17 19:52:11 CST"
  [5] "2021-02-17 19:52:21 CST" "2021-02-17 19:52:31 CST" "2021-02-17
19:52:41 CST" "2021-02-17 19:52:51 CST"
  [9] "2021-02-17 19:53:01 CST" "2021-02-17 19:53:11 CST" "2021-02-17
19:53:21 CST" "2021-02-17 19:53:31 CST"
 [13] "2021-02-17 19:53:41 CST" "2021-02-17 19:53:51 CST" "2021-02-17
19:54:01 CST" "2021-02-17 19:54:11 CST"
 [17] "2021-02-17 19:54:21 CST" "2021-02-17 19:54:31 CST" "2021-02-17
19:54:41 CST" "2021-02-17 19:54:51 CST"
```

And the series `Y` has the values of the variables which you can check printing a few rows

```
> Y[1:5,]
    V2    V3    V4
1 1.49 14.10 17.77
2 1.49 13.99 17.93
3 1.49 13.99 17.93
4 1.49 13.99 17.93
5 1.48 14.21 17.60
>
```

Or using `View()`.

Now, we learn how to visualize the data captured using as a time series. For this purpose, we use the R package `xts`, which we can install using Install packages from the Packages Menu of the RGui. Once installed, we load it using `library(xts)` or `require(xts)`. For example, we create an `xts` series, from the time base build by using strptime in the previous exercise, and the last three columns of data of X, bind as an xts object, give names, and plot

```
require(xts)
x.ts<- xts(Y,tt)
names(x.ts) <- c("Vout(V)","Rt(kΩ)","Temp(°C)")
plot(x.ts,ylab=names(x.ts),main="",multi.panel=TRUE,yaxis.same=FALSE)
```

Examples of resulting plots are shown in Figures 3.15 and 3.16.

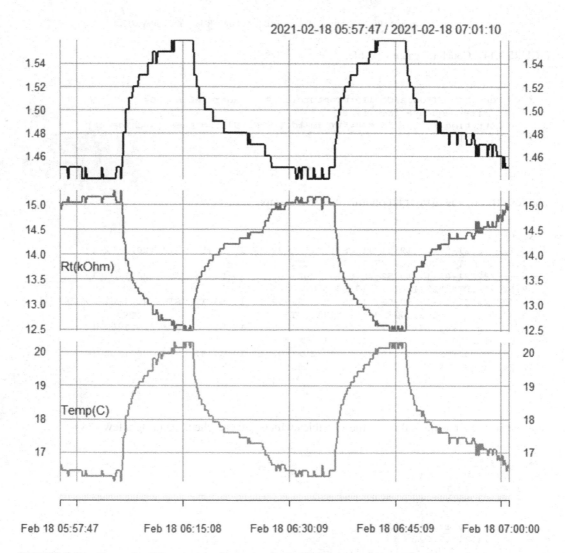

FIGURE 3.15   Example of time series of output voltage, thermistor resistances, and temperature.

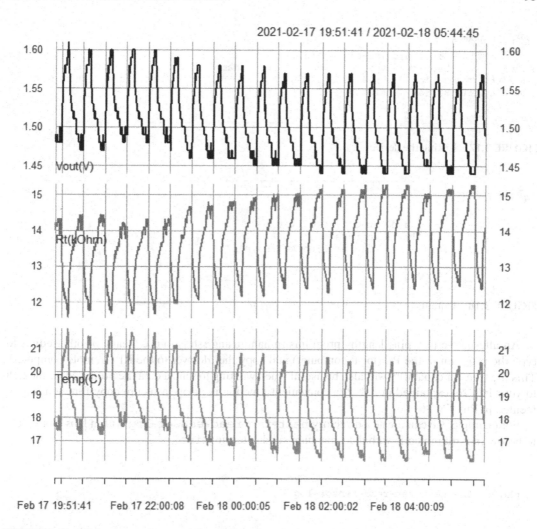

FIGURE 3.16   Time series of output voltage, thermistor resistances, and temperature.

USING ARDUINO CLI ON THE RPI AND ECHO OUTPUT USING PYTHON

Now, we are going to run the Arduino code on the RPi using the Command Line Interface or CLI, which we learned in Lab 2. First, connect the Arduino to the RPi using the micro-USB adapter or adapter cable. Now at the RPi ssh terminal using PuTTY we will type a command to identify the Arduino board connected to the RPi

```
arduino-cli board list
```

As shown in Figure 3.17, we have identified the board FQBN arduino:avr:uno and the core for this board arduino:avr, if you have not yet, install the core for this board as described in Lab 2. Verify installation (Figure 3.18).

```
arduino-cli core list
```

```
🐧 pi@raspberrypi: ~                                              —    □    ×
pi@raspberrypi:~ $ arduino-cli board list
Port            Protocol Type                Board Name  FQBN            Core
/dev/ttyACM0 serial    Serial Port (USB) Arduino Uno arduino:avr:uno arduino:avr
/dev/ttyAMA0 serial    Unknown

pi@raspberrypi:~ $
```

FIGURE 3.17 Listing the devices.

```
🐧 pi@raspberrypi: ~                                              —    □    ×
pi@raspberrypi:~ $ arduino-cli core list
ID           Installed Latest Name
arduino:avr 1.8.4     1.8.4  Arduino AVR Boards

pi@raspberrypi:~ $
```

FIGURE 3.18 Core list.

An alternative to typing lengthy programs in nano using ssh is to use Geany or other editor to type the program in the PC (local computer) and copy the file over to the RPi (remote computer). This file transfer capability will also be useful when wanting to send a data file produced on the RPi to your PC. As explained in Lab 2, a convenient function is `pscp`, a secure copy function implemented in PuTTY.

For example, we can create a directory `heat-fan-read-sensor-log` within labs on the RPi to house an Arduino code with the same name and extension `.ino`

```
mkdir labs/heat-fan-read-sensor-log
```

Assume we have a PC (local) `labs/arduino` directory with script `heat-fan-read-sensor-log`. We can use `pscp` to copy the `.ino` file within this local folder to the RPi. The asterisk `*` denotes a wildcard to catch any filename.

```
c:\labs\arduino>pscp heat-fan-read-sensor-log\*ino pi@192.168.1.132:labs/
heat-fan-read-sensor-log
pi@192.168.1.132's password:
heat-fan-read-sensor-log. | 2 kB |   2.3 kB/s | ETA: 00:00:00 | 100%
```

Alternatively, if using windows, we can use WinSCP to transfer the folder `heat-fan-read-sensor-log` to the RPi. To do this, download and open WinSCP, select File protocol as SCP. Then for hostname enter the RPi's IP address (or the Hostname you selected for the RPi), enter username, and password. You will see two panes. The left side is your Local directory (the PC in this case). The right side is the remote directory (the RPi in this case). WinSCP facilitates navigation within each side and copying the file from one side to the other. To facilitate repeated use of WinSCP, you can go to `Session|Save Workspace` and save the server information. For example, save as `rtem-labs`. The next time you open WinSCP, it will go directly to a login window. Press login and enter password.

Now at the ssh session, verify that the Arduino folder `heat-fan-read-sensor-log` was indeed transferred to labs. We are ready to compile the Arduino sketch

```
arduino-cli compile --fqbn arduino:avr:uno labs/heat-fan-read-sensor-log
```

Once it has been compiled, we can upload it

```
arduino-cli upload -p /dev/ttyACM0 --fqbn arduino:avr:uno labs/
heat-fan-read-sensor-log
```

We will use the python serial capture test_serial.py that we wrote in Lab 2

```
python3 labs/test_serial.py
```

and obtain the streaming data on the RPi (Figure 3.19).

Next, we will add a timestamp to the streaming data. The RPi Zero W does not have a real-time clock, but it will acquire the date and time information from the network. To make sure you use the correct time zone, you can adjust the configuration

```
sudo raspi-config
```

select localization and navigate to your time zone.

To add the timestamp to the data stream, we add a few lines to the `test _ serial.py` script. First, import the `datetime` module. Then within the loop, call `datetime.now()` and format the timestamp using `strftime()`. The format we will use is `%Y-%m-%d %H:%M:%S %Z`, which translates, for example, 2nd of February 2022 at 6:56:48 as the string 2022-02-02 06:56:48. The last `%Z` includes the time zone in the timestamp. Once we have a string for the timestamp, we concatenate (using "+") with the received string `linestr` to form a line of data.

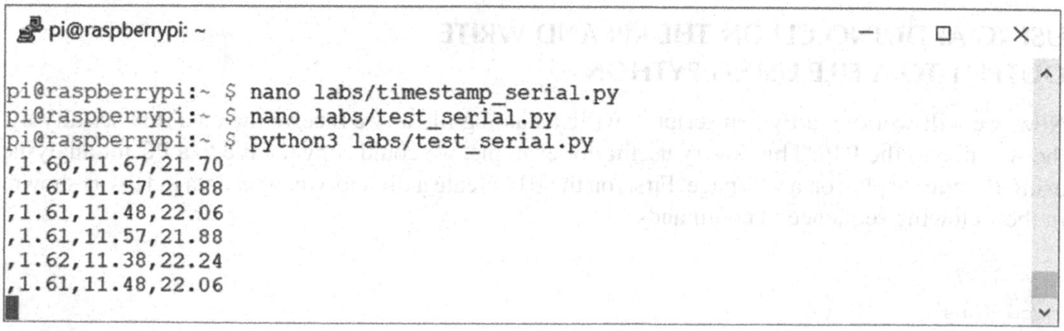

```
pi@raspberrypi: ~                                             —   □   ×

pi@raspberrypi:~ $ nano labs/timestamp_serial.py
pi@raspberrypi:~ $ nano labs/test_serial.py
pi@raspberrypi:~ $ python3 labs/test_serial.py
,1.60,11.67,21.70
,1.61,11.57,21.88
,1.61,11.48,22.06
,1.61,11.57,21.88
,1.62,11.38,22.24
,1.61,11.48,22.06
```

FIGURE 3.19   Check the test_serial output.

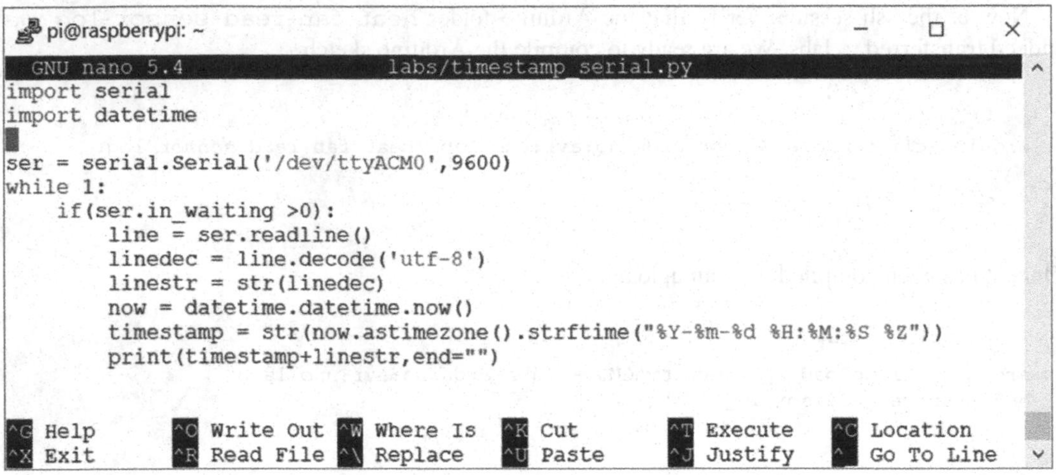

```
pi@raspberrypi: ~                                          —   □   ×
  GNU nano 5.4                    labs/timestamp_serial.py              ^
import serial
import datetime

ser = serial.Serial('/dev/ttyACM0',9600)
while 1:
    if(ser.in_waiting >0):
        line = ser.readline()
        linedec = line.decode('utf-8')
        linestr = str(linedec)
        now = datetime.datetime.now()
        timestamp = str(now.astimezone().strftime("%Y-%m-%d %H:%M:%S %Z"))
        print(timestamp+linestr,end="")



^G Help        ^O Write Out ^W Where Is  ^K Cut      ^T Execute  ^C Location
^X Exit        ^R Read File ^\ Replace   ^U Paste    ^J Justify     Go To Line  v
```

FIGURE 3.20   Python script to add timestamp to the data transmitted serially from the Arduino.

```
import serial
import datetime

ser = serial.Serial('/dev/ttyACM0',9600)
while 1:
    if(ser.in_waiting >0):
        line = ser.readline()
        linedec = line.decode('utf-8')
        linestr = str(linedec)
        now = datetime.datetime.now()
        timestamp = str(now.astimezone().strftime("%Y-%m-%d %H:%M:%S %Z"))
        print(timestamp+linestr,end="")
```

Save this new script in labs as `timestamp _ serial.py` (Figure 3.20) and run it

```
python3 labs/timestamp_serial.py
```

You will see that the timestamp has been added to each line of data (Figure 3.21).

USING ARDUINO CLI ON THE RPI AND WRITE OUTPUT TO A FILE USING PYTHON

Now, we will work on a Python script to write a datalog file as we receive the data sent serially by the Arduino to the RPi. This is very useful, for example, we could copy the file to a PC for analysis using R, or to display on a web page. First, on the RPi create a directory named data in labs as shown in the following sequence of commands

```
cd labs
mkdir data
cd ..
```

```
pi@raspberrypi: ~                                    —    □    ×
pi@raspberrypi:~ $ nano labs/timestamp_serial.py
pi@raspberrypi:~ $ python3 labs/timestamp_serial.py
2022-02-02 08:42:27 CST,1.60,11.67,21.70
2022-02-02 08:42:37 CST,1.60,11.67,21.70
2022-02-02 08:42:47 CST,1.60,11.67,21.70
2022-02-02 08:42:57 CST,1.60,11.67,21.70
2022-02-02 08:43:07 CST,1.60,11.67,21.70
2022-02-02 08:43:17 CST,1.60,11.77,21.52
2022-02-02 08:43:27 CST,1.61,11.57,21.88
```

FIGURE 3.21 The timestamp is now the first element of a line.

the last line would take you back to the home directory. Verify that you have the `data` directory by using `ls labs`. Next, use the nano editor and write a Python program `file _ serial.py` to write the data stream to a `datalog.csv` file

```
nano labs/file_serial.py
```

add content according to the next script (or download from the RTEM repository). After importing `os`, we import `datetime` and `timezone` from `datetime` (this allows to shorten code when invoking `datetime`). Then, define file name and the header. Package `os` is used to check whether the file exists or not and accordingly create the file and write the header (open with mode `"w"`) or just append records (open with mode `"a"`). The loop is like the previous script, but instead of the time zone code, we write UTC numeric offset for the time zone using `timezone.utc` and `%z` (note lowercase z). In the last two lines of the loop, we write a record to the file and flush.

```
import os
import serial
from datetime import datetime,timezone

logfile = "data/datalog.csv"
header = "TimeStamp,Vout,Rt,Temp\n"
if os.path.exists(logfile):
        print("File exists, appending records\n")
        file = open(logfile, "a")
else:
        print("File does not exists, writing header\n")
        print(header)
        file = open(logfile, "w")
        file.write(header)

ser = serial.Serial('/dev/ttyACM0',9600)
while 1:
    if(ser.in_waiting >0):
        line = ser.readline()
        linedec = line.decode('utf-8')
        linestr = str(linedec)
        now = datetime.now(timezone.utc)
        timestamp = str(now.astimezone().strftime("%Y-%m-%d %H:%M:%S %z"))
```

```
            print(timestamp+linestr,end="")
            file.write(timestamp+linestr)
            file.flush()
```

as shown in Figure 3.22.
    Run this python code

```
python3 labs/file_serial.py
```

the data stream will echo on the terminal and then written to a file (Figure 3.23). Without interrupting the process of writing the file, we can use an additional PuTTY session to examine the file with the nano editor (Figure 3.24)

```
nano labs/data/datalog.csv
```

In addition, we can copy it from the RPi to the PC using pscp

```
c:\labs>pscp pi@192.168.1.132:labs/data/datalog.csv data/datalog.csv
pi@192.168.1.132's password:
datalog.csv                    | 1 kB |    1.2 kB/s | ETA: 00:00:00 | 100%

c:\acevedo\courses\rtem\labs>
```

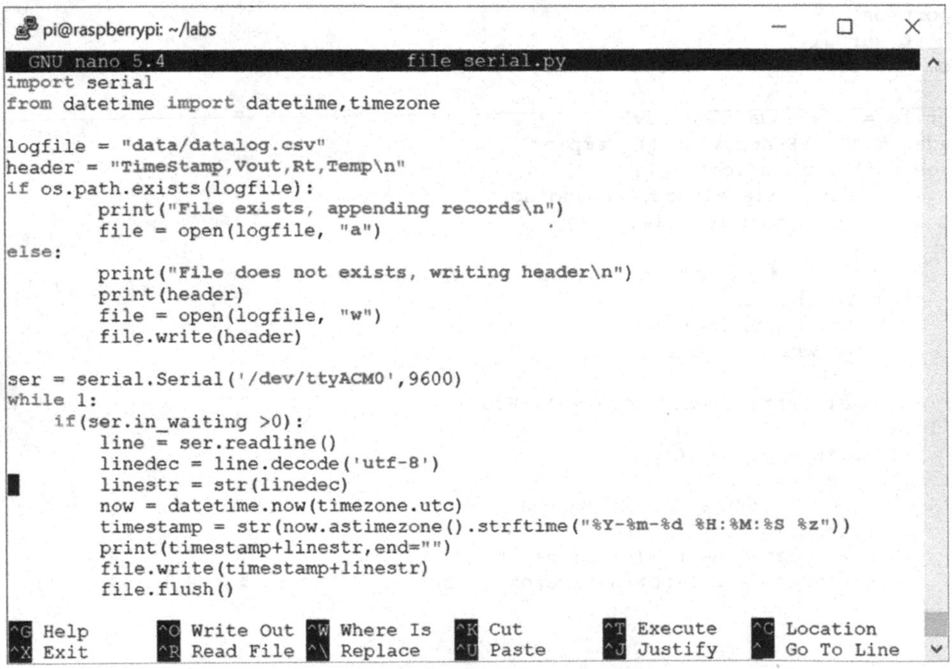

```
 pi@raspberrypi: ~/labs                                              —    □    ×
  GNU nano 5.4                       file_serial.py                        ^
import serial
from datetime import datetime,timezone

logfile = "data/datalog.csv"
header = "TimeStamp,Vout,Rt,Temp\n"
if os.path.exists(logfile):
        print("File exists, appending records\n")
        file = open(logfile, "a")
else:
        print("File does not exists, writing header\n")
        print(header)
        file = open(logfile, "w")
        file.write(header)

ser = serial.Serial('/dev/ttyACM0',9600)
while 1:
    if(ser.in_waiting >0):
        line = ser.readline()
        linedec = line.decode('utf-8')
        linestr = str(linedec)
        now = datetime.now(timezone.utc)
        timestamp = str(now.astimezone().strftime("%Y-%m-%d %H:%M:%S %z"))
        print(timestamp+linestr,end="")
        file.write(timestamp+linestr)
        file.flush()

^G Help       ^O Write Out  ^W Where Is   ^K Cut        ^T Execute    ^C Location
^X Exit       ^R Read File  ^\ Replace    ^U Paste      ^J Justify       Go To Line  v
```

FIGURE 3.22   Python script to obtain serial output and write a file.

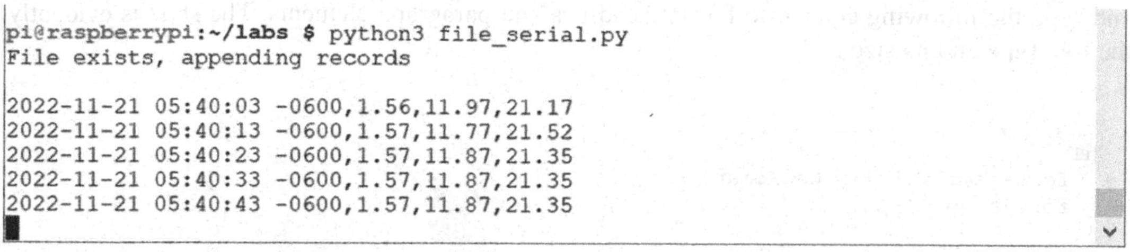

```
pi@raspberrypi:~/labs $ python3 file_serial.py
File exists, appending records

2022-11-21 05:40:03 -0600,1.56,11.97,21.17
2022-11-21 05:40:13 -0600,1.57,11.77,21.52
2022-11-21 05:40:23 -0600,1.57,11.87,21.35
2022-11-21 05:40:33 -0600,1.57,11.87,21.35
2022-11-21 05:40:43 -0600,1.57,11.87,21.35
```

FIGURE 3.23   Serial output using Python when the file exists. The time zone is CST expressed in offset -0600 with respect to UTC.

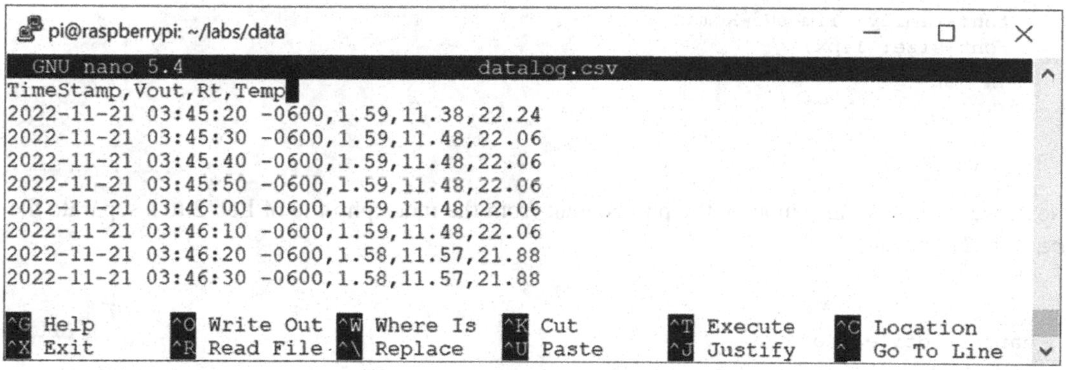

```
pi@raspberrypi: ~/labs/data                                    —    □    ×
  GNU nano 5.4                       datalog.csv                         ^
TimeStamp,Vout,Rt,Temp
2022-11-21 03:45:20 -0600,1.59,11.38,22.24
2022-11-21 03:45:30 -0600,1.59,11.48,22.06
2022-11-21 03:45:40 -0600,1.59,11.48,22.06
2022-11-21 03:45:50 -0600,1.59,11.48,22.06
2022-11-21 03:46:00 -0600,1.59,11.48,22.06
2022-11-21 03:46:10 -0600,1.59,11.48,22.06
2022-11-21 03:46:20 -0600,1.58,11.57,21.88
2022-11-21 03:46:30 -0600,1.58,11.57,21.88

^G Help       ^O Write Out ^W Where Is  ^K Cut     ^T Execute  ^C Location
^X Exit       ^R Read File ^\ Replace   ^U Paste   ^J Justify  ^  Go To Line  ∨
```

FIGURE 3.24   First few lines of file written using the serial streaming data using Python including the header.

or alternatively use WinSCP.

Likewise, we may want to copy the python programs developed on the RPi to our PC

```
c:\labs>pscp pi@192.168.1.132:labs/*py py/
pi@192.168.1.132's password:
test.py                   | 0 kB |   0.0 kB/s | ETA: 00:00:00 | 100%
test_serial.py            | 0 kB |   0.2 kB/s | ETA: 00:00:00 | 100%
timestamp_serial.py       | 0 kB |   0.4 kB/s | ETA: 00:00:00 | 100%
simpleloop.py             | 0 kB |   0.0 kB/s | ETA: 00:00:00 | 100%
file_serial.py            | 0 kB |   0.4 kB/s | ETA: 00:00:00 | 100%
```

HTML, CSS, PHP, AND JAVA SCRIPT

We will now continue learning how to display results and data on web pages by modifying the html and php documents we worked with in lab 2 and adding CSS to style them and JS to read data and plot it as SVG (Scalable Vector Graphics).

Let us create a stylesheet file style.css using nano

```
nano style.css
```

and type the following content to format headings and paragraph elements. The style is evidently the font type and its size.

```
h1 {
   font-family: TimesNewRoman;
   font-size: 18px;
}
h2 {
   font-family: TimesNewRoman;
   font-size: 16px;

}
p {
   font-family: TimesNewRoman;
   font-size: 14px;
}
```

Next, we will move and modify the php content from the index.php file of Lab 2 to a separate file readfile.php

```
nano readfile.php
```

with the following content

```
<?php
 $lines = file("data/datalog.csv");
 foreach ($lines as $line_num=> $line) {
    echo $line . "<br>";
 }
?>
```

We explained this segment of code in Lab 2 but note that now we are referencing the datalog file.

Next, we will download the lab3/plotd3.zip archive from the RTEM repository, extract its contents plotd3.html, and copy it to folder labs in the RPi. This html file has a java script containing D3.js code to plot a time series. Let us open it using nano

```
nano labs/plotd3.html
```

and examine some details

```
<!-- load the d3.js library -->
<script src="http://d3js.org/d3.v4.min.js"></script>
```

```
<!-- Create a div for the graph -->
<div id="graph"></div>
<script>
```

We are loading the `d3.js` library and creating a `div` element to contain the graph. It is followed by a `<script>` tag that will contain the JavaScript and will be closed at the end. The first part establishes the dimensions of the graph area and appends a SVG to generate the graph.

```
// dimensions
var margin = {top:10, right:20, bottom:40, left:60},
    width = 520-margin.left-margin.right,
    height = 380-margin.top-margin.bottom;
// svg
var svg = d3.select("#graph")
  .append("svg")
    .attr("width", width + margin.left + margin.right)
    .attr("height", height + margin.top + margin.bottom)
  .append("g")
    .attr("transform",
          "translate(" + margin.left + "," + margin.top + ")");
```

Very importantly, using `d3.csv` we read the data file and parse the entries of each record according to the header. Notice that the `TimeStamp` is interpreted according to the date and time format we have encountered previously in these lab guides. D3.js will interpret the `timezone` as an offset with respect to UTC (the same format we used to write the file in Python).

```
// read the file
d3.csv("data/datalog.csv",
  function(d){
    return {TimeStamp:d3.timeParse("%Y-%m-%d %H:%M:%S %Z")
    (d.TimeStamp), Vout:d.Vout, Rt:d.Rt, Temp:d.Temp}
  },

  // code for the graph
  function(data) {
```

This initiates the contents of a function that uses the data and is part of the `d3.csv` call. Consequently, we will have to close the function and the call, using "})" at the end when we get done with the D3 code.

This code consists of several parts, establishing the axes, the line, and the labels. We establish the scale for both axes, the x axis as time and y axis as linear scale. The plus sign "+" in front of `d.Temp` declares this to be a real numeric value, and not just a string.

```
    // x axis as timestamp
    var x = d3.scaleTime()
      .domain(d3.extent(data,function(d){return d.TimeStamp;}))
      .range([ 0, width ]);
```

```
    svg.append("g")
      .attr("transform", "translate(0," + height + ")")
      .call(d3.axisBottom(x));

    // y axis as linear
    var y = d3.scaleLinear()
      .domain([d3.min(data,function(d){return +d.Temp;}),
 d3.max(data,function(d){return +d.Temp;})])
        .range([height, 0]);
    svg.append("g")
      .call(d3.axisLeft(y));
```

Once this is complete, we draw the line

```
    // line
    svg.append("path")
      .datum(data)
      .attr("fill", "none")
      .attr("stroke", "steelblue")
      .attr("d", d3.line()
        .x(function(d)  {return x(d.TimeStamp)})
        .y(function(d)  {return y(d.Temp)})
        )
```

The final step is to add labels to the axis.

```
    // x axis label
    svg.append("text")
      .attr("transform",
            "translate(" + (width/2) + " ," +
                          (height + margin.top + 20) + ")")
      .style("text-anchor", "middle")
      .style("font-size", "12px")
      .text("TimeStamp");
    // y axis label
    svg.append("text")
      .attr("transform", "rotate(-90)")
      .attr("y", 0 - margin.left)
      .attr("x",0 - (height / 2))
      .attr("dy", "1em")
      .style("text-anchor", "middle")
      .style("font-size", "12px")
      .text("Temp(C)");
```

As we mentioned above, we close the function and the d3 call

```
    })
```

and finalize closing tags for script, body, and html.

```
</script>
</body>
</html>
```

We are ready to rewrite the index.php file that will handle all the calls to the CSS, PHP, and JS code.

```
nano index.php
```

and type the following content

```
<!DOCTYPE html>
<html>
<head>
  <title>Real-Time Environmental Monitoring</title>
  <link rel="stylesheet" href="css/style.css">
</head>
<body>
  <h1>RTEM Test Page</h1>
  <h2>Learning HTML, PHP, JS</h2>

  <p>Example of URL link: external web page
  <a href=https://github.com/mfacevedo1/rtem>RTEM GitHub Repository</a>
  </p>

  <p>Example of URL link: internal from the server file system
  <a href="data/datalog.csv">Download datalog file</a>
  </p>

  <p>Display datalog file contents using php</p>
  <iframe src="readfile.php" width=600 height=200></iframe>

  <p>Read datalog file and plot chart using d3.js</p>
  <iframe src="plotd3.html" width=600 height=400></iframe>

</body>
</html>
```

By now, you would understand most of this code. New elements are <iframe> which are containers for the separate components, displaying the file and the plot. Once you refresh index.php on the browser, the result will be as shown in Figure 3.25.

As time goes on and the Arduino keeps collecting data, streaming to the RPi, and appending datalog.csv, you can refresh the browser to see updates to the graph (Figure 3.26).

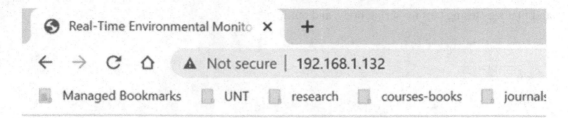

RTEM Test Page

Learning HTML, PHP, JS

Example of URL link: external web page <u>RTEM GitHub Repository</u>

Example of URL link: internal from the server file system <u>Download datalog file</u>

Display datalog file contents using php

```
2022-11-21 04:01:40 -0600,1.69,9.65,23.77
2022-11-21 04:01:51 -0600,1.68,9.74,25.57
2022-11-21 04:02:01 -0600,1.69,9.65,25.77
2022-11-21 04:02:11 -0600,1.69,9.65,25.77
2022-11-21 04:02:21 -0600,1.69,9.65,25.77
2022-11-21 04:02:31 -0600,1.69,9.65,25.77
2022-11-21 04:02:41 -0600,1.69,9.57,25.96
2022-11-21 04:02:51 -0600,1.69,9.57,25.96
2022-11-21 04:03:01 -0600,1.69,9.65,25.77
2022-11-21 04:03:11 -0600,1.69,9.57,25.96
2022-11-21 04:03:21 -0600,1.69,9.65,25.77
2022-11-21 04:03:31 -0600,1.69,9.65,25.77
```

Read datalog file and plot chart using d3.js

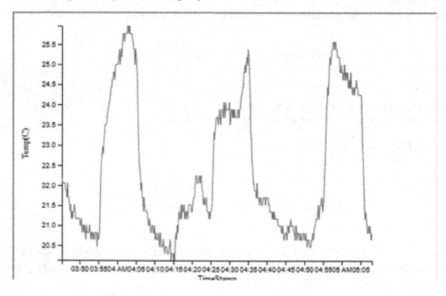

FIGURE 3.25   HTML example illustrating CSS, PHP, and JS.

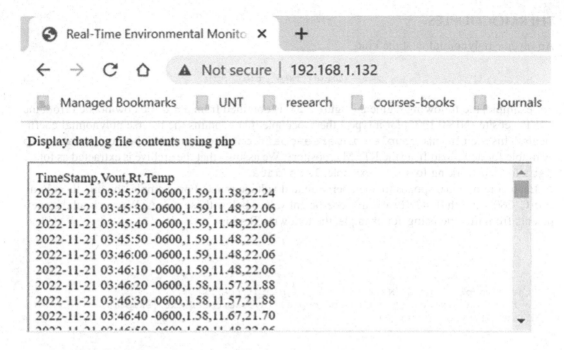

Display datalog file contents using php

```
TimeStamp,Vout,Rt,Temp
2022-11-21 03:45:20 -0600,1.59,11.38,22.24
2022-11-21 03:45:30 -0600,1.59,11.48,22.06
2022-11-21 03:45:40 -0600,1.59,11.48,22.06
2022-11-21 03:45:50 -0600,1.59,11.48,22.06
2022-11-21 03:46:00 -0600,1.59,11.48,22.06
2022-11-21 03:46:10 -0600,1.59,11.48,22.06
2022-11-21 03:46:20 -0600,1.58,11.57,21.88
2022-11-21 03:46:30 -0600,1.58,11.57,21.88
2022-11-21 03:46:40 -0600,1.58,11.67,21.70
2022-11-21 03:46:50 -0600,1.59,11.48,22.06
```

Read datalog file and plot chart using d3.js

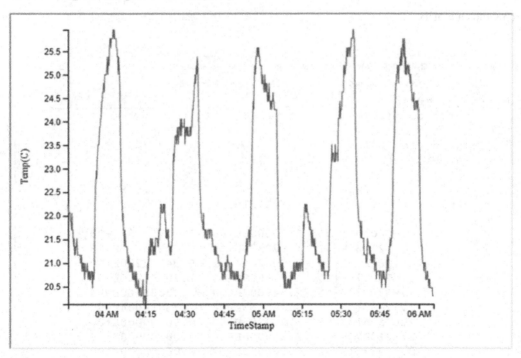

FIGURE 3.26   Updates to the data and graph as time unfolds.

THERMOCOUPLES

An inverse polynomial is of the kind

$$T = a_0 + a_1 V + a_2 V^2 + a_3 V^3 + \cdots \tag{3.2}$$

For example, the following segment (Figure 3.27) is obtained from a text file downloaded from the NIST website (NIST 1995) for a type J thermocouple that contains the inverse polynomial coefficients. This file `thermocoupleJ-inverse-coeff.txt` is contained in the archive `data.zip` available for download from the RTEM repository. We assume that the archive is extracted as folder `data` to your working folder, for example, `labs/data`.

Each column corresponds to a temperature and voltage range. For example, the middle column is 0°C–760°C, with 0–42.919 mV, and coefficient values below those ranges. We can read the coefficients from this file using. for example, the following R script

```
#thermocouple type J (NSIT inverse polynomial coeffs)
coeff <- matrix(scan("data/thermocoupleJ-inverse-coeff.txt",skip=8,
nlines=9),byrow=T,ncol=3)
```

which yields a matrix

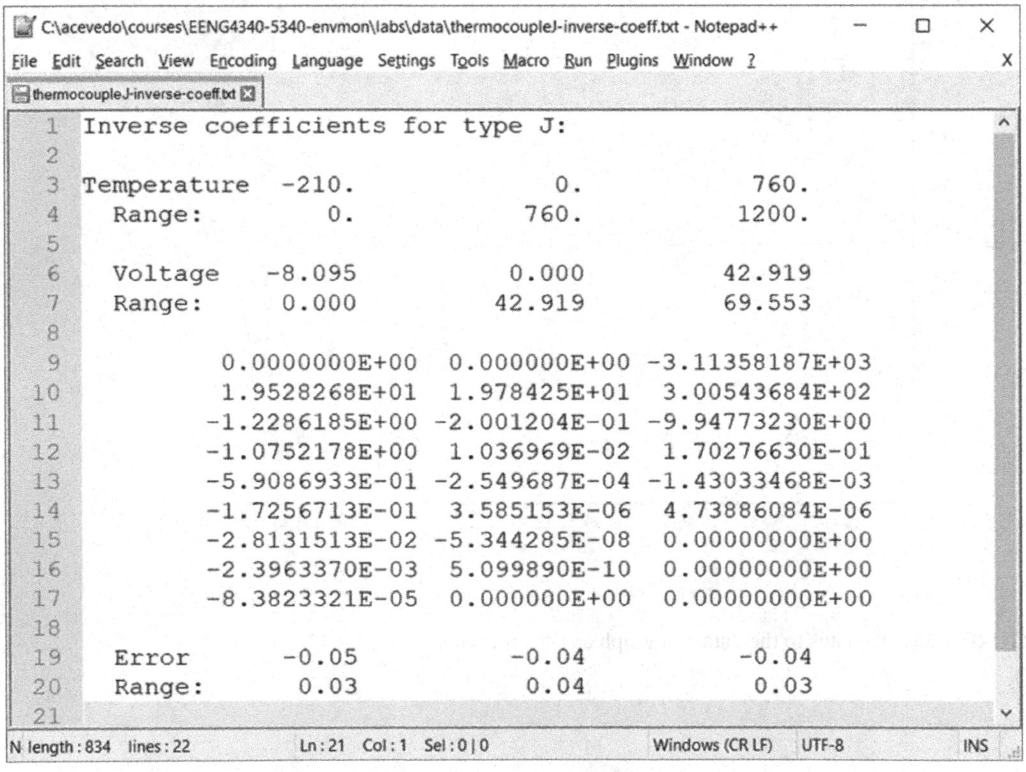


```
Inverse coefficients for type J:

Temperature   -210.            0.            760.
  Range:         0.          760.           1200.

  Voltage     -8.095          0.000          42.919
  Range:       0.000         42.919          69.553

         0.0000000E+00  0.000000E+00 -3.11358187E+03
         1.9528268E+01  1.978425E+01  3.00543684E+02
        -1.2286185E+00 -2.001204E-01 -9.94773230E+00
        -1.0752178E+00  1.036969E-02  1.70276630E-01
        -5.9086933E-01 -2.549687E-04 -1.43033468E-03
        -1.7256713E-01  3.585153E-06  4.73886084E-06
        -2.8131513E-02 -5.344285E-08  0.00000000E+00
        -2.3963370E-03  5.099890E-10  0.00000000E+00
        -8.3823321E-05  0.000000E+00  0.00000000E+00

  Error        -0.05          -0.04          -0.04
  Range:        0.03           0.04           0.03
```

FIGURE 3.27 Text file listing inverse polynomial coefficients.

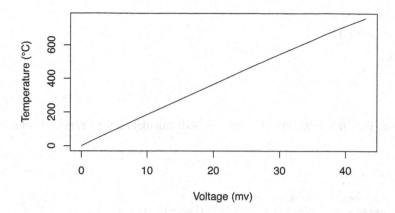

FIGURE 3.28 Thermocouple *J*, temperature calculated from mV for the range 0°C–760°C using NSIT inverse polynomial.

```
> coeff
                  [,1]            [,2]            [,3]
 [1,]   0.000000e+00    0.000000e+00   -3.113582e+03
 [2,]   1.952827e+01    1.978425e+01    3.005437e+02
 [3,]  -1.228619e+00   -2.001204e-01   -9.947732e+00
 [4,]  -1.075218e+00    1.036969e-02    1.702766e-01
 [5,]  -5.908693e-01   -2.549687e-04   -1.430335e-03
 [6,]  -1.725671e-01    3.585153e-06    4.738861e-06
 [7,]  -2.813151e-02   -5.344285e-08    0.000000e+00
 [8,]  -2.396337e-03    5.099890e-10    0.000000e+00
 [9,]  -8.382332e-05    0.000000e+00    0.000000e+00
>
```

We can select the 0°C–760°C range

```
# select middle column range 0- 270
> a <- coeff[,2]
> a
[1]   0.000000e+00  1.978425e+01 -2.001204e-01  1.036969e-02 -2.549687e-04
[6]   3.585153e-06 -5.344285e-08  5.099890e-10  0.000000e+00
>
```

A value of temperature can be obtained from a value of voltage, say for 5 mV, by calculating a polynomial of voltage and summing the product of coefficient values times the polynomial terms. This is the linear combination given in equation (3.2).

```
mv <- 5
# polynomial of mV
mv.pol <- mv^(0:(length(a)-1))
# calculating temp from linear combination
```

```
Temp <- sum(c*mv.pol)
> Temp
[1] 95.0655
>
```

The temperature for 5 mV is 95.0655°C. Now, we will calculate it for a range of voltage

```
#select full range 0-760
mv <- seq(0,42.919,0.1)
# form a polynomial for each value of mv
mv.pol <- sapply(mv, function(x) x^(0:(length(a)-1)))
# calculate Temp as linear combination for each value of mV
Temp <- apply(mv.pol, MARGIN=2, function(x) sum(a*x))
```

Now, we can plot

```
# plot
plot(mv,Temp,type="l",ylab="Temperature (°C)",xlab="Thermoelectric Voltage
(mv)")
```

to obtain the plot shown in Figure 3.28.
    To calculate sensitivity

```
# using endpoints
FS <- mv[length(mv)]
sens <- (mv[length(mv)]-mv[1])/(Temp[length(mv)]-Temp[1])
# in uV
> round(sens*1000,2)
[1] 56.47
>
```

The sensitivity is 56.47 µV/°C in the full range of 0 to FS=42.919 mV. Calculate linearity error in % of FS for the range 0–FS

```
mv.lin <- sens*(Temp-Temp[1]) + mv[1]
# max deviance of response with respect to linear
dev <- abs(mv-mv.lin); max.dev <- max(dev)
lin.err <- 100*max.dev/FS
> round(lin.err,2)
[1] 1.97
>
```

which means that the linearity error is 1.97% of FS.

USING A THERMOCOUPLE

We will use the ADM02 DMM or CAMWAY DMM 6000 with a K-type thermocouple temperature probe. Connect the thermocouple probe to the banana plug, temp input jacks, observing correct polarity, COM (common) and positive. Use it to measure changes of temperature as we did for the thermistor resistance but use the thermocouple instead of the thermometer. Compare measurements to those obtained by the Arduino. Pictures are available in the supplementary material.

EXERCISES

Exercise 3.1 Thermistor resistance and B parameter estimate

Setup your Arduino, heater, fan, and thermistor to measure thermistor resistance and temperature, record these data, and use linear regression to obtain an estimate of B for your thermistor. Show your data and prediction using a figure similar to Figure 3.4. You will need the B estimate result for the remaining exercises. Keep your breadboard and devices setup for the remaining exercises.

Exercise 3.2 Transducer from voltage divider

Using the R functions you have built as part of the transducer-functions.R code, determine the expected response of a voltage divider transducer (Figure 3.5) for $R_f = 10$ kΩ, $R_1 = 10$ kΩ, and $V_s = 5$ V. Use a temperature range from 20°C to 30°C. Use the B parameter value of the thermistor that you measured in the previous exercise. Plot the response including sensitivity and linearity error for the voltage divider transducer. Plot graphs to evaluate self-heating.

Exercise 3.3 Implement your transducer

Follow all the steps to implement your temperature transducer. Show your implementation, the data collected, and response on the serial monitor. Compare serial monitor readings to the one done by thermometer and DMM.

Exercise 3.4 Displaying output in the Serial Plotter

Run Arduino script `heat-fan-read-sensor-plot` and display on the Serial Plotter. Run at least one cycle of heating and cooling. Plot V_{out}, R_t, and Temp. Take a screenshot to record the resulting graph.

Exercise 3.5 Capturing output with CoolTerm (or RealTerm)

Use Arduino script `heat-fan-read-sensor-log`. Generate a file `capture.txt` using CoolTerm. Examine the file to verify its contents. Apply the `Rstrptime` function to your timestamp. Create `xts` series and display it as graphical output.

Exercise 3.6 Arduino and RPi.

Connect the Arduino to the RPi Zero W. Secure copy (pscp) the heat-fan-read-sensor-log from the PC to the RPi. Delete printing the comma at the beginning of the record. Compile and upload the Arduino code using the Arduino CLI software installed on the RPi Zero W. Establish serial communication between Arduino and the RPi Zero W in order to write a datalog.csv file on the RPi. Copy the file to the PC.

Exercise 3.7 Web page.

Create directories and generate code for HTML, CSS, PHP, JS as indicated in the guide to produce a web page similar to Figure 3.26.

Exercise 3.8 Calculating Thermocouple specifications
   Use R as given in the guide to calculate sensitivity, linearity error, of the J thermocouple discussed in the guide.

Exercise 3.9 Using a thermocouple
   Use the DMM with the thermocouple to measure changes of temperature as we did to determine thermistor resistance but use the thermocouple instead of the thermometer. Compare measurements with those obtained by the Arduino.

REFERENCES

Acevedo, M.F. 2024. *Real-Time Environmental Monitoring: Sensors and Systems - Textbook, Second Edition*. Boca Raton, FL: CRC Press, Taylor & Francis Group, 392 pp.
Amazon. 2023. *KETOTEK Digital Thermometer Temperature Meter Gauge*. Accessed January 2023. https://www.amazon.com/KETOTEK-Thermometer-Temperature-Waterproof-Fahrenheit/dp/B01DVSIYN2/ref=asc_df_B01DVSIYN2?tag=bngsmtphsnus-20&linkCode=df0&hvadid=79989588513714&hvnetw=s&hvqmt=e&hvbmt=be&hvdev=c&hvlocint=&hvlocphy=&hvtargid=pla-4583589114951828&th=1.
Arduino. 2016. *Download the Arduino Software*. Accessed August 2016. https://www.arduino.cc/en/Main/Software.
Arduino. 2023. *Arduino-CLI*. Accessed January 2023. https://github.com/arduino/arduino-cli.
Geany. 2023. *Geany - The Flyweight IDE*. Accessed January 2023. https://www.geany.org/.
Meier, R. 2023. *Roger Meier's Freeware*. Accessed January 2023. https://freeware.the-meiers.org/.
NIST. 1995. *Tables of Thermoelectric Voltages and Coefficients for Download NIST*. Accessed 2014. http://srdata.nist.gov/its90/download/download.html.
Posit. 2023. *RStudio IDE*. Accessed January 2023. https://posit.co/downloads/.
PuTTY. 2023. *Download PuTTY*. https://www.putty.org/.
Python. 2022. *Python*. Accessed January 2022. https://www.python.org/.
R Project. 2023. *The Comprehensive R Archive Network*. Accessed January 2023. http://cran.us.r-project.org/.
WinSCP. 2023. *Free Award-Winning File Manager*. Accessed January 2023. https://winscp.net/eng/index.php.

4 Bridge Circuits and Signal Conditioning

INTRODUCTION

We analyze bridge circuits, bipolar signals, and signal conditioning. For this purpose, you will conduct exercises based on R: balanced-source voltage divider, quarter-bridge circuit, and linearized bridge. Then, we learn how to make a unipolar (i.e., always positive) voltage from a bipolar signal. We study signal conditioning to improve the bridge circuit response. For this purpose, we will use the linearized bridge and connect it to an instrumentation amplifier. Bridge circuits produce bipolar signals and require powering op-amps by means of split rails (e.g., −12 and +12 V or −5 and +5 V). For this purpose, we will implement a rail splitter. Then, we learn how to convert a bipolar signal to unipolar (i.e., always positive) so that it can be input to an Arduino.

MATERIALS

READINGS

For theoretical background, you can use Chapter 4 of Acevedo, M.F. 2024. *Real-Time Environmental Monitoring: Sensors and Systems Second Edition - Textbook*, which is a companion to these guides (Acevedo 2024). Other bibliographical references are cited throughout the guide.

COMPONENTS

- LM 358 Dual operational amplifier through hole (Texas Instruments 2023)
- INA126P Instrumentation Amplifier
- MT3608 DC Boost Converter 2–24 V to 5–28 V 2A (with a USB connector) (Components Info 2023)
- Protoboard and wires
- Resistors (four 10 kΩ) to be used in the voltage divider
- Thermistor (10 kΩ nominal) to be used as a sensor
- Resistors (two 100 kΩ for the rail splitter)
- Resistors (two 33 kΩ for the unipolar conversion to Arduino)
- Higher power 50 Ω resistor (1 W) to be used as a heater
- Small cooling fan to be used to cool the sensor
- Potentiometer 15 or 20 kΩ (to be used as a gain resistor for the In-amp)

MAJOR COMPONENTS AND INSTRUMENTS

- Arduino UNO
- Raspberry Pi Zero W.
- Digital temperature probes, e.g., KETOTEK (Amazon 2023) or KEYNICE Digital Thermometer
- Digital multimeter ADM02 DMM or CAMWAY DMM 6000 (with thermocouple temperature probe)

DOI: 10.1201/9781003184362-4

TOOLS (RECOMMENDED)

- Long nose pliers
- Wire strippers and clippers
- Small Screwdriver

SOFTWARE

- R, for data analysis (R Project 2023)
- RStudio, an IDE to use R (Posit 2023)
- PuTTY, for Secure Shell (PuTTY 2023)
- WinSCP, to transfer files from the Raspberry Pi (RPi) to a Windows PC (WinSCP 2023)
- Python (installed from RPi)
- Arduino IDE, Interface to Arduino (Arduino 2016)
- Geany, IDE to edit programs as well as data (Geany 2023)
- Arduino-CLI, Arduino software to run using commands from the RPi (Arduino 2023)
- PHP, for RPi.

SCRIPTS AND DATA FILES

All are available from the RTEM GitHub repository https://github.com/mfacevedol/rtem

- R scripts contained in archive R-bridge-transducers.zip
- Arduino scripts contained in archive `arduino-bridge-transducers.zip`
- Python scripts `python-rpi.zip`

SUPPLEMENTARY SUPPORT MATERIAL

Supplementary support material including additional screenshots, images, and procedures are available from the publisher's eResources web page provided for this book.

LINEARIZED THERMISTOR: SMALL VARIATION ANALYSIS

As explained in Chapter 4 of the companion textbook, when variations of the measurand are small, it is possible to simplify the analysis of the sensor response by assuming a linear response around the operating value of the measurand. We will investigate this concept further using R. Create a folder `labs/R/lab4` to contain the files for this lab session and start a script file with the name `bridge-functions.R` (or use the one available from the RTEM GitHub repository).

Start by developing a function `B.param.der` to calculate the derivative of the B parameter model. We can type this function in the file `bridge-functions.R`

```
B.param.der <- function(Rt.nom,T.C,B){
 T.K <- T.C + 273; T0.K <- Rt.nom$T0.C+273
 dRt.k <- B*B.param(Rt.nom,T.C,B)$Rt.k*(-1/T.K^2)
 return(list(dRt.k=round(dRt.k,5),T.C=T.C))
}
```

Now, start a `labs/R/lab4/lab4.R` file and source the functions file along with the thermistor functions we developed in Lab 3,

```
source("R/lab3/thermistor-functions.R")
source("R/lab4/bridge-functions.R")
```

In the `lab4.R` script file, type code to use the function `B.param.der` to calculate derivative of the B model as a function of an array `T.C` in the range 20°C–30°C.

```
# thermistor slope by derivative
Rt.nom <- list(T0.C = 25, R0.k=10)
T.C=seq(20,30,0.01); B=4116
# derivative in the range
dRt.k <- B.param.der(Rt.nom,T.C,B)
```

The result of the function has two components: `dRt.k` has the value of the derivative whereas the other component repeats the values of `T.C` for easy reference. We are interested in the value of the derivative at nominal `T0` and thus we calculate it and store as object `S`.

```
# derivative at nominal
S <- B.param.der(Rt.nom,Rt.nom$T0.C,B)$dRt.k
```

Verify that `S` has the value −0.46349

```
> S
[1] -0.46349
>
```

This is in kΩ/°C. Next, we plot the derivative as a function of `T.C`

```
plot(dRt.k$T.C,dRt.k$dRt.k,type="l",xlab="Temp (°C)", ylab="dR/dT (kΩ/°C)")
abline(v=Rt.nom$T0.C,lty=2,col="grey")
abline(h=S,lty=2,col="grey")
```

See the result in Figure 4.1.

Let us add a function to file `bridge-functions.R` that evaluates the linearized response based on the slope at a nominal value.

```
therm.lin <- function(Rt.nom,T.C,S){
 # linear model
 # Rt.nom is list(T0.C, R0.k)
 T.K <- T.C + 273; T0.K <- Rt.nom$T0.C+273
 Rt.k <- Rt.nom$R0.k-S*(T.C-Rt.nom$T0.C)
 return(list(Rt.k=round(Rt.k,2),T.C=T.C))
}
```

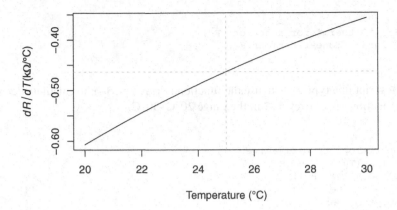

FIGURE 4.1   Derivative of the *B* model for *B*=4100 and nominal 10 kΩ at 25°C.

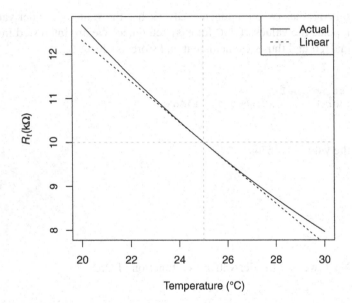

FIGURE 4.2   Linearized thermistor response.

Source the updated `bridge-functions.R` and type the following code to obtain Figure 4.2.

```
# nonlinear with slope
T.C=seq(20,30,0.01);B=4116
S <- B.param.der(Rt.nom,Rt.nom$T0.C,B)$dRt.k
#linearized using S
RT.lin <- therm.lin(Rt.nom,T.C,S)
RT.B <- B.param(Rt.nom,T.C,B)

# plot together curve and linearized
plot(RT.B$T.C, RT.B$Rt.k, type="l", xlab="Temp(°C)", ylab="Rt(kΩ)")
lines(RT.B$T.C, RT.lin$Rt.k,lty=2)
legend("topright",leg=c("Actual","Linear"),lty=1:2,col=1)
abline(v=Rt.nom$T0.C,lty=2,col="grey")
abline(h=Rt.nom$R0.k,lty=2,col="grey")
```

Add the following function to the `bridge-functions.R` file or download from the RTEM repository. This function calculates the voltage divider based on the linearized thermistor response

```
ckt.div.lin <- function(T.C,Rt.nom,S,Rf.k,R1.k,Vs){
  Rt.k <- therm.lin(Rt.nom,T.C,S)$Rt.k
  R <- R1.k + Rt.k + Rf.k
  Vout <- Vs*Rf.k/R
  Pow <- (Vs/R)^2*Rt.k
  return(list(T.C=T.C,Vout=round(Vout,3),Pow.mW=round(Pow,3)))
}
```

Source the `bridge-functions.R` file to update the functions and use `ckt.div.lin` by writing the following code in `lab4.R` where we use a voltage supply of 5 V.

```
Rf.k=10; R1.k=10; Vs=5;kt=1.5
# using linearized
x<- ckt.div.lin(T.C,Rt.nom,S,Rf.k,R1.k,Vs)
X<- specs.transd(x)
specs.plot(X)
```

The divider response is given in Figure 4.3, which indicates that the linearized thermistor response in a voltage divider yields a nonlinear transducer response because of the voltage divider formula itself.

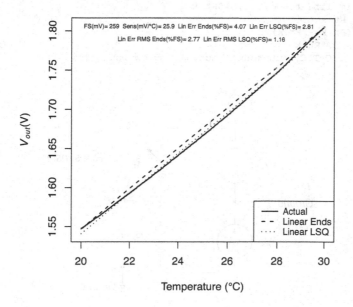

FIGURE 4.3 Transducer response in a voltage divider is nonlinear.

BALANCED SOURCE VOLTAGE DIVIDER

Consider a voltage divider circuit but with a balanced source (Figure 4.4). We assume the sensor is a thermistor. As explained in Chapter 4, the output voltage of the transducer is

$$V_{\text{out}} = V_s \left(\frac{R_f - R_t}{R_t + R_f} \right) \tag{4.1}$$

and that assuming variations are sufficiently small $2R_f \gg dR$ we can simplify the output to

$$V_{\text{out}} \approx V_s \left(\frac{-dR}{2R_f} \right) = -dR \frac{V_s}{2R_f} \tag{4.2}$$

As well as the sensitivity of the transducer as

$$\frac{dV}{dT} \approx -\frac{V_s}{2R_f} S \tag{4.3}$$

where S is the slope of the thermistor at the nominal temperature $S = \left(\dfrac{dR}{dT} \right) \bigg|_{T=T0}$. As a numeric example, using $V_s = 5\,\text{V}$, $R_f = 10\,\text{k}\Omega$, $S = -0.46\,\text{k}\Omega/°\text{C}$, we obtain a transducer sensitivity of

$$\frac{dV}{dT} \approx \left(-\frac{V_s}{2R_f} \right) S = \left(-\frac{5\,\text{V}}{2 \times 10\,\text{k}\Omega} \right) \left(-0.46 \frac{\text{k}\Omega}{°\text{C}} \right) = 115 \frac{\text{mV}}{°\text{C}}.$$

A more exact calculation of equation (4.1) can be accomplished using the following R function. Add this function to the `bridge-functions.R` file or download it from the RTEM repository.

```
ckt.bal.div <- function(T.C,Rt.nom,S,Rf.k,Vs){
 Rt.k <- therm.lin(Rt.nom,T.C,S)$Rt.k
 R <- Rf.k + Rt.k
 Vout <- Vs*(Rf.k - Rt.k)/R
 Pow <- (Vs/R)^2*Rt.k
 return(list(T.C=T.C,Vout=round(Vout,3),Pow.mW=round(Pow,3)))
 }
```

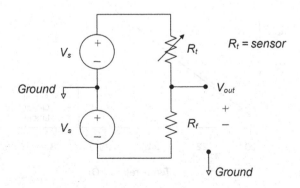

FIGURE 4.4   Balanced source voltage divider.

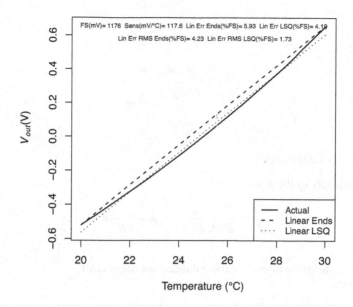

FIGURE 4.5 Output voltage of the balanced source voltage divider.

Source `bridge-functions.R` and use `ckt.bal.div` as follows to obtain Figure 4.5.

```
# balanced source divider
Rf.k=10; Vs=5; B=4100;kt=1.5
x<- ckt.bal.div(T.C,Rt.nom,B,Rf.k,Vs)
X<- specs.transd(x)
specs.plot(X)
```

Note that now the output voltage can swing positive or negative depending on the sign of the departure from the zero point where both resistors are the same, which corresponds to the nominal temperature. The sensitivity is 117.6 mV/°C, which is slightly larger than the value of 115 mV/°C we obtained by the simplified equation (4.3), and as expected we have non-linear effects due to the nature of the circuit itself.

WHEATSTONE BRIDGE

Now consider a Wheatstone bridge where the upper left branch is a sensor (e.g., thermistor), subject to a changing variable (e.g., temperature). See Figure 4.6. We select all resistances R_1, R_2, and R_3 equal to a fixed resistor, which has a value equal to R_f which has a value equal to the nominal value of the resistive sensor at the nominal value of the measurand. For example, using a thermistor, the nominal value R_f is the value of the sensor resistance R_t at the nominal operating value of temperature, say T_0.

In Chapter 4, we demonstrated that the transducer output voltage is

$$V_{out} = V_s \frac{R_f - R_t}{2 \times (R_t + R_f)} \tag{4.4}$$

and that assuming variations are sufficiently small $2R_f \gg dR$ we can simplify this expression to

$$V_{out} \approx -\frac{V_s}{4R_f} \times dR \tag{4.5}$$

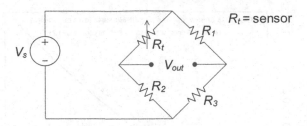

FIGURE 4.6   Quarter-bridge circuit.

As well as the sensitivity of the transducer

$$\frac{dV}{dT} = \frac{dV}{dR}\left(\frac{dR}{dT}\right)\Bigg|_{T=T0} \approx -\frac{V_s}{4R_f} \times S$$

where $S = \left(\dfrac{dR}{dT}\right)\Bigg|_{T=T0}$ as given above for the balanced source divider.

This circuit is called a "quarter-bridge" because the Wheatstone bridge contains one sensor out of the possible four resistances. Note also that the sensitivity has the constant 1/4 corresponding to a quarter of the total V_s/R_f. One could substitute R_1 for another sensor ("reference" sensor), which will remain undisturbed by variations of the variable being measured. This way we will be sensing differences of a variable with respect to the reference.

As a numerical example, consider a thermistor with a nearly linear relationship of slope $S = -0.46$ kΩ/°C at a nominal $R_{0=}10$ kΩ at $T_{0=}25$°C. We build a quarter-bridge with a reference resistance of $R_{f=}10$ kΩ and a voltage source of $V_{s=}5$ V. The sensitivity of the transducer is

$$\frac{dV}{dT} \approx \left(-\frac{V_s}{4R_f}\right)S = \left(-\frac{5\text{ V}}{4\times10\text{ k}\Omega}\right)\left(-0.46\,\frac{\text{k}\Omega}{^\circ\text{C}}\right) = 57.5\,\frac{\text{mV}}{^\circ\text{C}}$$

which is half of the balanced source divider. A more exact calculation can be accomplished using R functions.

We will write a function to calculate equation (4.4) and add it to the file `bridge-functions.R` or use the one downloaded from the RTEM repository

```
ckt.quart.bridge <- function(T.C,Rt.nom,S,Rf.k,Vs){
 Rt.k <- therm.lin(Rt.nom,T.C,S)$Rt.k
 R <- 2*(Rf.k + Rt.k)
 Vout <- Vs*(Rf.k - Rt.k)/R
 Pow <- (Vs/R)^2*Rt.k
 return(list(T.C=T.C,Vout=round(Vout,3),Pow.mW=round(Pow,3)))
}
```

Use this function in `lab4.R` to analyze the bridge voltage response.

```
# quarter bridge
Rf.k=10; Vs=5;kt=1.5
x<- ckt.quart.bridge(T.C,Rt.nom,S,Rf.k,Vs)
X<- specs.transd(x)
specs.plot(X)
```

Figure 4.7 shows the graphical response. Note that the sensitivity is 58.8 mV/°C close to the simple approximation 57.5 mV/°C and the linearity error indicates non-linear behavior due to the circuit equation.

LINEARIZED BRIDGE BY OP-AMP

In Chapter 4, we discussed the circuit of Figure 4.8, which is a differential amplifier with the resistive sensor in the feedback branch and a common supply to both inputs. It forces the difference at the midpoint of the bridge to be zero when the bridge is balanced as demonstrated in the previous section. Small changes of the sensor should produce an output voltage that departs from zero. The op-amp linearizes the output of the bridge, but not the sensor response, i.e., not the change in resistance vs. the measurand.

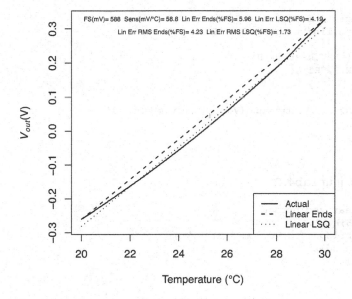

FIGURE 4.7 Results for the quarter-bridge circuit.

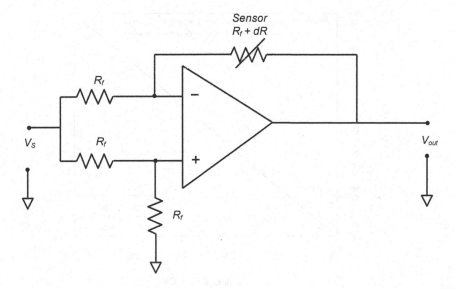

FIGURE 4.8 Linearized bridge circuit.

As given in Chapter 4, the output is

$$V_{out} = -\frac{V_s}{2}\frac{dR}{R_f} = -\frac{V_s}{2}\frac{R_t - R_f}{R_f} \tag{4.6}$$

Assuming R_f is equal to the nominal resistance of the thermistor. This is very useful because the response is linear with respect to dR even for large values of dR and the denominator has been reduced by half, i.e., from 4 to 2. In other words, we get the same magnitude of the response as a two-sensor bridge (half-bridge), but it has the advantage of being linear and of using only one sensor.

We can analyze this transducer using the following R function. Type it or use it from the archive downloaded from the RTEM repository.

```
ckt.lin.bridge <- function(T.C,Rt.nom,S,Rf.k,Vs){
Rt.k <- therm.lin(Rt.nom,T.C,S)$Rt.k
R <- 2*Rf.k
Vout <- Vs*(-(Rt.k-Rf.k)/R)
Pow <- (Vs/R)^2*Rt.k
```

```
return(list(T.C=T.C,Vout=round(Vout,3),Pow.mW=round(Pow,3)))
```

```
}
```

Source and use it in `lab4.R`

```
# op amp linearized bridge
Rf.k=10; Vs=5;kt=1.5
x<- ckt.lin.bridge(T.C,Rt.nom,S,Rf.k,Vs)
X<- specs.transd(x)
specs.plot(X)
```

We obtain Figure 4.9 showing a sensitivity of ~115 mV/°C and low linearity error.

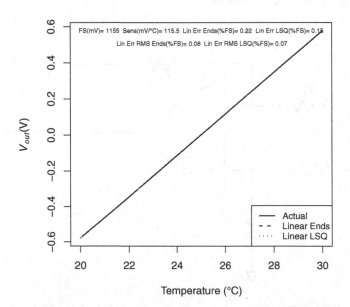

FIGURE 4.9   Linearized bridge output.

The op-amp will require both negative and positive power rails because the output must be able to go negative. Because the output is that of an op-amp, it has low impedance, and this is convenient for the subsequent processing of the signal. As we discussed in Chapter 4 of the textbook it is important to realize that the op-amp only linearizes the circuit and that therefore if we were to consider the full non-linear response of the thermistor, the output would be non-linear.

DC–DC CONVERSION AND SPLIT RAIL

BOOST OR STEP-UP DC–DC CONVERTER

The simplest DC–DC converters are the *buck* or step-down type of converter (Figure 4.10, left) and the *boost* or step-up type of converter (Figure 4.10, right). The switch utilized in both circuits represents a pulse waveform produced by pulse width modulation (PWM). In simple terms, the periodic PWM waveform is composed of a pulse in the on state that lasts a fraction D of the period T. For the remainder time $(1-D)T$, the signal is in the off state. The fraction D is called the *duty cycle* of the pulse wave.

For a boost converter, when the switch is closed the input source charges the coil and when the switch opens, the coil injects its current into the capacitor and thus charges it. Then when the switch closes again, the capacitor holds the charge, which keeps building up, thus achieving a higher voltage at the output. In this case, $\langle V_{\text{out}} \rangle = \dfrac{1}{1-D} V_{\text{in}}$ and the current is $\langle I_{\text{out}} \rangle = (1-D)\langle I_{\text{in}} \rangle$. For example, what would be the duty cycle for a boost converter to step up from 5 to 24 V? $\langle V_{\text{out}} \rangle = \dfrac{1}{1-D} V_{\text{in}}$ then $D = 1 - \dfrac{V_{\text{in}}}{\langle V_{\text{out}} \rangle} = 1 - \dfrac{5}{24} = 0.791$ or the pulse should be on for 79% of the period.

In this lab session, we will use an adjustable Step-up Converter MT3608 DC Boost Converter 2–24 V to 5–28 V 2A (with a USB connector) (Components Info 2023; Autodesk Instructables 2023). Figure 4.11 shows the back of the component indicating the direction of voltage conversion and the front of the component with a USB connector for power.

The easiest way to work with this component is to solder header pins to the output and insert these pins on the breadboard. Power it with the USB cable and charging block, and verify that you have 5 V at the input terminals using the digital multimeter (DMM). Measure the output voltage with the DMM while turning the potentiometer clockwise to decrease the voltage and clockwise to increase it. Adjust the output voltage to as close as 12 V as possible, it may be a few tenths of volts off.

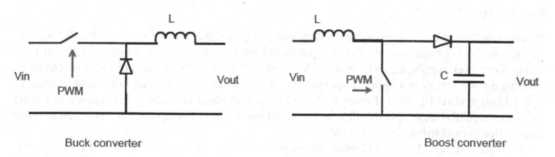

FIGURE 4.10 DC–DC converter.

FIGURE 4.11   MT3608 DC Boost Converter 2–24 V to 5–28 V 2A (with USB connector).

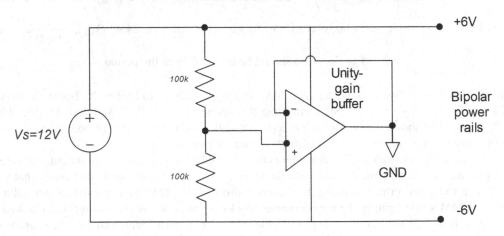

FIGURE 4.12   Rail splitter using an op-amp.

RAIL SPLITTER

When using bridge circuits, the output voltage can swing positive or negative, i.e., a bipolar voltage. Amplifying this voltage requires that the operational amplifier be biased or powered by a *split-rails* or bipolar power supply, e.g., +12 and −12 V, or +6 and −6 V. A practical and low cost circuit to obtain a bipolar supply is to split a unipolar rail voltage, say 12 V into +6 and −6 V, using an op-amp to establish a virtual ground (Figure 4.12). Here, the two equal resistors (1% tolerance or better) form a voltage divider to split the 12 V in two equal parts of 6 V. The op-amp is connected as unity gain buffer to establish a virtual ground.

Build the circuit of Figure 4.12 using the output of the step-up converter as the 12 V supply, two 100 kΩ resistors (1%) for voltage divider, one half of an LM358 (Texas Instruments 2023) as unity-gain buffer. See Figure 4.13. Test your circuit and verify the new +6, −6 V power rails.

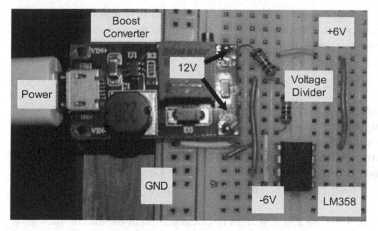
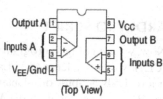

FIGURE 4.13   Rail splitter implementation.

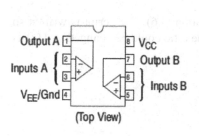
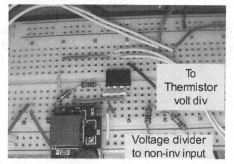

FIGURE 4.14   Linearized bridge: breadboard implementation.

LINEARIZED BRIDGE IMPLEMENTATION

Implement the circuit of Figure 4.8 using the second half of the LM358, with power rails, produced by the rail splitter, and use $R_{f=}$ 10 kΩ resistors in all branches except the feedback resistor which will be the thermistor. Connect the power supply of V_s=5V at the bridge input (see Figure 4.14). Do not connect the output to an Arduino port yet. When the temperature is approximately that of the nominal thermistor resistance, the output should be ~0V. When we run the heater and fan as in Lab 3, the temperature will change above and below the nominal, and the output V_{out} should vary linearly around zero volts, both negative and positive. We will cover the range from 20°C to 30°C.

READING BIPOLAR VOLTAGE INTO AN ARDUINO

Before we connect the output of the bridge to an Arduino analog port, we need to build a voltage divider to convert bipolar voltage to unipolar (i.e., always positive) and change the code accordingly. Whenever an ADC converter accepts only a positive voltage and converts it to unsigned integers, as is the case for the Arduino analog ports that can read 0–5V, we can bias the input to half the maximum voltage, i.e., 2.5V so that excursions below and above this value become negative and positive readings. The required bias can be implemented with a voltage divider as shown in Figure 4.15. We would have to subtract the digital value corresponding to 2.5 V to obtain the values simulating the bipolar signal. The following Arduino script demonstrates how this works.

ARDUINO SCRIPT

Now, we are going to program and run Arduino code on the RPi using the Command Line Interface (CLI), which we learned in Lab sessions 2 and 3. First, connect the Arduino to the RPi using the micro-USB adapter or adapter cable. We will write a script `heat-fan-read-bridge` and use the RPi to interface with the Arduino; for this purpose, the script will reside in the `labs` folder of the RPi. We can write the script using nano on the RPi or write it in Geany using a PC and copy it to the RPi using SCP or use the script `heat-fan-read-bridge` available from the RTEM repository.

First in the declarations (Figure 4.16), we add new lines to calculate half of the maximum digital number (1023)

```
float dHalf = 1023/2.00;
```

then declare a ratio of 2.00 for the linearized bridge as in equation (4.6). The script is written so that is applicable to a Wheatstone quarter-bridge. In this case, the ratio could be changed to 4.00 as given in equation (4.5)

```
// change this ratio to 4.00 for quarter bridge
float ratio = 2.00;
```

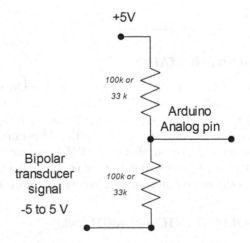

FIGURE 4.15   Bipolar input to Arduino analog port.

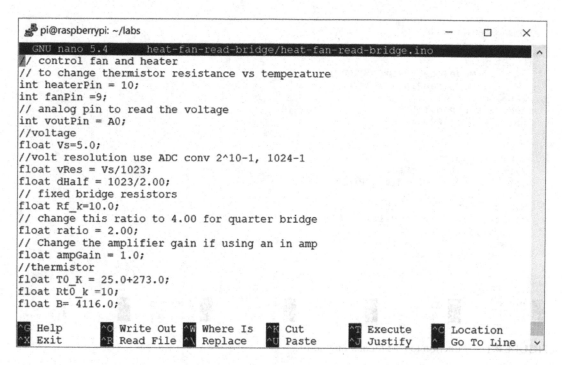

FIGURE 4.16 Script to read bridge output. Declarations.

For generality, the script is written to account for an amplifier and thus we declare a variable to give value to the amplifier gain. For now, leave it as $G = 1.00$, but we will be able to change it later when we use an amplifier.

```
// Change the amplifier gain if using an in amp
float ampGain = 1.0;
```

Next, we write the setup (same as in the scripts we have used before) and a function to read bipolar voltage (Figure 4.17). This function subtracts half of the maximum digital number from the ADC reading, and then multiplies by 2.00 to obtain what would be the transducer output.

```
float readBipolar(){
    //read digital number
    int din = analogRead(voutPin);
    //convert to voltage
    // subtract half point
    float din1 = din - dHalf;
    //convert to voltage
    float vout= 2.00*din1*vRes;
    return vout;
}
```

Then, we build a function to read the bridge output (Figure 4.18) where we convert voltage to temperature using the inverse relation for the bridge. Recall from Chapter 4 of the textbook that for a bridge circuit the thermistor resistance is calculated from

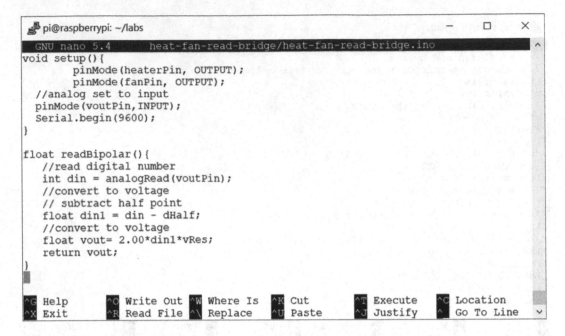

```
pi@raspberrypi: ~/labs                                         −   □   ×

  GNU nano 5.4          heat-fan-read-bridge/heat-fan-read-bridge.ino
void setup(){
        pinMode(heaterPin, OUTPUT);
        pinMode(fanPin, OUTPUT);
  //analog set to input
  pinMode(voutPin,INPUT);
  Serial.begin(9600);
}

float readBipolar(){
  //read digital number
  int din = analogRead(voutPin);
  //convert to voltage
  // subtract half point
  float din1 = din - dHalf;
  //convert to voltage
  float vout= 2.00*din1*vRes;
  return vout;
}


^G Help       ^O Write Out  ^W Where Is  ^K Cut      ^T Execute   ^C Location
^X Exit       ^R Read File  ^\ Replace   ^U Paste    ^J Justify   ^  Go To Line
```

FIGURE 4.17   Function to read bipolar output.

$$R_t \simeq R_0 - \frac{bV_{\text{out}}R_f}{V_s} \tag{4.7}$$

Here, we have generalized using b which takes the value 2 for a linearized bridge and 4 for a quarter bridge. In cases when we have an amplifier of gain G

$$R_t \approx R_0 - \frac{bV_{\text{out}}R_f}{V_s \times G} \tag{4.8}$$

Once we calculate the thermistor resistance, we obtain temperature from the inverse of the B model

$$T = \left(T_0^{-1} + B^{-1} \ln\left(\frac{R_t}{R_0}\right) \right)^{-1} \tag{4.9}$$

These equations are applied within the `readBridge` function (Figure 4.18) using ratio for b and `ampGain` for G. Recall that both variables were declared at the outset of the script.

```
    float voutRaw= readBipolar();
    float vout= voutRaw+calBias;
    // use inverse equation
    // ratio=2 for linear bridge and ratio=4 for quarter bridge
    float Rt_k = Rt0_k - ratio*Rf_k*vout/(Vs*ampGain);
    float T_K =1/(1/T0_K+(1/B)*log(Rt_k/Rt0_k));
    float T_C = T_K - 273.0;
```

Finally, we program the loop function (Figure 4.19) in a similar fashion to other cases but change the call to read the transducer to function `readBridge`.

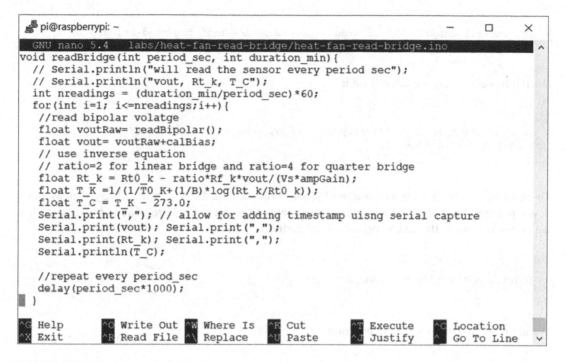

```
pi@raspberrypi: ~                                              —    □    ×
  GNU nano 5.4    labs/heat-fan-read-bridge/heat-fan-read-bridge.ino    ^
void readBridge(int period_sec, int duration_min){
  // Serial.println("will read the sensor every period sec");
  // Serial.println("vout, Rt_k, T_C");
  int nreadings = (duration_min/period_sec)*60;
  for(int i=1; i<=nreadings;i++){
  //read bipolar volatge
  float voutRaw= readBipolar();
  float vout= voutRaw+calBias;
  // use inverse equation
  // ratio=2 for linear bridge and ratio=4 for quarter bridge
  float Rt_k = Rt0_k - ratio*Rf_k*vout/(Vs*ampGain);
  float T_K =1/(1/T0_K+(1/B)*log(Rt_k/Rt0_k));
  float T_C = T_K - 273.0;
  Serial.print(","); // allow for adding timestamp uisng serial capture
  Serial.print(vout); Serial.print(",");
  Serial.print(Rt_k); Serial.print(",");
  Serial.println(T_C);

  //repeat every period_sec
  delay(period_sec*1000);
  }
}

^G Help      ^O Write Out  ^W Where Is  ^K Cut     ^T Execute  ^C Location
^X Exit      ^R Read File  ^\ Replace   ^U Paste   ^J Justify     Go To Line  v
```

FIGURE 4.18 Function to invert the bridge output to temperature.

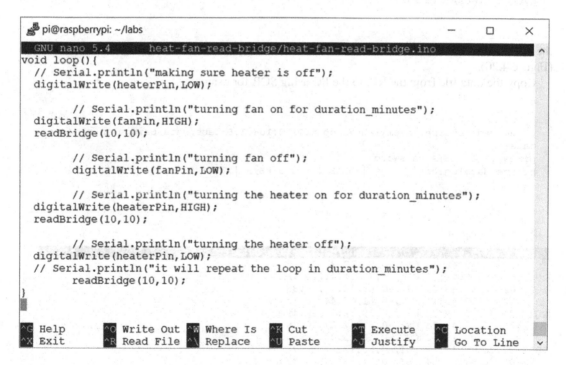

```
pi@raspberrypi: ~/labs                                        —    □    ×
  GNU nano 5.4      heat-fan-read-bridge/heat-fan-read-bridge.ino    ^
void loop(){
  // Serial.println("making sure heater is off");
  digitalWrite(heaterPin,LOW);

       // Serial.println("turning fan on for duration_minutes");
  digitalWrite(fanPin,HIGH);
  readBridge(10,10);

       // Serial.println("turning fan off");
       digitalWrite(fanPin,LOW);

       // Serial.println("turning the heater on for duration_minutes");
  digitalWrite(heaterPin,HIGH);
  readBridge(10,10);

       // Serial.println("turning the heater off");
  digitalWrite(heaterPin,LOW);
  // Serial.println("it will repeat the loop in duration_minutes");
       readBridge(10,10);
}

^G Help      ^O Write Out  ^W Where Is  ^K Cut     ^T Execute  ^C Location
^X Exit      ^R Read File  ^\ Replace   ^U Paste   ^J Justify     Go To Line  v
```

FIGURE 4.19 Loop function.

Now we will use Arduino-cli on the RPi to compile and run the script. First, verify that the Arduino folder `heat-fan-read-bridge` and its file `*.ino` is indeed in the labs folder of the RPi. Then, compile the Arduino sketch

```
arduino-cli compile --fqbn arduino:avr:uno labs/heat-fan-read-bridge
```

Once it has compiled, we can upload it

```
arduino-cli upload -p /dev/ttyACM0 --fqbn arduino:avr:uno labs/
heat-fan-read-bridge
```

To examine the output, we will use the Python serial capture file _ serial.py that we have developed previously but edit it to change the name of the file to bridge-datalog.csv so that we do not overwrite the datalog.csv used earlier.

```
logfile = "labs/data/bridge-datalog.csv"
```

We are ready to run this Python program

```
python3 labs/file_serial.py
```

the data stream will echo on the terminal and then written to file bridge-datalog.csv (Figure 4.20).
Copy the data file from the RPi to the PC using SCP, for example

```
c:\acevedo\courses\rtem\labs>pscp pi@192.168.1.5:labs/data/bridge-datalog.csv
data/
pi@192.168.1.5's password:
bridge-datalog.csv          | 9 kB |    9.3 kB/s | ETA: 00:00:00 | 100%
```

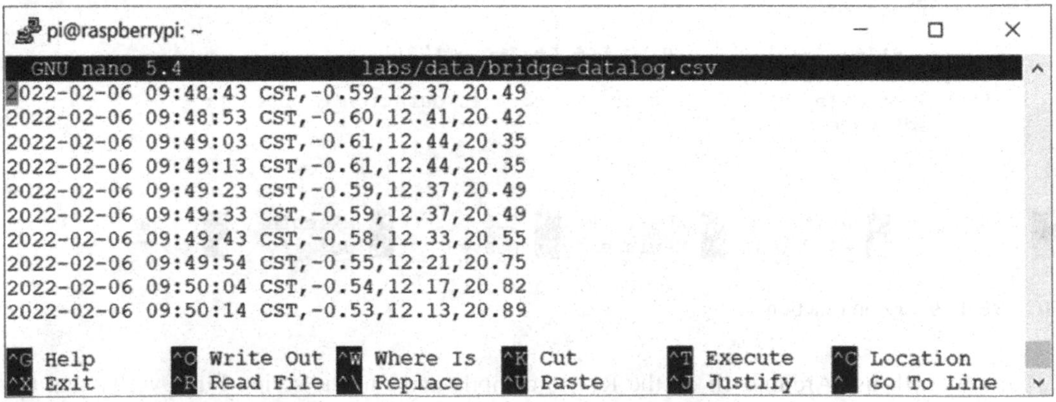

FIGURE 4.20 Captured file using the Python program on the RPi.

PLOT THE TIME SERIES USING R PACKAGE XTS

Now, using R, we read this `bridge-datalog.csv` file and plot. For this purpose, we use the R package `xts`, which we install using Install packages from the Packages Menu of the GUI, or simply type

```
install.packages('xts')
```

Once installed, you can load it using `library(xts)`.

```
library(xts)
```

We read the file, use `strptime` to build a time base from column 1 which has the time stamp

```
# read data from Arduino convert timestamp
X <- read.table("data/bridge-datalog.csv", header=TRUE, sep=",")
# read time sequence from file
tt <- strptime(X[,1], format="%Y-%m-%d %H:%M:%S %z")
```

select the variables, which are in columns 2–4, bind them with the time base as an `xts` object, and give names to the variables

```
Y <- X[,2:4]
x.ts<- xts(Y,tt)
names(x.ts) <- c("Vout(V)","Rt(kOhm)","Temp(C)")
```

There are two functions to plot `xts` objects, one is `plot.xts` (the default) and the other is `plot.zoo`, and there are pros and cons to each. Using all default argument values, we can simply type `plot` to use the default function

```
plot(x.ts)
```

you would notice that because the units are different, we cannot appreciate the details of each trace. To improve this, we set arguments `multi.panel=TRUE` and `yaxis.same=FALSE`

```
plot(x.ts, multi.panel=TRUE,yaxis.same=FALSE)
```

Now, we have separate traces and can observe the details of each. By default, the line width is `lwd=2`, which may look too thick in some graphs, we can change here to `lwd=1`; also, by default the title `main` is the label for the time series object, we can change to `main="Bridge"`

```
plot(x.ts,main="Bridge",multi.panel=TRUE,yaxis.same=FALSE,lwd=1)
```

We obtain a time series plot like that shown in Figure 4.21.

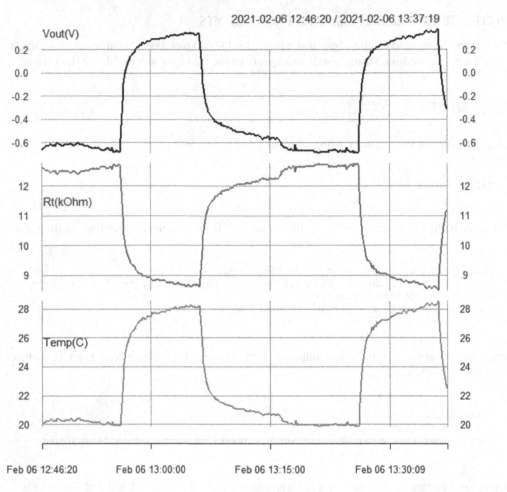

FIGURE 4.21   Linearized bridge output: time series plot using plot.xts.

Alternately, if we use `plot.zoo`, by default

```
plot.zoo(x.ts)
```

it would separate the graphs in separate windows, it would use `Index` for the x-axis label and the object name in the main title. We can change arguments `main="Bridge"`, and `xlab="Hour"`

```
plot.zoo(x.ts,main="Bridge",xlab="Hour")
```

and obtain a plot like the one shown in Figure 4.22.

Bridge

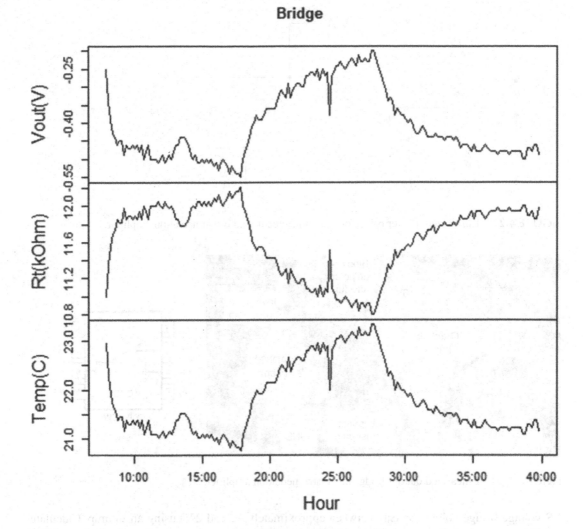

FIGURE 4.22   Linearized bridge output: time series plot using plot.zoo.

BRIDGE AND INSTRUMENTATION AMPLIFIER

Small deviations of the variable (e.g., temperature) produce small voltage output of the bridge; we can amplify the bridge output to obtain greater FS and sensitivity. An instrumentation amplifier (or in-amp) is a good choice for amplifying the bridge output. We will use the in-amp AD 620 or the In-amp INA126 that comes in a single 8-pin IC and its gain G can be programmed using a single resistor R_G. For the AD620, the gain is $G = \dfrac{49.4\,\text{k}\Omega}{R_G} + 1$ and for the INA126 is $G = 5 + \dfrac{80\,\text{k}\Omega}{R_G}$. See datasheets in RTEM repository for the pin-out and detailed specifications of these devices. We will continue using the linearized bridge already implemented and take its output to the in-amp as shown in Figure 4.23 so that we preserve linearity as provided by the linearized bridge and increase sensitivity using the in-amp.

Disconnect the bipolar-to-unipolar voltage divider from the output of the bridge and connect it to the output of the in-amp; now, the output of the bridge is connected to the input of the in-amp. The midpoint of the bipolar-to-unipolar voltage divider is connected to port A0 of the Arduino (Figure 4.24). Using the knowledge of your bridge output, calculate the gain G needed to bring the

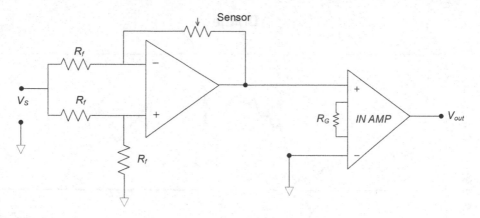

FIGURE 4.23 Linearized quarter bridge output connected to and instrumentation amplifier.

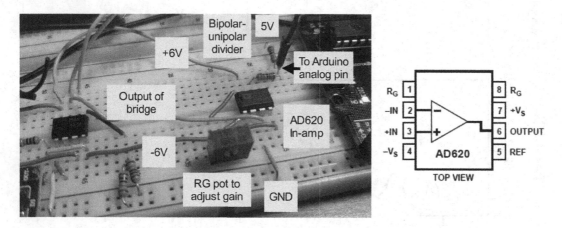

FIGURE 4.24 Linearized quarter bridge and instrumentation amplifier.

FS voltage bridge voltage up (say between approximately −2 and 2 V) using an in-amp. Calculate the R_G needed to attain this gain G. Use a 15 or 20 kΩ potentiometer for the R_G and vary it to obtain the desired larger signal and adjust the gain (Figure 4.24).

Let us change the gain in the Arduino script, e.g., to 10.0 in the declaration area of the script `heat-fan-read-bridge.ino`

```
// Change the amplifier gain if using an in amp
float ampGain = 10.0;
```

This small edit can be easily done using nano on the RPi. Now, we will use Arduino-cli on the RPi to compile and run the script.

```
arduino-cli compile --fqbn arduino:avr:uno labs/heat-fan-read-bridge
arduino-cli upload -p /dev/ttyACM0 --fqbn arduino:avr:uno labs/heat-fan-
read-bridge
```

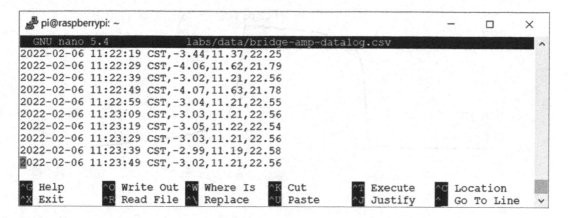

FIGURE 4.25   Captured file.

To examine the output, we will use the Python serial capture `file _ serial.py` that we have developed previously but edit it to change the name of the file to `bridge-amp-datalog.csv` so that we do not overwrite the files used earlier. The outcome will be an updated file (Figure 4.25).

Copy the data file from the RPi to the PC using SCP, for example,

```
c:\acevedo\courses\rtem\labs>pscp pi@192.168.1.5:labs/data/bridge-amp-
datalog.csv data/
pi@192.168.1.5's password:
bridge-amp-datalog.csv        | 8 kB |   8.7 kB/s | ETA: 00:00:00 | 100%
```

PLOT THE TIME SERIES USING R PACKAGE XTS

Now, in a similar way to the linearized bridge activity, we employ R to read this `bridge-amp-datalog.csv` file, convert to time series using `xts`, and plot; the only change needed is the name of the file to `bridge-amp-datalog.csv`. The code is given here for easy reference

```
# read data from Arduino convert timestamp
X <- read.table("data/bridge-amp-datalog.csv", header=TRUE,sep=",")
# read time sequence from file
tt <- strptime(X[,1], format="%Y-%m-%d %H:%M:%S %z")
Y <- X[,2:4]
x.ts<- xts(Y,tt)
names(x.ts) <- c("Vout(V)","Rt(kOhm)","Temp(C)")
plot(x.ts,main="Bridge Amp",multi.panel=TRUE,yaxis.same=FALSE,lwd=1,col=1)
```

We obtain the time series plot like that shown in Figure 4.26 using plot for default of `plot.xts`, or you can also use `plot.zoo` as explained previously.

```
plot.zoo(x.ts,main="Bridge Amp",xlab="Hour")
```

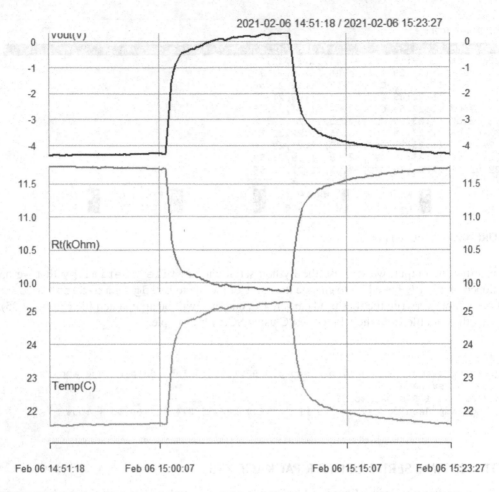

FIGURE 4.26 Output from linearized quarter bridge and instrumentation amplifier.

DISPLAYING TRANSDUCER OUTPUT ON A WEB PAGE

In this part, we display the results of the bridge transducer plus instrumentation amplifier on a web page using HTML, PHP, and JS in a similar manner to that explained in Lab session 3; however, this time we will learn how to plot more than one line on the SVG graph and two y-axes, one on the left and one on the right. For this purpose, we will download the lab4/multiplotd3.zip from the RTEM repository and store its contents multiplotd3.html in the working directory labs of the RPi. Let us open this html file using nano

```
nano labs/multiplotd3.html
```

and go to the d3.csv statement and see that the data file name is bridge-amp-datalog.csv

```
d3.csv("data/bridge-amp-datalog.csv",
   function(d){
     return {TimeStamp:d3.timeParse("%Y-%m-%d %H:%M:%S %Z")(d.TimeStamp),
Vout:d.Vout, Rt:d.Rt, Temp:d.Temp}
   },
```

The next changes relate to the y-axis. First, scale the y-axis considering that the minimum of the domain would be given by the Rt value since it is smaller than Temp.

```
   // y axis as linear
   var y = d3.scaleLinear()
     .domain([d3.min(data,function(d){return +d.Rt;}),
d3.max(data,function(d){return +d.Temp;})])
     .range([height, 0]);
   svg.append("g")
     .call(d3.axisLeft(y));
```

Second, create a right y-axis to plot the voltage

```
   // y right axis as linear
   var y1 = d3.scaleLinear()
     .domain([d3.min(data,function(d){return +d.Vout;}),
d3.max(data,function(d){return +d.Vout;})])
     .range([height,0]);
   svg.append("g")
        // note that we need to translate to the right
        .attr("transform","translate(" + width + ",0)")
     .call(d3.axisRight(y1));
```

Once this is complete, we add lines for R_t and V_{out}

```
   svg.append("path")
     .datum(data)
     .attr("fill", "none")
     .attr("stroke", "blue")
     .attr("d", d3.line()
       .x(function(d) { return x(d.TimeStamp)})
       .y(function(d) { return y(d.Rt)})
       )
   svg.append("path")
     .datum(data)
     .attr("fill", "none")
     .attr("stroke", "red")
     .attr("d", d3.line()
       .x(function(d) { return x(d.TimeStamp)})
       .y(function(d) { return y1(d.Vout)})
       )
```

The final step is to add labels to the axes. The labels for the x-axis and left y-axis are done as in the `plotd3.html` of Lab 3; we just relabel it to `"Rt(kohm)  and  Temp(C)"`. However, we must add labels to the right y-axis, which requires using `width` in the y coordinate instead of 0.

```
// y right axis label
svg.append("text")
  .attr("transform", "rotate(-90)")
  .attr("y", width + (margin.right/2))
  .attr("x", 0 - (height/2))
  .attr("dy", "1em")
  .style("text-anchor", "middle")
  .style("font-size", "10px")
  .text("Vout(V)");
```

We must also edit the `readfile.php` script so that it refers to the `bridge-amp-datalog.csv` file

```
nano readfile.php
```

as shown here

```
<?php
$lines = file("data/bridge-amp-datalog.csv");
foreach ($lines as $line_num=> $line) {
    echo $line . "<br>";
}
?>
```

Finally, we are ready to edit the `index.php` file so that it refers to the `bridge-amp-datalog.csv` file and `multiplotd3.html`. Use nano

```
nano index.php
```

and edit accordingly

```
<!DOCTYPE html>
<html>
<head>
 <title>Real-Time Environmental Monitoring</title>
 <link rel="stylesheet" href="css/style.css">
</head>
<body>
 <h1>RTEM Test Page</h1>
 <h2>Learning HTML, PHP, JS</h2>

 <p>Example of URL link: external web page
 <a href=https://github.com/mfacevedol/rtem>RTEM GitHub Repository</a>
 </p>
 <p>Example of URL link: internal from the server file system
```

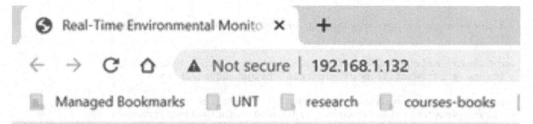

RTEM Test Page

Learning HTML, PHP, JS

Example of URL link: external web page RTEM GitHub Repository

Example of URL link: internal from the server file system Download bridge amp datalog file

Display bridge amp datalog file contents using php

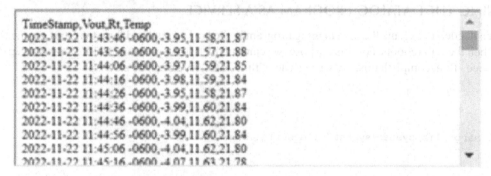

```
TimeStamp,Vout,Rt,Temp
2022-11-22 11:43:46 -0600,-3.95,11.58,21.87
2022-11-22 11:43:56 -0600,-3.93,11.57,21.88
2022-11-22 11:44:06 -0600,-3.97,11.59,21.85
2022-11-22 11:44:16 -0600,-3.98,11.59,21.84
2022-11-22 11:44:26 -0600,-3.95,11.58,21.87
2022-11-22 11:44:36 -0600,-3.99,11.60,21.84
2022-11-22 11:44:46 -0600,-4.04,11.62,21.80
2022-11-22 11:44:56 -0600,-3.99,11.60,21.84
2022-11-22 11:45:06 -0600,-4.04,11.62,21.80
2022-11-22 11:45:16 -0600,-4.07,11.63,21.78
```

Read bridge amp datalog file and plot multi chart using d3.js

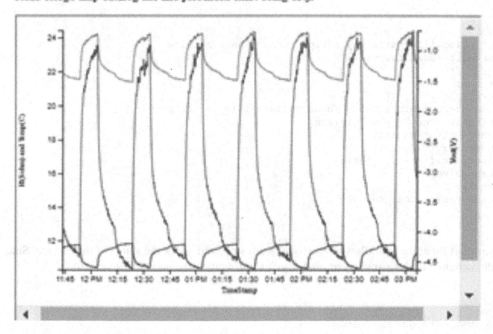

FIGURE 4.27   HTML example illustrating CSS, PHP, and JS.

```
<a href="data/bridge-amp-datalog.csv">Download bridge amp datalog file</a>
</p>

<p>Display bridge amp datalog file contents using php</p>
<iframe src="readfile.php" width=600 height=200></iframe>

<p>Read bridge amp datalog file and plot multi chart using d3.js</p>
<iframe src="multiplotd3.html" width=600 height=400></iframe>

</body>
</html>
```

Once you refresh `index.php` on the browser, the result will be as shown in Figure 4.27. As time goes on and the Arduino keeps collecting data, streaming to the RPi, and appending `bridge-amp-datalog.csv`, you can refresh the browser to see updates to the graph.

RUNNING THE DATALOG PROGRAM AS A SERVICE

It is often needed to execute the data logging program in the background and for it to restart automatically when the RPi reboots. For this purpose, we can run the Python program `file _ serial.py` as a service. To accomplish this, we write a file within /etc/system/system

```
sudo nano /lib/system/system/file_serial.service
```

with the following content

```
[Unit]
Description=file_serial.py script running as a service
After=multi-user.target

[Service]
Type=simple
ExecStart=/usr/bin/python3 /home/pi/labs/file_serial.py
WorkingDirectory= /home/pi/labs
StandardOutput=inherit
StandardError=inherit
SyslogIdentifier=file_serial
Restart=always
[Install]
WantedBy=multi-user.target
```

The command `systemctl` allows to activate, start, stop, and check the status of the service. Since we made a change in the /etc/systemd/system directory, we need to reload with

```
sudo systemctl daemon-reload
```

Then, we enable the service

```
sudo systemctl enable file_serial.service
```

and start it

```
sudo systemctl start file_serial.service
```

We can verify that the file `bridge-amp-datalog.csv` is being written to the data directory. Also, to verify the service we can use status

```
sudo systemctl status file_serial.service
```

The service will keep a log that can be read with

```
sudo journalctl -f -u file_serial.service
```

We want to verify that the service will restart after an RPi reboot. Test it with

```
sudo reboot
```

You would have to start PuTTY again, and check that we are writing to the file again. Alternatively, just refresh your browser and verify that new records are being recorded and plotted.

EXERCISES

Exercise 4.1 Writing datalog file to RPi.
Connect the Arduino to the RPi Zero W. Secure copy (pscp) the heat-fan-read-sensor-log from the PC to the RPi. Compile and upload the Arduino code using the Arduino CLI software installed on the RPi Zero W. Write the python code to establish serial communication between the Arduino and the RPi Zero W in order to write a `datalog.csv` file that includes the timestamp. Secure copy the file to the PC.

Exercise 4.2 Bridge Analysis
Analyze a balanced source voltage divider, a quarter bridge, and a linearized bridge using R. Plot the output voltage and calculate sensitivity. Employ the same circuit and thermistor parameter values given in the lab activity.

Exercise 4.3 Linearized Bridge Implementation
Implement the linearized bridge following the steps provided in the lab activity. The implementation includes rail splitter and linearizing op-amp.

Exercise 4.4 Arduino Reading Linearized Bridge Output.
Implement reading the linearized bridge output using the Arduino and Raspberry Pi Zero W into a datalog file. Use R code to produce time series graphs.

Exercise 4.5 Instrumentation Amplifier Implementation.

Add the instrumentation amplifier to the linearized bridge. Modify the Arduino code to include the gain and use the Raspberry Pi Zero W to write a datalog file. Use R code to produce time series graphs.

Exercise 4.6 Displaying Transducer Output on a Web Page

Implement the web page to download and display your datalog file. Refresh the browser to verify that new data are recorded and plotted as time progresses.

Exercise 4.7 Running the Datalog Program as a Service

Implement the service to run the `file _ serial.py` program in the background. Test that it works after a reboot to the RPi.

REFERENCES

Acevedo, M.F. 2024. *Real-Time Environmental Monitoring: Sensors and Systems - Textbook, Second Edition.* Boca Raton, FL: CRC Press, Taylor & Francis Group, 392 pp.

Amazon. 2023. *KETOTEK Digital Thermometer Temperature Meter Gauge.* Accessed January 2023. https://www.amazon.com/KETOTEK-Thermometer-Temperature-Waterproof-Fahrenheit/dp/B01DVSIYN2/ref=asc_df_B01DVSIYN2?tag=bngsmtphsnus-20&linkCode=df0&hvadid=79989588513714&hvnetw=s&hvqmt=e&hvbmt=be&hvdev=c&hvlocint=&hvlocphy=&hvtargid=pla-4583589114951828&th=1.

Arduino. 2016. *Download the Arduino Software.* Accessed August 2016. https://www.arduino.cc/en/Main/Software.

Arduino. 2023. *Arduino-CLI.* Accessed January 2023. https://github.com/arduino/arduino-cli.

Autodesk Instructables. 2023. *DC-DC Boost Converter MT3608.* Accessed January 2023. https://www.instructables.com/DC-DC-Boost-Converter-MT3608/.

Components Info. 2023. *MT3608 DC To DC Step Up Boost Converter Module Pinout, Datasheet, Specs.* Accessed January 2023. https://www.componentsinfo.com/mt3608-module-pinout-datasheet/.

Geany. 2023. *Geany - The Flyweight IDE.* Accessed January 2023. https://www.geany.org/.

Posit. 2023. *RStudio IDE.* Accessed January 2023. https://posit.co/downloads/.

PuTTY. 2023. *Download PuTTY.* https://www.putty.org/.

R Project. 2023. *The Comprehensive R Archive Network.* Accessed January 2023. http://cran.us.r-project.org/.

Texas Instruments. 2023. *LM358-N.* Accessed January 2023. https://www.ti.com/product/LM358-N.

WinSCP. 2023. *Free Award-Winning File Manager.* Accessed January 2023. https://winscp.net/eng/index.php.

5 Dataloggers and Sensor Networks

INTRODUCTION

This lab session covers aspects of datalogging, digital interface types, and sensor networks. We start by making a datalogger from an Arduino conducting several activities in sequence: adding a real-time clock (RTC) to an Arduino while learning the principles of I2C interface, pulsing and reading a sensor, combining sensor reading and RTC, interfacing an SD card with an Arduino while learning the Serial Peripheral Interface (SPI) interface, and combining the RTC and SD card. We then add an Ethernet module to an Arduino and learn how to combine sensor readings with an interface based on TCP/IP that places information on a web server. We take the time in this lab session to learn about reducing program space and dynamic memory when writing Arduino sketches. This is important because the Arduino Uno has limited space and memory. Since some environmental sensors come equipped with a digital interface such as RS-485, UART, IC-2, or SDI-12, we practice one example, interfacing RS-485 to a Raspberry Pi (RPi). The chapter ends with comments on Arduino datalogger and ethernet shields.

MATERIALS

READINGS

For theoretical background, you can use Chapter 5 of Acevedo, M.F. 2024. *Real-Time Environmental Monitoring: Sensors and Systems Second Edition - Textbook*, which is a companion to these guides (Acevedo 2024). Other bibliographical references are cited throughout the guide.

COMPONENTS

- Protoboard and wire jumpers
- 10 kΩ resistor to be used in the voltage divider
- Potentiometer 15 or 20 kΩ (to be used to simulate a resistive sensor)
- Ethernet cable CAT-5

MAJOR COMPONENTS AND INSTRUMENTS

- Arduino Uno
- Raspberry Pi Zero W.
- Micro SD adapter (e.g., HiLetgo Micro SD TF Card Adapter Reader Module 6Pin SPI Interface Driver Module with chip Level Conversion for Arduino)
- Micro SD and adapter (e.g., 16 GB)
- Tiny RTC (WINGONEER Tiny RTC I2C DS1307 AT24C32 Real-Time Clock Module for Arduino)
- ENC28J60 Module (Ethernet LAN Network Module SPI Interface 3.3 V for Arduino)
- RS-485 Interface Module (based on MAX485 TTL to RS-485)
- Network Switch 5-port if needed

DOI: 10.1201/9781003184362-5

TOOLS (RECOMMENDED)

- Long nose pliers
- Wire strippers and clippers
- Small Screwdriver

SOFTWARE

- R, for data analysis (R Project 2023)
- RStudio, an IDE to use R (Posit 2023)
- PuTTY, for Secure Shell (PuTTY 2023)
- WinSCP, to transfer files from the RPi to a windows PC (WinSCP 2023)
- Python (installed from RPi)
- Arduino IDE, Interface to Arduino (Arduino 2016)
- Geany, IDE to edit programs as well as data (Geany 2023)
- Arduino-CLI, Arduino software to run using commands from the RPi (Arduino 2023a)
- PHP (installed from RPi) for RPi.
- Arduino library RTClib (Adafruit 2023)
- Arduino SPI library (Arduino 2023b)
- Arduino SD library (Arduino 2022c)
- Arduino SoftwareSerial library (Arduino 2022d)
- Arduino EthernetENC library (Arduino 2022a)

SCRIPTS

All available from the RTEM GitHub repository https://github.com/mfacevedol/rtem

- Arduino scripts contained in archive `Arduino-datalogger.zip`
- Python script contained in archive `rs485 _ logger.zip`
- JS D3 script `multiplotd3-rs485.zip`

SUPPLEMENTARY SUPPORT MATERIAL

Supplementary support material including additional screenshots, images, and procedures are available from the publisher eResources web page provided for this book.

REAL-TIME CLOCK AND THE I2C INTERFACE

As discussed in Chapter 5 of the companion textbook, an RTC is a circuit that keeps track of current time, even if the processor is powered off. It is based on counters for tenths/hundredths of seconds, seconds, minutes, hours, days, date, month, year, and century. RTCs are very important in monitoring, since dataloggers must provide a timestamp to each data record. An RTC can be implemented using an RTC integrated circuit (IC). A low-cost example of an RTC IC is the DS1307, which can count seconds, minutes, hours, date of the month, month, day of the week, and year with leap-year compensation (Maxim Integrated 2022). Registers for time and calendar are in BCD format; the register banks have addresses 00–07 hours, each register contains two nibbles or one byte (Table 5.1).

Take for example the seconds register, the low nibble or least significant (LS) nibble has the LS digit of seconds in BCD from 0 to 59; that is from 0 to 9. The high nibble contains the tens of seconds or from 1 to 5 by using bits 6 to 5 (Figure 5.1).

TABLE 5.1
Register Bank of DS1307

00H	Seconds
01H	Minutes
02H	Hours
03H	Day
04H	Date
05H	Month
06H	Year
07H	Control

	Bit7	Bit6	Bit5	Bit4	Bit3	Bit2	Bit1	Bit0
	Clock Halt	10's of seconds			Seconds			
	0=enable	3 LS bits of high nibble			Low nibble			
Examples	53	1	0	1	0	0	1	1
	15	0	0	1	0	1	0	1

FIGURE 5.1 RTC seconds' register.

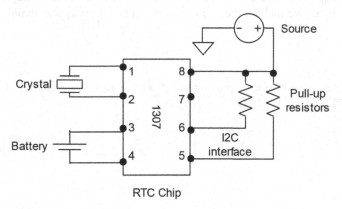

FIGURE 5.2 Circuit schematic for an RTC based on the DS1307.

Figure 5.2 shows a simple circuit to build an RTC using the DS1307. Here, the source is 5 V, connected to pin 8, the pull-up resistors are 1 kΩ, pins 6 and 5 are the SCL (clock line) and SDA (data line) serial lines of the I2C interface that allow connecting to a microcontroller unit (MCU). The crystal, for example, 32.768 kHz, is connected between pins 1 and 2, and the battery (3.3 V) is connected between pins 3 and 4. Instead of wiring it from scratch, we will use the Tiny RTC, a practical low-cost circuit based on DS1307 (ElectroPeak 2022; Instructables Circuits 2022).

As mentioned above, the SCL and SDA lines of the serial I2C interface are used to transfer time and calendar information to an MCU. The DS1307 operates as a slave device; register contents are sent on the SDA line, with transmission using START and STOP. We use analog pins A4 and A5 of the Arduino, connected to SDA and SCL, respectively, to communicate with the RTC via the I2C interface. Use jumper wires to implement the circuit given in Figure 5.3.

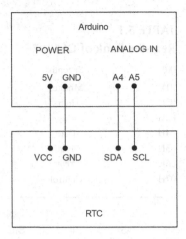

FIGURE 5.3   Connecting the Tiny RTC circuit to an Arduino.

Now we write the code to make the Arduino communicate with the RTC circuit using the
RTClib library, which contains convenient functions to use the DS1307 with Arduino and the
Wire library which manages I2C serial communication. The RTClib is available online; for
example (Adafruit 2023), and you can also use the Arduino IDE to install it, via the menu Tools|
Manage Libraries. Search for RTClib and install.

Type the following code, save as rtc.ino, or use the one provided in the download from the
RTEM repository. First, we include the libraries and declare rtc as shorthand to use the library
functions.

```
// Date and time functions using a DS1307 RTC connected via I2C and Wire lib
#include <Wire.h>
#include "RTClib.h"

RTC_DS1307 rtc;
```

Second, the setup

```
void setup () {
  Serial.begin(9600);
  if (! rtc.begin()) {
    Serial.println("No RTC");
    while (1);
  }
  delay(1000);
  if (! rtc.isrunning()) {
    Serial.println("Set the time as needed");
    // remove comment to one of two lines
    // This line sets the RTC to the date & time the sketch is compiled
```

```
   // rtc.adjust(DateTime(F(__DATE__), F(__TIME__)));
   // This line sets a date & time, e.g.,12/07/2022 23:00 would be:
   // rtc.adjust(DateTime(2022, 12, 7, 17, 23, 0));
  }

}
```

You can set the date and time according to the compilation time

```
rtc.adjust(DateTime(F(__DATE__), F(__TIME__)));
```

or manually

```
rtc.adjust(DateTime(YEAR, MONTH, DAY, HOUR , MINUTE, SECOND));
```

You need to set up date and time only once, and after that, comment out with "//" or delete the related line from your code. Otherwise, date and time will be set again any time you run the script. If you opt to set it manually, type the time a few seconds earlier before uploading the code to compensate for the delay to upload the code.

If you have trouble changing the time, then copy the `rtc.adjust` line outside the `if (!rtc.isrunning())` control block. It could be placed just after the `delay(1000)` line and before the `if` line. Once it works then comment out that line afterwards so that it does not execute the next time you compile.

Finally, we write the loop function which gets current time and streams it to the serial monitor every five seconds, using the format `YY/MM/dd  HH:MM:SS`. For that purpose, we use the print functions of single characters '/', '', and ':'.

```
void loop () {
    DateTime now = rtc.now();
    Serial.print(now.year(), DEC); Serial.print('/');
    Serial.print(now.month(), DEC); Serial.print('/');
    Serial.print(now.day(), DEC); Serial.print(' ');
    Serial.print(now.hour(), DEC); Serial.print(':');
    Serial.print(now.minute(), DEC); Serial.print(':');
    Serial.print(now.second(), DEC); Serial.println();
    //repeat every 5 sec
    delay(5000);
}
```

Compile, upload, open the serial monitor window, and verify the result (Figure 5.4).

FIGURE 5.4 RTC script serial monitor output.

READING SENSORS BY PULSING POWER

As an example, consider reading the signal from a simulated water electrical conductivity (EC) transducer made as voltage divider using a fixed resistor and a resistive sensor (Figure 5.5). For simplicity, we will use a potentiometer to simulate the sensor so that we can focus on the datalogging aspects rather than the details of the sensors as we covered in Labs 3 and 4. The excitation to the voltage divider will be a pulse, using `A1` as an output pin, instead of a constant voltage source to the probe using the 5 V pin of the Arduino. This helps save power consumed by the sensor. In this configuration, we use analog `A0` as input to read the transducer output voltage.

Next, we will write a program based on a program by Gertz (2014) that would read EC from this sensor (Gertz and Di Justo 2012). Type this script and save it as `pulsed _ sensor` or use it from the RTEM repository. Since the Arduino Uno has only 32 kB of space and 2 kB of dynamic memory, at this point, we start introducing some strategies that would help save space and memory. It is not critical for this script, but during this lab session we will keep adding other devices and requiring more memory (Arduino 2022b).

The global variables declarations and setup are as follows:

```
// Simulated Water Electrical Conductivity (EC) Monitor
// Sensor simulated by a potentiometer in a voltage divider

// sensor circuit parameter
float Vs = 5.0;
float Rf = 10.0; // in kohm
//sensor pins
int pinPulse=A1;
int pinSignal=A0;
// logical to print or skip header
bool flagHeader= false;

void setup() {
```

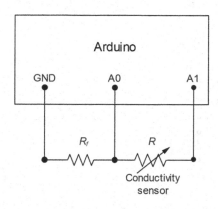 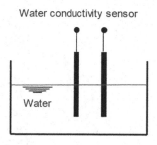

FIGURE 5.5 Arduino and simulated sensor.

```
Serial.begin(9600);
//sensor pins
pinMode(pinPulse,OUTPUT);
pinMode(pinSignal,INPUT);
}
```

In the above segment, we limited the global variables as much as possible to reduce the usage of dynamic memory, selecting only those that would allow us to reassign circuit parameters and reconfigure pins easily. Now, we write a function to read the sensor, thinking also of reducing storage by printing to the serial monitor optional and to be used only for verification, and we return the EC value. This avoids making EC a global variable and allowing it to be passed to other segments of the code when calling the function. By working in $k\Omega$ and mS units, we reduce significant digits by being able to use a float. However, for better precision in we could opt for more significant digits using EC as a double.

```
float readSensor(){
  digitalWrite(pinPulse,HIGH);
  delay(1000); // allow transients to stop
  int din = analogRead(pinSignal);
  digitalWrite(pinPulse,LOW);
  float Vout = din*Vs/1023;
  float Resist = Vs*Rf/Vout - Rf; //in kohm
  //can remove comment next lines to verify calculations
  //Serial.print(din);Serial.print(',');
  //Serial.print(Vout);Serial.print(',');
  //Serial.println(Resist);
  float EC = 1.0/Resist; // in mS
  return EC;
}
```

In the loop shown below, we print a header that could be skipped using the boolean flag. Note that we start using F("some string literal") in the stream print because string literals consume storage and memory. Finally, repetitively every five seconds we call readSensor and use the returned value in mS to calculate µS and TDS base 500.

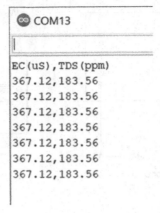

```
⊚ COM13

|
EC(uS),TDS(ppm)
367.12,183.56
367.12,183.56
367.12,183.56
367.12,183.56
367.12,183.56
367.12,183.56
367.12,183.56
```

FIGURE 5.6 Serial monitor.

```
void loop() {
 //prints output header
 if(flagHeader==false){
  // Discard first reading of sensor, print header, and set flag
  readSensor();
  Serial.print(F("EC(uS)"));Serial.print(F(","));  Serial.println(F("TDS(ppm)"));
  flagHeader=true;
 }
 // Calculate EC in uS and TDS in ppm (base 500) and print values in csv
 float mS = readSensor();
 Serial.print(1000.0*mS);Serial.print(F(","));Serial.println(500.0*mS);
 //repeat every 5 sec
 delay(5000);
}
```

Once compiled, you can see the storage and memory usage in the bottom banner of the IDE. In this case, we see that we use 5568 bytes of storage space (17% of 32 KB) and 317 bytes for global variables (15% of the 2048 bytes of memory),

```
Sketch uses 5568 bytes (17%) of program storage space. Maximum is 32256
bytes.
Global variables use 317 bytes (15%) of dynamic memory, leaving 1731 bytes
for local variables. Maximum is 2048 bytes.
```

You can confirm the script output opening the serial monitor (Figure 5.6).

COMBINING SENSOR READING AND RTC

Our next step toward building a datalogger is to combine sensor reading with an RTC to provide a timestamp for each reading as shown in the diagram of Figure 5.7. This system now uses four analog pins of the Arduino, two for the sensor and two for the I2C communication with the RTC.

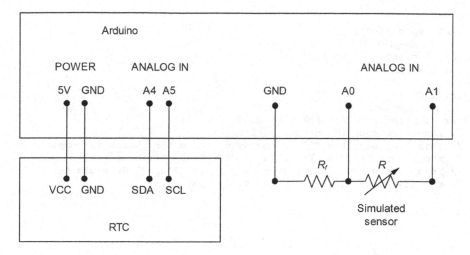

FIGURE 5.7 Combining sensor reading and RTC.

Combine the scripts RTC and pulsed _ sensor into a new script sensor _ timestamp, which is also available in the download from the RTEM repository. We have already discussed the libraries for RTC in the RTC script and the global variable declarations in the pulse _ sensor script. We just put them together.

```
// Simulated Water Electrical Conductivity (EC) Monitor
// Sensor simulated by a potentiometer in a voltage divider
// read sensor and print together with timestamp using an RTC

// Date and time functions using a DS1307 RTC connected via I2C and Wire lib
#include <Wire.h>
#include "RTClib.h"
RTC_DS1307 rtc;

// sensor circuit parameter
float Vs = 5.0;
float Rf = 10.0; // in kohm
//sensor pins
int pinPulse=A1;
int pinSignal=A0;
// logical to print ot skip header
bool flagHeader= false;
```

Similarly, in the setup we put together the contents of the RTC script with the pulse_sensor one.

```
void setup () {
  Serial.begin(9600);
  //sensor pins
  pinMode(pinPulse,OUTPUT);
  pinMode(pinSignal,INPUT);
```

```
  Serial.begin(9600);
  if (! rtc.begin()) {
    Serial.println(F("No RTC"));
    while (1);
  }
  delay(1000);
  if (! rtc.isrunning()) {
    Serial.println(F("Set the time as needed"));
    // remove comment to one of two lines
    // This line sets the RTC to the date & time the sketch is compiled
    // rtc.adjust(DateTime(F(__DATE__), F(__TIME__)));
    // This line sets a date & time, e.g.,12/07/2022 23:00 would be:
    // rtc.adjust(DateTime(2022, 12, 7, 17, 23, 0));
  }
}
```

The readSensor function remains the same. Finally, within the loop after writing the header, we write the timestamp first and then the EC and TDS values; please note that the line printing the second has a comma "," character at the end so that we have a full csv record.

```
// timestamp
DateTime now = rtc.now();
Serial.print(now.year(), DEC); Serial.print('/');
Serial.print(now.month(), DEC); Serial.print('/');
Serial.print(now.day(), DEC); Serial.print(' ');
Serial.print(now.hour(), DEC); Serial.print(':');
Serial.print(now.minute(), DEC); Serial.print(':');
Serial.print(now.second(), DEC); Serial.print(',');

// Calculate EC in uS and TDS in ppm (base 500) and print values in csv
float mS = readSensor();
Serial.print(1000.0*mS);Serial.print(F(","));Serial.println(500.0*mS);
//repeat every 5 sec
delay(5000);
```

Once compiled, we see that usage of storage space and dynamic memory has increased with this more complete script. You should be able to verify script output opening the serial monitor (Figure 5.8). It is opportune to note that even though it may seem logical to concatenate the strings to form the timestamp, instead of multiple Serial.print calls, it is not better because it would take more program space and memory (Arduino 2022b).

DATALOGGER BASED ON ARDUINO, RTC, AND SD CARD

Now that we have timestamped data, we will add SD (Security Digital) card storage to complete a datalogger, which can be implemented using SD cards. We will use a microSD Card Adapter (writer/reader) that communicates with the Arduino using the SPI. A convenient one is the HiLetgo Micro SD TF Card Adapter Reader Module 6Pin SPI Interface Driver Module with chip Level Conversion for Arduino. Recall from Chapter 5 of the companion textbook that SPI is synchronous and uses clock (SCK), MOSI (Master Output/Slave Input) used by the master to send data to slave,

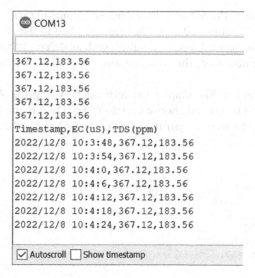

FIGURE 5.8   Serial output showing timestamp and values in CSV.

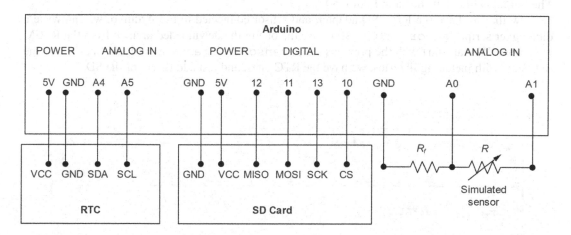

FIGURE 5.9   Adding an SD card adapter to an Arduino to complete a datalogger.

MISO (Master Input/Slave Output) used by a slave to send data to the master, and device ID (Chip Select – CS) lines. It is the clock line that keeps devices synchronized, by telling the receiver when to sample the bits on the data line. Wire the SPI lines of the SD adapter (VCC i.e., power, GND i.e., ground, MISO, MOSI, SCK, and CS) to the Arduino pins as given in Figure 5.9.

We will use a new SD card for the work with the Arduino. Do not use the one you have configured to run the RPi because we will still use the RPi in subsequent exercises. You first format the new card using a PC equipped with an SD writer/reader. When this adapter is for a large format SD, you will need an adapter for size. There are several programs available to format an SD card, for example, the SD Memory Card Formatter (SD Association 2022), or you can use the file manager, find the SD card, right click and select format. You can use quick format, verify the capacity of the card (e.g., 16 GB), and label the card (e.g., ARDUINO LAB) so that you can identify it later.

To write a script controlling the SD adapter, we use additional libraries, the SPI library (Arduino 2023b) and the SD library (Arduino 2022c). These are bundled in the IDE and should be ready to use. The SD library contains several script examples, such as `Card Info` (Get info about your SD card), `Datalogger` (Log data from three analog sensors to an SD card), and `Files` (Create and destroy an SD card file).

Insert the micro card in the SD adapter (writer/reader). Using the Arduino IDE navigate to `File|Examples`, go down to SD and choose `Card Info`. Declare the CS pin to 10 if it happens to be set to 4 or 8. You do this in line `const int chipSelect =10` as shown in this segment of the script

```
// change this to match your SD shield or module;
// Arduino Ethernet shield: pin 4
// Adafruit SD shields and modules: pin 10
// Sparkfun SD shield: pin 8
// MKRZero SD: SDCARD_SS_PIN
const int chipSelect =10;
```

The output is like that shown in Figure 5.10.

Now that we have the RTC, SD adapter, and sensor connected to the Arduino, we can write a datalogger script `sensor _ RTC _ SD` or use the one in the downloaded archive from the RTEM repository. We can start with the previous script `sensor _ timestamp` and add lines to use the SD. Start with including libraries, we have the RTC ones, and just add those for the SD.

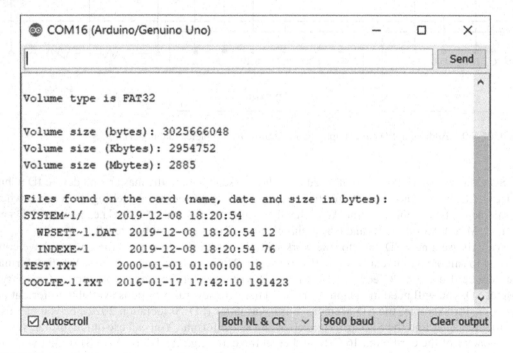

FIGURE 5.10   Serial monitor output for CardInfo.

```
// SD card using SPI and SD libraries
#include <SPI.h>
#include <SD.h>
```

Next, declare variables, we have everything but add the pin number, 10, for CS

```
// SPI bus: MOSI-pin11, MISO-pin12, CLK-pin13
// SD CS on pin10
int pinCS =10;
```

In the setup function, we add a few lines to initialize the SD card.

```
  // Initializing SD card at pin 10
  if (!SD.begin(pinCS)) {
    Serial.println(F("SD failed"));
    while (1);
  }
  Serial.println(F("SD initialized"));
}
```

Now in the loop function, we add lines to write the header to file only one time (as controlled by the header flag), forming a string separated by commas (CSV) and write it to the SD.

```
  // writes output header
  if(flagHeader==false){
    // Discard first reading of sensor, write header, and set flag
    readSensor();
    // open the file
    File dataFile = SD.open("datalog.txt", FILE_WRITE);
    // write header to file if available echo to serial monitor
    if (dataFile) {
      dataFile.print(F("Timestamp"));dataFile.print(F(","));
      dataFile.print(F("EC(uS)"));dataFile.print(F(",")); dataFile.
println(F("TDS(ppm)"));
      dataFile.close();
    }
    // else report error
    else {
      Serial.println(F("error opening file"));
    }
    // echo to serial monitor
    Serial.print(F("Timestamp"));Serial.print(F(","));
    Serial.print(F("EC(uS)"));Serial.print(F(",")); Serial.
println(F("TDS(ppm)"));
    flagHeader=true;
  }
```

Now, we continue the loop function by reading the time and sensor value and forming a record for the file with time and data to write it to the card. Again, we use multiple calls to `Serial.print` rather than string concatenation to save storage and memory (Arduino 2022b).

```
//time and sensor value
DateTime now = rtc.now();
float mS = readSensor();

// open the file
File dataFile = SD.open("datalog.txt", FILE_WRITE);
// write to file if available
if (dataFile) {
  // timestamp
  dataFile.print(now.year(), DEC); dataFile.print('/');
  dataFile.print(now.month(), DEC); dataFile.print('/');
  dataFile.print(now.day(), DEC); dataFile.print(' ');
  dataFile.print(now.hour(), DEC); dataFile.print(':');
  dataFile.print(now.minute(), DEC); dataFile.print(':');
  dataFile.print(now.second(), DEC); dataFile.print(',');
  // Calculate and write EC in uS and TDS in ppm (base 500)
  dataFile.print(1000.0*mS);dataFile.print(',');dataFile.println(500.0*mS);
  dataFile.close();
}
// else report error
else {
 Serial.println(F("error opening file"));
}
```

The last part of the loop would be to echo to the serial monitor and use delay to control the sampling time.

```
// echo timestamp and sensor reading to serial monitor
Serial.print(now.year(), DEC); Serial.print('/');
Serial.print(now.month(), DEC); Serial.print('/');
Serial.print(now.day(), DEC); Serial.print(' ');
Serial.print(now.hour(), DEC); Serial.print(':');
Serial.print(now.minute(), DEC); Serial.print(':');
Serial.print(now.second(), DEC); Serial.print(',');
Serial.print(1000.0*mS);Serial.print(',');Serial.println(500.0*mS);

// sample every 5 sec
delay(5000);
```

Once compiled, we check that this larger more complete sketch uses about half of program storage space and that global variables also use about a half of dynamic memory. Keep in mind that including libraries contributes to the use of storage and memory in addition to the code in your own script. To finalize the exercise, verify the output on the serial monitor (Figure 5.11) and that indeed file `datalog.csv` is stored in the SD card. You can do this by using the scripts `Files` and `ReadWrite` of the SD library that you can access via `File|Examples` in the Arduino IDE.

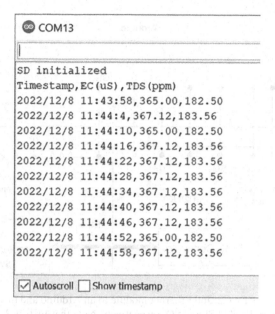

```
● COM13

SD initialized
Timestamp,EC(uS),TDS(ppm)
2022/12/8 11:43:58,365.00,182.50
2022/12/8 11:44:4,367.12,183.56
2022/12/8 11:44:10,365.00,182.50
2022/12/8 11:44:16,367.12,183.56
2022/12/8 11:44:22,367.12,183.56
2022/12/8 11:44:28,367.12,183.56
2022/12/8 11:44:34,367.12,183.56
2022/12/8 11:44:40,367.12,183.56
2022/12/8 11:44:46,367.12,183.56
2022/12/8 11:44:52,365.00,182.50
2022/12/8 11:44:58,367.12,183.56

☑ Autoscroll  ☐ Show timestamp
```

FIGURE 5.11   Datalogger output.

ADDING ETHERNET TO AN ARDUINO AND TCP/IP

To provide Ethernet capabilities to the Arduino, we will use the ENC28J60 Ethernet module based on the SPI. Since we have the SD card connected to the SPI lines of the Arduino, when we add the Ethernet SPI device, we would have a configuration based on two slaves and one master, recall from Chapter 5 that we could have the two devices on the same clock and data lines (SCK, MOSI, MISO) and use separate CS lines to address a particular device. For simplicity of this exercise, instead of handling two devices and needing a larger script, we will disconnect the SD card device from the SPI bus and connect the ENC28J60 module by itself (Figure 5.12).

This would also allow us to focus on the TCP/IP aspects of the communication and configure a script that would fit in the limited storage and memory resources of the Arduino Uno. When wiring the module to the Arduino using Figure 5.12 note that power VCC is 3.3 V and the SDI bus lines have a shorthand of SI and SO. For CS, we use pin 9 of the Arduino so that we reserve pin 10 for an SD adapter if we wish to use it later. Even though the power is 3.3 V, the pins are 5 V tolerant and thus we can connect it to the Arduino pins. In addition, we need to plug an Ethernet cable to the RJ45 connector on the ENC28J60 module and the other end of the cable to a port of the LAN router or to a network switch connected to the router.

We will use libraries SPI, as before to handle the SPI, and `EthernetENC` (Arduino 2022a) to handle the Ethernet functions of the module. You can install this library using the Arduino Library Manager. In this exercise, we will use the TCP/IP to run a web server on the Arduino that can be accessed by a client using the HTPP protocol. First, some default settings of this library may make the script too large for the storage and memory available in the Arduino Uno. Use the file manager to browse to the folder named `utility` under library `EthernetENC` in your libraries folder. Find file `uipethernet-conf.h` and edit a setting to reduce the use of dynamic memory. `UIP _ CONF _ MAX _ CONNECTIONS` defines the maximum number of client connections; the default value is 4. Let us set it to 1, assuming you will be using one browser to access the server. Edit this line

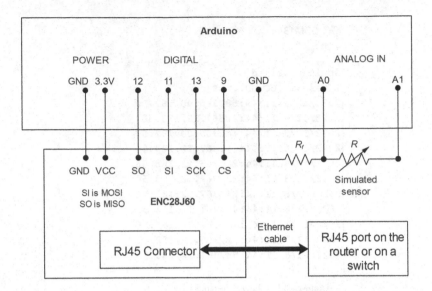

FIGURE 5.12 Connecting the ENC28J60 Ethernet module to an Arduino and to a LAN. On the ENC28J60, SDI pins MOSI and MISO are labeled SI and SO, respectively. Note that VCC is 3.3 V.

```
#define UIP_CONF_MAX_CONNECTIONS 1
```

and save this file.

We are ready to write the sketch. Start with new lines in the declaration area to include the library, and define the MAC, IP address, and port of the server we are going to run on the Arduino. The MAC is arbitrary and not defined by the manufacturer, and the IP depends on your LAN. Here, we are selecting 192.168.1.133, which is in the static address sector of this router. Recall that in lab session 2 we used 192.168.1.132 for the static address of the RPi. This may be a good time to look at that section of lab 2 if you want to refresh those concepts. For simplicity, we use the standard port 80 for HTTP.

```
// SPI library for ethernet module
#include <SPI.h>
// Ethernet library for ENC28J60
#include <EthernetENC.h>
//ENC28J60 on SPI bus: MOSI-pin11, MISO-pin12, CLK-pin13, CS-pin 9

// ethernet mac and ip
byte mac[] = {0xDE, 0xAD, 0xBE, 0xEF, 0xFE, 0xED};
IPAddress ip(192, 168, 1, 133);
// web port
EthernetServer server(80);
```

After defining the parameters and pins of the sensor as before, we define pin 9 as CS for this device.

```
// sensor circuit parameter
float Vs = 5.0;
float Rf = 10.0; // in kohm
//sensor pins
int pinPulse=A1;
int pinSignal=A0;
// SPI bus: MOSI-pin11, MISO-pin12, CLK-pin13
// Ethernet CS on pin9
int pinEtherCS =9;
```

In the setup function, we initialize the serial port and mode of the sensor pins as before, plus new lines to initialize the ethernet connection. Note that we use the macro F() in Serial.print to reduce memory use by string literals.

```
void setup () {
  Serial.begin(9600);

  //sensor pins
  pinMode(pinPulse,OUTPUT);
  pinMode(pinSignal,INPUT);

  // Initializing Ethernet module
  Ethernet.init(pinEtherCS);
  Ethernet.begin(mac, ip);
  if (Ethernet.hardwareStatus() == EthernetNoHardware) {
    Serial.println(F("No ENC28J60"));
    return;
  }
  Serial.println(F("Ethernet initialized"));

  // start the server
  server.begin();
  Serial.print(F("server is at IP "));
  Serial.println(Ethernet.localIP());
}
```

Insert the readSensor function as before and we are ready to write the loop() function, which we will base on the WebServer example of the EthernetENC library. The code we will use is reproduced here in entirety for an overview and then we will clarify specific aspects of this code.

```
void loop () {
  // listen for incoming clients
  EthernetClient client = server.available();
  if (client) {
    Serial.println(F("new client"));
    // an http request ends with a blank line
    boolean currentLineIsBlank = true;
    while (client.connected()) {
```

```
        if (client.available()) {
          char c = client.read();
          //can remove comment to see details of client
          // Serial.write(c);
          // if newline and blankline the http request ended
          // send a reply
          if (c == '\n' && currentLineIsBlank) {
            // send a standard http response header
            client.println("HTTP/1.1 200 OK");
            client.println("Content-Type: text/html");
            client.println("Connection: close");  // connection closed after
response
            client.println("Refresh: 5");  // refresh the page automatically every
5 sec
            client.println();
            client.println("<!DOCTYPE HTML>");
            client.println("<html>");
            client.println("<head><title>RTEM Arduino Ethernet</title></head>");
            client.println("<body>");
            client.println("<h1>Arduino Pulsed Sensor WebServer</h1>");
            client.println("<h2>Using Ethernet module ENC28J60 and library
EthernetENC</h2>");
            client.println("<p>Readings of EC, and TDS</p>");
            // Calculate EC in uS and TDS in ppm (base 500)
            float mS = readSensor();
            client.print(1000.0*mS);client.print(F(","));client.println(500.0*mS);
            client.println("<br />");
            client.println("</body>");
            client.println("</html>");
            break;
          }
          if (c == '\n') {
            // you're starting a new line
            currentLineIsBlank = true;
          } else if (c != '\r') {
            // you've gotten a character on the current line
            currentLineIsBlank = false;
          }
        }
      }
      // give the web browser time to receive the data
      delay(1);
      // close the connection:
      client.stop();
      Serial.println(F("client disconnected"));
  }

}
```

In the first part, we check that the server is available, listen for clients (your browser pointing to the IP 192.168.1.133), and get ready to respond when the client is available. By removing the comment of the Serial.write line, we could see the details of the client displayed on the serial monitor. After that, the server will send a reply to your browser, which is HTML code written in multiple client.println lines, instead of a file. The line client.println("Refresh: 5") will refresh the browser every five seconds. It is opportune to mention that we cannot run php scripts on the Arduino, and thus we write the sensor data dynamically from the sketch itself.

Once compiled, we see that the sketch uses 16,042 bytes (49%) of program storage space and global variables use 946 bytes (46%) of dynamic memory. Since we saved memory by adjusting the default settings of the library, we could add more content to the web page or alternatively, we can allow more HTTP connections (other browser) by re-editing the UIP _ CONF _ MAX _ CONNECTIONS definition.

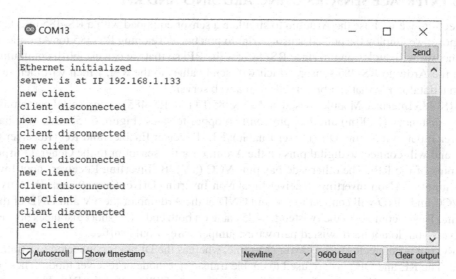

FIGURE 5.13 Serial monitor showing the exchange between the server (on the Arduino) and the client (your browser).

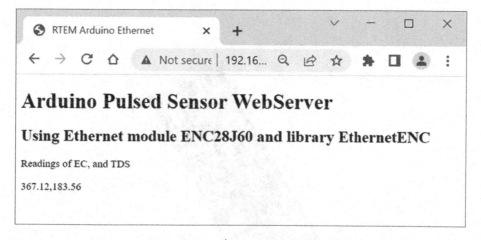

FIGURE 5.14 Browser displaying dynamic sensor readings on the web page served by the Arduino.

Run the server script on the Arduino, open a browser and point to 192.168.1.133 to request connection to the server. Verify the repetitive client connects and disconnects every five seconds displayed on the serial monitor (Figure 5.13), and the output on the browser displaying the HTML and refreshing the readings every five seconds (Figure 5.14).

RS-485 INTERFACE SENSORS USING ARDUINO AND RPi

In this section, we will use the Arduino to simulate a sensor equipped with a RS-485 interface. For this purpose, we employ the pulsed sensor script to practice developing RS-485 for Arduino. At the same time, we will deploy an interface RS-485 for the RPi so that we can establish communication between the Arduino RS-485 sensor, which will send values to the RPi, which will then store the values in a datalog file that can be served from a web server.

The RS-485 Interface Module (based on MAX485 TTL to RS-485) allows serial communication over long distances (1200 m) and has pins on two opposite sides (Figure 5.15). One side has pins DI (Driver Input or TX pin), DE (Driver Enable), RE (Receiver Enable), and RO (Receiver Out or RX pin) and will connect to digital pins on the Arduino for the sensor or to the General Purpose IO (GPIO) pins of the RPi. The other side has pins VCC (5 V), B (Inverting Receiver Input Inverting Driver Output), A (Non-inverting Receiver Input Non-Inverting Driver Output), and GND (0V). The power VCC and GND will connect to 5 V and GND of the Arduino or the 5 V and GND of the RPi. Pins A and B will connect to the twisted RS485 cable on both ends. For short distances between the transceivers you do not need twisted pair wires; jumper wires would suffice.

The RO pin is connected to a receiver (Rx) line, whereas the DI pin is connected to a transmitter (Tx) line. The RE and DE pins are used to enable transmitter mode or receiver mode. These can be jumped and connected to a digital pin of the Arduino or to a GPIO pin of the RPi. These can be set low for a receiver or set high for a transmitter.

FIGURE 5.15   RS-485 module.

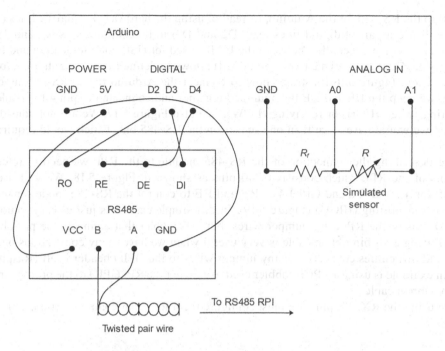

FIGURE 5.16 Connecting the RS-485 module to an Arduino.

FIGURE 5.17 RS-485 module and Arduino. Shows twisted pair.

To set up the RS-485 for the Arduino, instead of using the hardware Rx and Tx lines (D0 and D1), we will use a pair of digital lines (e.g., D2 and D3) to be configured as Rx and Tx using software. This will avoid conflict between the UART used for USB communication and RS-485. The library `SoftwareSerial` (Arduino 2022d) allows to define these separate pins for serial communication. Figure 5.16 illustrates how to connect the Arduino to the RS-485 module. In this case, we keep the RE and DE lines jumped together and connected to pin 4 to enable either transmitting (when HIGH) or receiving (LOW). See also Figure 5.17. We are not considering a case when we would have more than one sensor on the RS-485 bus, which would require an ID for each.

Before describing the connection of the RS-485 module to the RPi, we need to refer to the GPIO pins of the RPi, which consists of 40 pins as shown in Figure 5.18. We will use GPIO 14 and 15 for Tx and Rx, and GPIO 4 to RE and DE to control the RS-485 mode for receiving (LOW) or transmitting (HIGH) (Figure 5.19). In this simple case, it is just as easy to connect a few GPIO pins of the RPi using jumper wires, but a T-cobbler that connects the protoboard to the GPIO using a 40-pin ribbon cable is very useful when we have many connections because it would avoid difficulties connecting many jumper wires to the GPIO header itself (Raspberry Pi 2022). An example is using a RPi T-cobbler used to connect the RPi GPIO to the protoboard using a 40-wire ribbon cable.

Before using the RX, TX pins, we must configure them for serial communication. This is done using

```
sudo raspi-config
```

Select interface option, then Serial. Answer "No" to the question on a login shell accessible over serial, finish configuration, and accept to reboot or do it manually by

Pin Numbers

+3V3	1	2	+5V
GPIO2/SDA1	3	4	+5V
GPIO3 /SCL1	5	6	GND
GPIO4	7	8	GPIO14 / TX
GND	9	10	GPIO15 / RX
GPIO17	11	12	GPIO18
GPIO27	13	14	GND
GPIO22	15	16	GPIO23
+3V3	17	18	GPIO24
GPIO10 / MOSI	19	20	GND
GPIO9 /MISO	21	22	GPIO25
GPIO 11 /SCLK	23	24	GPIO8 CE0#
GND	25	26	GPIO7 /CE1 #
GPIO0 /ID_SD	27	28	GPIO1 /ID_SC
GPIO5	29	30	GND
GPIO6	31	32	GPIO12
GPIO13	33	34	GND
GPIO19 / MISO	35	36	GPIO16 /CE2#
GPIO26	37	38	GPIO20 /MOSI
GND	39	40	GPIO21 /SCLK

FIGURE 5.18 Raspberry Pi GPIO.

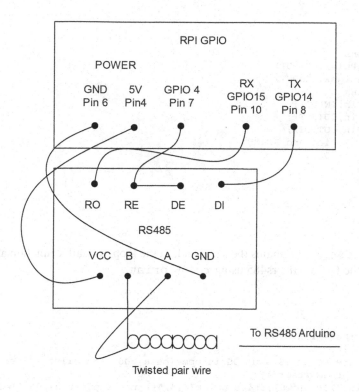

FIGURE 5.19   Connecting the RPi and RS485 modules.

```
sudo reboot
```

Now, we are ready to write an Arduino script for the sensor and a Python script on the RPi for data-logging. Let us first discuss the Arduino code rs485sensor.ino. This file is available from the RTEM repository. We include the library SoftwareSerial, define pins for RO, DI, DE/RE.

```
#include <SoftwareSerial.h>
int rxPin = 2;
int txPin = 3;
int enPin = 4;

// sensor circuit parameter
float Vs = 5.0;
float Rf = 10.0; // in kohm
//sensor pins
int pinPulse=A1;
int pinSignal=A0;

//RS485 on rxpin and txpin
SoftwareSerial rs485(rxPin,txPin);
```

Then, in the setup() function, we define the mode of pins and initialize serial communication

```
void setup() {
 //sensor pins
 pinMode(pinPulse,OUTPUT);
 pinMode(pinSignal,INPUT);
 // rs485 pins
 pinMode(rxPin,INPUT);
 pinMode(txPin,OUTPUT);
 pinMode(enPin,OUTPUT);
 // baud for Serial and RS485
 Serial.begin(9600);
 rs485.begin(9600);
 delay(1000);
}
```

The function `readSensor` remains the same and, in the loop, we call it function and transmit the sensor values to the RPi via the rs485 using `rs485.print`.

```
void loop() {
    digitalWrite(enPin,HIGH);
    delay(100);
    // Calculate EC in uS and TDS in ppm (base 500) and print values in csv
    float mS = readSensor();
    rs485.print(1000.0*mS);rs485.print(F(","));rs485.println(500.0*mS);
    Serial.print(F("sent by RS485 = "));
    Serial.print(1000.0*mS);Serial.print(F(","));Serial.println(500.0*mS);
    delay(5000);
}
```

An alternative to implement RS485 on Arduino is to use a RS485 library, for example (Tillaart 2022), which employs the hardware serial.

Now, we turn to the RPi and write a Python script `rs485_logger.py`. This file is available from the RTEM repository as well. We use the opportunity to learn about GPIO as well as the `try` method in Python. In addition to the import statements that we have already used we include others we will need

```
import os
from datetime import datetime,timezone
import serial
import RPi.GPIO as GPIO
import time
from time import sleep
```

we define the serial stream for RS485

```
# serial channel for RS-485
rs485  = serial.Serial(
        port='/dev/serial0',
        baudrate = 9600,
```

```
          parity=serial.PARITY_NONE,
          stopbits=serial.STOPBITS_ONE,
          bytesize=serial.EIGHTBITS,
          timeout=0
)
```

And define file name, header, and checking if the file exists as we have done previously

```
logfile = "data/rs485-datalog.csv"
header = "TimeStamp,EC,TDS\n"

if os.path.exists(logfile):
        print("File exists, appending records\n")
        file = open(logfile, "a")
else:
        print("File does not exists, writing header\n")
        print(header)
        file = open(logfile, "w")
        file.write(header)
```

Now, we use `try` with an exception to catch those instances when we interrupt the program, and a `finally` option to clean up the GPIO.

```
try:
        GPIO.setmode(GPIO.BOARD)
        # GPIO number in setup refers to pin number on the header
        # in this example pin number 7 is GPIO4
        GPIO.setup(7, GPIO.OUT)
        print("Listen for input")
        GPIO.output(7, GPIO.LOW)
        time.sleep(0.5)
```

Within the `try:` we insert the main loop like what we have used before but adding the `rs485.write`,

```
while 1:
if(rs485.in_waiting >0):
        line = rs485.readline()
        linedec = line.decode('utf-8')
        linestr = str(linedec)
        now = datetime.now(timezone.utc)
        timestamp = str(now.astimezone().strftime("%Y-%m-%d %H:%M:%S %z")+',')
        print(timestamp,linestr,end="")
        file.write(timestamp+linestr)
        file.flush()
time.sleep(0.5)
```

as well as the `except` and the `finally`

```
#print this if CTRL + C is used to stop python script
except KeyboardInterrupt:
        print("Program interrupted")
finally:
        GPIO.cleanup()
```

Once the Python script is ready, we run it using python3. At this point, we should test both programs running together. On the Serial Monitor of the Arduino IDE, we would see a sequence of lines indicating the ongoing process (Figure 5.20). Whereas on the PuTTY monitor, we will see a sequence of lines indicating the progress of the communication from the RPi side (Figure 5.21). As the process unfolds, the `rs485 _ logger.py` program writes a file `data/rs485-datalog.csv`.

As a final step of this exercise, we can modify the `index.php`, `readfile.php`, and `multiplotd3.html` scripts to display the results on a web page. In file `index.php`, we edit the name of the file to be `data/rs485-datalog.csv`

```
<p>Example of URL link: internal from the server file system
 <a href="data/rs485-datalog.csv">Download RS485 datalog file</a>
 </p>
```

and change the name of `multiplotd3.html` which we will edit later

```
<p>Read RS485 sensor datalog file and plot multi chart using d3.js</p>
 <iframe src="multiplotd3-rs485.html" width=600 height=400></iframe>
```

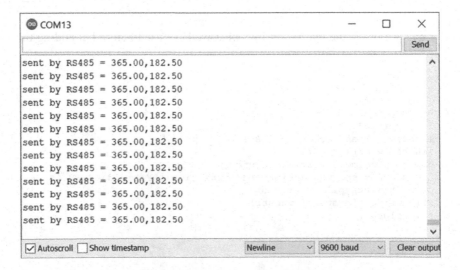

FIGURE 5.20   Serial monitor showing communication from the Arduino side.

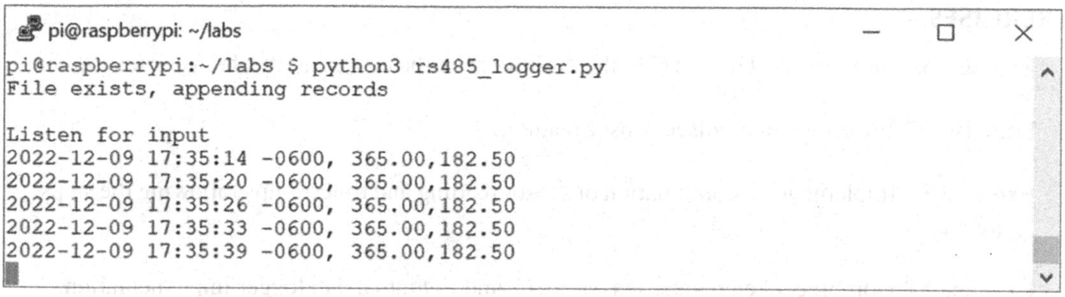

```
pi@raspberrypi: ~/labs                                    —    □    ×
pi@raspberrypi:~/labs $ python3 rs485_logger.py
File exists, appending records

Listen for input
2022-12-09 17:35:14 -0600, 365.00,182.50
2022-12-09 17:35:20 -0600, 365.00,182.50
2022-12-09 17:35:26 -0600, 365.00,182.50
2022-12-09 17:35:33 -0600, 365.00,182.50
2022-12-09 17:35:39 -0600, 365.00,182.50
```

FIGURE 5.21 PuTTY screen showing communication from the RPi side.

and make other edits to adapt the text to be displayed on the browser.

In the `readfile.php` file, we simply change the name of the file

```php
<?php
 $lines = file("data/rs485-datalog.csv");
 foreach ($lines as $line_num=> $line) {
    echo $line . "<br>";
 }
?>
```

Finally, in the file `multiplotd3.html`, we edit the name of the file and variables in the header. A complete file `multiplotd3-rs485.html` is available from the RTEM repository. We now point the browser to the RPi IP and can see the results of the web page.

SDI-12 SENSORS

As we leave this chapter, we mention that we did not exercise topics related to other interfaces such as SDI-12. In Lab 13, we will work with soil moisture data collected from a sensor network on an SDI-12 bus.

SHIELDS

There add-on boards to complement the capabilities of the Arduino. These add-on circuits are called "shields" because they typically fit on top of the Arduino resembling a protective shield. There are hundreds of shield makers and shields available (Shieldlist 2014) that enable an Arduino to perform a variety of functions. For example, an Ethernet connection shield allows an Arduino to connect to the Internet; a GPS shield allows it to obtain location, and relevant to environmental monitoring, a datalogger shield makes an Arduino into a datalogger. The latter provides an RTC, an SD card interface, and area for prototyping additional circuitry and sensors. Once we connect it to the Arduino, we would have a datalogger that would perform as we developed it in the previous activities. The Adafruit Assembled Datalogging Shield for Arduino (Adafruit 2014) provides an RTC to keep time even when the Arduino is unplugged using a battery backup. It has an SD card interface that works. It has an onboard 3.3 V regulator that can be used as a reference and power the SD card. Adafruit provides libraries and example code for both SD and RTC. Connect the shield to the sensor using the same configuration employed in previous activities. The Arduino pins are extended on the shield. Use the same scripts as above. Upload and once running check output on the serial monitor.

EXERCISES

Exercise 5.1 Connect the Tiny RTC to the Arduino, run the code, and display your results.

Exercise 5.2 Implement the pulsed sensor reading.

Exercise 5.3 Implement the combination of sensor reading and timestamps following the steps above.

Exercise 5.4 Combine all elements, sensor, RTC, and SD into a datalogger implementation.

Exercise 5.5 Displaying transducer output on a web page using the ENC28J60 Ethernet module.

Exercise 5.6 Displaying transducer output on a web page using the RS-485 interface between Arduino and RPi.

REFERENCES

Acevedo, M.F. 2024. *Real-Time Environmental Monitoring: Sensors and Systems - Textbook, Second Edition.* Boca Raton, FL: CRC Press, Taylor & Francis Group, 392 pp.

Adafruit. 2014. *Logger Shield.* Adafruit Industries. Accessed October. http://shieldlist.org/adafruit/logger.

Adafruit. 2023. *RTClib.* Accessed January 2023. https://github.com/adafruit/RTClib.

Arduino. 2016. *Download the Arduino Software.* Accessed August 2016. https://www.arduino.cc/en/Main/Software.

Arduino. 2022a. *EthernetENC.* Accessed December 2022. https://reference.arduino.cc/reference/en/libraries/ethernetenc/.

Arduino. 2022b. *Reduce the Size and Memory Usage of Your Sketch.* Accessed December 2022. https://support.arduino.cc/hc/en-us/articles/360013825179-Reduce-the-size-and-memory-usage-of-your-sketch.

Arduino. 2022c. *SD.* Accessed December 2022. https://www.arduino.cc/reference/en/libraries/sd/.

Arduino. 2022d. *SoftwareSerial Library.* Accessed December 2022. https://docs.arduino.cc/learn/built-in-libraries/software-serial.

Arduino. 2023a. *Arduino-CLI.* Accessed January 2023. https://github.com/arduino/arduino-cli.

Arduino. 2023b. *SPI.* Accessed January 2023. https://www.arduino.cc/en/Reference/SPI.

ElectroPeak. 2022. *Interfacing DS1307 RTC Module with Arduino & Make a Reminder.* Accessed February 2022. https://create.arduino.cc/projecthub/electropeak/interfacing-ds1307-rtc-module-with-arduino-make-a-reminder-08cb61.

Geany. 2023. *Geany - The Flyweight IDE.* Accessed January 2023. https://www.geany.org/.

Gertz, E. 2014. *EMWA/Chapter-5/Water Conductivity.* Accessed 2014. https://github.com/ejgertz/EMWA/blob/master/chapter-5/WaterConductivity.

Gertz, E., and P. Di Justo. 2012. *Environmental Monitoring with Arduino: Building Simple Devices to Collect Data about the World Around Us.* Sebastopol, CA.: O'Reilly Media Inc. pp.

Instructables Circuits. 2022. *Interfacing DS1307 I2C RTC with Arduino.* Accessed February 2022. https://www.instructables.com/Interfacing-DS1307-I2C-RTC-With-Arduino/.

Maxim Integrated. 2022. *DS1307 64 x 8, Serial, I2C Real-Time Clock.* Accessed December 2022. https://www.analog.com/media/en/technical-documentation/data-sheets/ds1307.pdf.

Posit. 2023. *RStudio IDE.* Accessed January 2023. https://posit.co/downloads/.

PuTTY. 2023. *Download PuTTY.* https://www.putty.org/.

R Project. 2023. *The Comprehensive R Archive Network.* Accessed January 2023. http://cran.us.r-project.org/.

Raspberry Pi. 2022. *Adafruit's New Raspberry Pi Breakout Kit: The Pi T-Cobbler.* Accessed December 2022. https://www.raspberrypi.com/news/adafruits-new-raspberry-pi-breakout-kit-the-pi-t-cobbler/.

SD Association. 2022. *SD Memory Card Formatter.* Accessed December 2022. https://www.sdcard.org/downloads/formatter/.

Shieldlist. 2014. *Arduino Shield List.* Accessed October. http://shieldlist.org/.

Tillaart, R. 2022. *RS485.* https://github.com/RobTillaart/RS485.

WinSCP. 2023. *Free Award-Winning File Manager.* Accessed January 2023. https://winscp.net/eng/index.php.

6 Telemetry and Wireless Sensor Networks

INTRODUCTION

Radio links are important to enable telemetry with remote environmental monitoring stations. In this context, we analyze radio links using R scripts and the Radio Mobile program emphasizing links at 2.4 GHz and 915 MHz. In subsequent exercises, we use both frequency bands to implement wireless sensor networks (WSNs). The first one, 2.4 GHz, is employed in Wi-Fi and we use it to implement a WSN in which each sensor node runs in a NodeMCU based on ESP8266. The second frequency, 915 MHz, allows for a longer transmission range and we use it to implement a WSN in which each node uses a Moteino to drive a RFM69 radio. In both cases, the nodes collect data from a simple transducer, made of a thermistor in a voltage divider circuit. For the Wi-Fi WSN, we use the MQ Telemetry Transport (MQTT) protocol, and each node publishes data to a Raspberry Pi (RPi) Zero W where we run a broker and a subscriber client. For the Moteino/RFM69 WSN, the nodes send data to a gateway (or base station) also consisting of a Moteino/RFM69, which streams data serially to a RPi Zero W. For both networks, we learn how to datalog network data and make available from a web server installed on the RPi.

MATERIALS

READINGS

For theoretical background, you can use Chapter 6 of Acevedo, M.F. 2024. *Real-Time Environmental Monitoring: Sensors and - Textbook, Second Edition* which is a companion to these guides (Acevedo 2024). Other bibliographical references are cited throughout the guide.

COMPONENTS

- Resistances: 10 kΩ (Qty 2)
- Thermistor (10 kΩ nominal at 25°C) (Quantity 2) to be used as a sensor
- Protoboard and jumper wires

EQUIPMENT AND NETWORK RESOURCES

- Laptop or PC with Wi-Fi and SD card reader/writer, as well admin rights to install software. We assume a windows PC but it can be adapted to Linux or Mac.
- Access to a small network (class C) provided by
  - Wireless router
  - Hotspot enabled from a smartphone (this will consume data from the phone plan)

MAJOR COMPONENTS AND INSTRUMENTS

- Raspberry Pi Zero W (RPi for short)
- Micro SD and adapter (SanDisk Mobile Class4 MicroSDHC Flash Memory Card-SDSDQM-B35A with Adapter 16 GB)

DOI: 10.1201/9781003184362-6

- NodeMCU 8266 – ESP-12 module with CP2102 (Qty 3)
- Moteino USB with RFM69CW radio (Quantity 3) (or Moteino+Future Technology Devices International (FTDI) Adapter)
- USB cable adapter USB regular female to micro-USB male
- USB cables regular USB to micro-USB (Quantity 3)

TOOLS AND INSTRUMENTS

- Long nose pliers
- Wire strippers and clippers
- Small Screwdriver

SOFTWARE (LINKS PROVIDED IN THE REFERENCES)

- R, for data analysis (R Project 2023)
- RStudio, an IDE to use R (Posit 2023)
- PuTTY, for Secure Shell (SSH) (PuTTY 2023)
- WinSCP, to transfer files from the RPi to a windows PC (WinSCP 2023)
- Python (installed from RPi)
- Arduino IDE, Interface to Arduino (Arduino 2016)
- Geany, IDE to edit programs as well as data (Geany 2023)
- Arduino-CLI, Arduino software to run using commands from the RPi (Arduino 2023a)
- PHP (installed from RPi) for RPi.
- Arduino library RTClib (Adafruit 2023)
- Arduino SPI library (Arduino 2023b)
- Arduino SD library (Arduino 2022b)
- Arduino SoftwareSerial library (Arduino 2022c)
- Arduino EthernetENC library (Arduino 2022a)
- Radio Mobile, to simulate radio links (Coudé 2022)
- Advanced IP scanner, to scan your network and find the IP address of the RPi

SCRIPTS AND DATA FILES

RTEM GitHub repository https://github.com/mfacevedol/rtem.
- R functions in `radiolink.zip`
- Scripts to program the NodeMCU WSN `SensorNodeMQTT _ Publish.zip`
- Scripts to program the Moteino WSN `Moteino _ Scripts.zip`
- Python code to program the RPi in the WSNs `RPiSubscribe.zip` and `RPiMoteino.zip`

SUPPLEMENTARY SUPPORT MATERIAL

Supplementary support material including additional screenshots, images, and procedures are available from the publisher eResources web page provided for this book.

TELEMETRY: RADIO LINK USING R

Download the archive `radiolink.zip`, available from the RTEM GitHub repository, which contains R functions `radiolink-functions.R`. We will use these functions to analyze radio links.

FREE SPACE MODEL

Function `fsl` calculates FSL in dB as a function of frequency in MHz and distance in km using the practical equation $\text{FSL}_{\text{dB}} = 20\log(f_{\text{MHz}}) + 20\log(d_{\text{km}}) + 32.45$ and then uses antenna gain (by default unity or 0 dB) and transmitted power (30 dBm by default) to calculate received power.

```
fsl <- function(f.MHz,d.km,G.dB=0,Pt.dBm=30){
  # PL for unity gain
  FSL.dB <- 20*log10(f.MHz)+20*log10(d.km)+32.45
  # received power in dBm
  rxp.dBm <- Pt.dBm + G.dB - FSL.dB
  return(list(fMHz.dkm.GdB.PtdBm=c(f.MHz,d.km,G.dB,Pt.dBm),
             FSL.dB=round(FSL.dB,2),rxp.dBm=round(rxp.dBm,2)))
}
```

We start a script `lab6.R` to source the functions

```
source("R/lab6/radiolink-functions.R")
```

Now, we calculate the path loss (PL) and received power for 2.4 GHz, over a 2 km link, with default gain and transmit power

```
# WiFi at 2 km
f.MHz <- 2400;d.km <- 2
fsl(f.MHz,d.km)
```

resulting in

```
> f.MHz <- 2400;d.km <- 2
> fsl(f.MHz,d.km)
$fMHz.dkm.GdB.PtdBm
[1] 2400    2    0   30

$FSL.dB
[1] 106.07

$rxp.dBm
[1] -76.07

>
```

which means that at 2.4 GHz the FSL would be ~106 dB and received power −76 dBm, which is low. We can increase the gain to 20 dB to increase received power, say by 20 dB, and therefore receive above the acceptable−60 dBm point.

```
> fsl(f.MHz,d.km,G.dB=20)
$fMHz.dkm.GdB.PtdBm
[1] 2400    2   20   30

$FSL.dB
[1] 106.07

$rxp.dBm
[1] -56.07

>
```

Function `fsl.set` (not shown here for the sake of space but can be examined opening `radio-link-functions.R`) uses results from `fsl` to calculate FSL and received power for a set of frequencies and distances. By default, it will also plot FSL and power vs. distance. We use this function `fsl.set` to compare the loss as we double the frequency and distance

```
> f.MHz <- c(900,1800,3600);d.km <- c(2,4,8)
> fsl.set(f.MHz,d.km)
$f.MHz
[1]   900 1800 3600

$d.km
[1] 2 4 8

$FSL.dB
      900    1800    3600
2   97.56 103.58 109.60
4  103.58 109.60 115.62
8  109.60 115.62 121.64

$rxp.dBm
      900    1800    3600
2 -67.56 -73.58 -79.60
4 -73.58 -79.60 -85.62
8 -79.60 -85.62 -91.64

>
```

We have confirmed that there is a ~6 dB increment in loss as we double the distance or as we double the frequency. The function `fsl.set` also would have plotted `FSL.dB` vs. distance and Rx power `rxp.dBm`.

Now, we will calculate the PL and power for a set of frequencies that are of interest for telemetry and WSN and compare the loss over a range of distances,

```
f.MHz <- c(915, 2400,5000)
d.km <- seq(1,10,0.1)
x <- fsl.set(f.MHz,d.km)
```

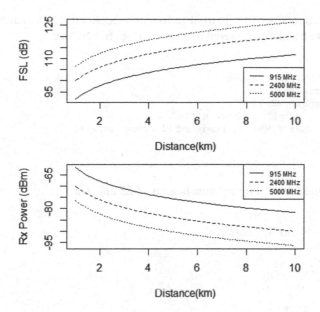

FIGURE 6.1 FSL as a function of link distance and frequency.

We obtain the plot shown in Figure 6.1. Clearly, FSL for 915 MHz is lower across all distances than 2.4 GHz, and the latter is lower across all distances than 5 GHz. Note that FSL is ~6 dB higher for 5 GHz compared to 2.4 GHz and that FSL is ~8 dB larger for 2.4 GHz compared to 915 MHz.

TWO-RAY MODEL

To apply the two-ray model, we first check the crossover distance, which requires to input the wavelength. Function `wlength` calculates wavelength in meters (m) given frequency in MHz

```
wlength <- function(f.MHz){
  c = 3 *10^8 # speed of light
  wl.m <- c/(f.MHz*10^6) # wavelength in m
  return(list(f.MHz=f.MHz,wl.m=round(wl.m,3)))
}
```

For example, for the set under consideration

```
> wlength(f.MHz)
$f.MHz
[1]   915 2400 5000

$wl.m
[1] 0.328 0.125 0.060

>
```

Note that the wavelength for 915 MHz is about a third of a meter, and for 2.4 GHz is 0.125 m or about an eight of one meter. This is relevant to the antenna height of a Moteino, which we will use later in the WSN hands-on activity.

The result of function `wlength` is used in function `dc.km` that calculates crossover distance given antenna heights `ht` and `hr` entered as a product `ht.hr=ht*hr`

```
dc.km <- function(f.MHz,ht.hr){
  # crossover distance
  wl.m <- wlength(f.MHz)$wl.m
  d.cross.km <-  (4*pi*ht.hr/wl.m)*10^-3
  return(list(f.MHz=f.MHz,dc.km=round(d.cross.km,2)))
}
```

For example, apply the function for antenna heights of 25 and 9 m.

```
> ht=20; hr=10; ht.hr <- ht*hr
> dc.km(f.MHz,ht.hr)
$f.MHz
[1]   915 2400 5000

$dc.km
[1]   7.66 20.11 41.89

>>
```

We get ~8, 20, and 42 km for 915, 2.4, and 5 GHz, respectively.

Function `two.ray` applies the two-ray model to calculate PL as a function of distance and the product of antenna heights

```
two.ray <- function(d.km,ht.hr,G.dB=1,Pt.dBm=30){
  # Calculating PL two ray
  PL.dB <- 40*log10(d.km)+120-20*log10(ht.hr)
  rxp.dBm <- Pt.dBm + G.dB - PL.dB
  return(list(d.km=d.km,PL.dB=round(PL.dB,2),rxp.dBm=round(rxp.dBm,2)))
}
```

For example, using the same antenna heights, a set of distances from 50 to 100 km that would be much larger than the crossover for 915 MHz,

```
> d.km <- seq(50,100,10)
> two.ray(d.km,ht.hr)
$d.km
[1]   50  60  70  80  90 100

$PL.dB
[1] 141.94 145.11 147.78 150.10 152.15 153.98

$rxp.dBm
[1] -110.94 -114.11 -116.78 -119.10 -121.15 -122.98

>
```

which shows that PL ranges from ~142 to ~154dB, with received power from ~111 to 123 dBm. These power values are too low; to increase them we can increase antenna height, say ht=hr=20, antenna gain to 40 dB, and transmit power to 40 dBm.

```
> ht=20; hr=20; ht.hr <- ht*hr
> d.km <- seq(50,100,10)
> two.ray(d.km,ht.hr,G.dB=40,Pt.dBm=40)
$d.km
[1]   50  60  70  80  90 100

$PL.dB
[1] 135.92 139.08 141.76 144.08 146.133 147.96

$rxp.dBm
[1] -55.92 -59.08 -61.76 -64.08 -66.13 -67.96
```

Function two.ray.set (not shown here for the sake of space, but you can examine it opening radiolink-functions.R) applies the two-ray model to a set of distances and antenna heights. For example,

```
> d.km =c(50,100,200,400)
> ht.hr= c(10,20,40,80,160)
> two.ray.set(d.km,ht.hr)
$d.km
[1]   50 100 200 400

$ht.hr
[1]  10  20  40  80 160

$PL.dB
         10     20     40     80    160
50   167.96 161.94 155.92 149.90 143.88
100  180.00 173.98 167.96 161.94 155.92
200  192.04 186.02 180.00 173.98 167.96
400  204.08 198.06 192.04 186.02 180.00

$rxp.dBm
          10      20      40      80     160
50   -136.96 -130.94 -124.92 -118.90 -112.88
100  -149.00 -142.98 -136.96 -130.94 -124.92
200  -161.04 -155.02 -149.00 -142.98 -136.96
400  -173.08 -167.06 -161.04 -155.02 -149.00
```

We note that PL increases by 12 dB as we double the distance and decreases by 6 dB as we double the product ht × hr.

FRESNEL ZONES CLEARANCE

Function fresnel.clear (not shown here for the sake of space, but you can examine it opening radiolink-functions.R) produces percent clearance of the first Fresnel zone. We will apply it to a 2.4 GHz link, over 2 km, with Line of Sight (LoS) at 10 m above the ground, and an obstacle at 1 km from the transmitter.

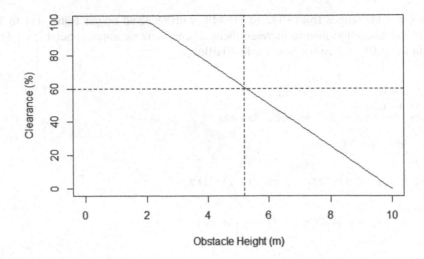

FIGURE 6.2   First Fresnel zone clearance.

```
f.MHz <- 2400;d.km <- 2
# height LoS in m
h.los = 10
wl.m <- wlength(f.MHz)$wl.m
# obstacle distance
d1.km = 1
fresnel.clear(wl.m,h.los,d.km,d1.km,d2.km)
```

and a graph like the one shown in Figure 6.2. The results indicate that when the obstacle is ~5.3 m or taller we clear <60% of the first zone.

TELEMETRY: RADIO LINK USING RADIO MOBILE

A useful program available on the web is RadioMobile (Coudé 2022). It can be used to perform a radio link study setting parameters for the two sites. As an example, enter DP (UNT Discovery Park campus) and EESAT (UNT Main campus) as sites. See Figure 6.3 for coordinates and elevation, which you will enter to create your sites.

Then from the main menu create a new link using the sites you created. Enter 25 m for the antenna height at EESAT and 2 m for the antenna height at DP. After setting parameters for the link, frequency 915 MHz, antenna gains Tx 6 dBi, RX 2 dBi, TX power 20 W, TX line loss 3 dB, RX line loss 0.5 dB, RX threshold (sensitivity) 0.5 μV. You obtain an image like the one shown in Figure 6.4. Below the image, you will see the information of each site and the link, for both sites, similar to Figure 6.5. Below those tables, you see the link displayed in Google maps as shown in Figure 6.6.

WIRELESS SENSOR NETWORKS

For hands-on activities in this lab session, we will use ESP8266-based NodeMCU and Moteino devices for WSN nodes, and RPi Zero W acting as server. The NodeMCU will illustrate the use of Wi-Fi 2.4 GHz for a WSN node, whereas the Moteino will illustrate the use of 915 MHz for a WSN node. In the NodeMCU example, the RPi will act as a MQTT broker, whereas in the Moteino example, the RPi will act to collect information from the WSN Moteino Gateway.

Radio Mobile		Radio Mobile	
My sites(2)		**My sites(2)**	
DP	▼	EESAT	▼
Latitude	33.25620250	Latitude	33.21425828
Longitude	-97.14971818	Longitude	-97.15133321
Zoom	15	Zoom	16
Elevation (m)	216.70	Elevation (m)	217.70
Description	TEO	Description	TEO1
Group		Group	
Latitude	33° 15' 22.33"N	Latitude	33° 12' 51.33"N
Longitude	097° 08' 58.99"W	Longitude	097° 09' 04.80"W
QRA	EM13KG	QRA	EM13KF
UTM (WGS84)	14S E672356 N3681217	UTM (WGS84)	14S E672288 N3676563
See on Google Maps		See on Google Maps	
Modify		Modify	
Define as Home in my settings		Define as Home in my settings	
Delete this site		Delete this site	

FIGURE 6.3 UNT sites: DP and EESAT.

FIGURE 6.4 Example of radio link study in Radio Mobile.

Radio Mobile		Pat By Roger Coudé VE2DBE		Information
Radio link study DP-EESAT				
DP (1)				**(2) EESAT**
Latitude	33.256203 °	Latitude		33.214258 °
Longitude	-97.149718 °	Longitude		-97.151333 °
Ground elevation	216.7 m	Ground elevation		217.7 m
Antenna height	2.0 m	Antenna height		2.0 m
Azimuth	181.85 TN | 178.50 MG °	Azimuth		1.84 TN | 358.49 MG °
Tilt	-0.01 °	Tilt		-0.03 °
Radio system				**Propagation**
TX power	43.01 dBm	Free space loss		89.07 dB
TX line loss	3.00 dB	Obstruction loss		12.41 dB
TX antenna gain	6.00 dBi	Forest loss		1.00 dB
RX antenna gain	2.00 dBi	Urban loss		11.47 dB
RX line loss	0.50 dB	Statistical loss		5.62 dB
RX sensitivity	-113.02 dBm	Total path loss		119.57 dB
Performance				
Distance				4.666 km
Precision				10.0 m
Frequency				146.000 MHz
Equivalent Isotropically Radiated Power				30.905 W
System gain				160.53 dB
Required reliability				70.000 %
Received Signal				-72.06 dBm
Received Signal				55.83 µV
Fade Margin				40.96 dB

FIGURE 6.5 Information about the simulated link.

FIGURE 6.6   Link displayed in Google maps.

WSN USING ESP8266 DEVICES AND MQTT

In this part of the lab session, we will use the NodeMCU module based on the ESP8266, which is a development board useful for WSN and Internet of Things (IoT), based on the ESP8266 Wi-Fi SoC and the ESP-12 module (Espressif Systems). The NodeMCU will communicate via Wi-Fi to a RPi Zero that will act as a server (Figure 6.7). In the exercise, we will learn the MQTT protocol that allows multiple devices to exchange information and is currently a popular protocol for IoT. The exercise will use two NodeMCUs as sensor nodes acting as MQQT publishers and a RPi acting as a MQTT broker using Python. A subscriber client will also run on the RPi (Figure 6.7).

The steps required are as follows: install and test the broker, install boards and libraries in the Arduino IDE to program the nodes, and program the subscriber in Python (Varnish 2022).

MQQT Broker: Raspberry pi

We first install the `mosquitto` broker in the RPi using `PuTTY` and connecting to the RPi. As we know, before installing a new package, it is good practice to update the RPi.

```
sudo apt-get update
sudo apt-get upgrade
```

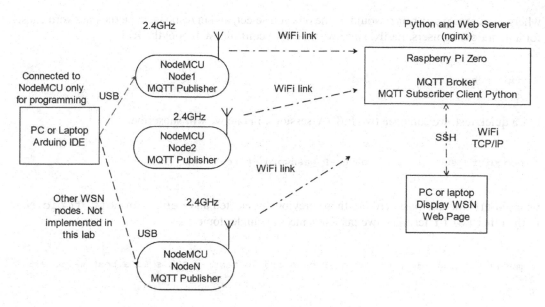

FIGURE 6.7   WSN using NodeMCUs (based on ESP8266) and MQTT protocol.

When finished updating, we install the `mosquitto` and `mosquitto-clients` packages.

```
sudo apt-get install mosquitto -y
sudo apt-get install mosquitto-clients -y
```

Once that is complete, edit the configuration file located at `/etc/mosquitto/mosquitto.conf`, by opening it with nano

```
sudo nano /etc/mosquitto/mosquitto.conf
```

Substitute the last line of this file,

```
include_dir /etc/mosquitto/conf.d
```

by two lines that will make `mosquitto` require a valid username and password

```
allow_anonymous false
password_file /etc/mosquitto/pwfile
```

and a third line telling `mosquitto` to listen for messages on port number 1883

```
listener 1883
```

Once edited, save the file, and set the username by typing

```
sudo mosquitto_passwd -c /etc/mosquitto/pwfile username
```

where of course username would be the one you select, and in response type the password chosen for authenticating users. Lastly, since we modify a conf file we reboot the RPi.

```
sudo reboot
```

For a quick test, we can open two PuTTY sessions, in one we will subscribe

```
mosquitto_sub -d -u username -P password -t test
```

where username and password are those previously selected and test is an mqtt topic. Now, in the other PuTTY terminal, we publish a message under topic test

```
mosquitto_pub -d -u username -P password -t test -m "This is a test message!"
```

where username and password are those previously selected and verify that the message is printed in the subscriber PuTTY terminal. Upon satisfactory test results, we know the broker is installed correctly.

SENSOR NODES: NODEMCU

The NodeMCU has a small size and includes a Tensilica 32-bit RISC CPU Xtensa LX106 and is equipped with Wi-Fi following 802.11b/g/n with +25 dBm (Figure 6.8). Its specifications and pinout can be consulted on many websites, for example (Components 101 2022), and more specifically in the datasheet for the ESP8266 (Espressif 2022). The operating voltage is 3.3 V, the clock speed is 80 MHz, and it has 16 digital I/O pins that include interfaces we have discussed previously UART, SPI, I2C, as well as GPIO; it also has one pin for ADC (0–3.3 V). Storage includes flash memory (4 MB) and SRAM (64 KB). It is equipped with USB-TTL based on CP2102, and PCB Antenna.

As we have studied in previous chapters, a simple transducer built from a thermistor sensor is a voltage divider circuit where the sensor is connected to a fixed resistor R_f (Figure 6.9). The voltage supply can be implemented as power provided by the NodeMCU 3.3 V pin (Figure 6.10).

The relevant NodeMCU pins of this exercise are 3.3 V power, Ground, and A0. Let us wire a temperature transducer to pins 3.3 V, GND and A0 of each one of the two NodeMCU modules as

FIGURE 6.8   NodeMCU on a protoboard wired as needed for this exercise.

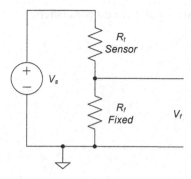

FIGURE 6.9   Voltage divider circuit with sensor.

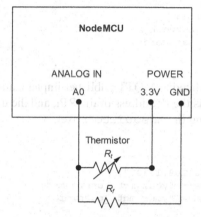

FIGURE 6.10   Temperature transducer using pin A0 of NodeMCUs (based on ESP8266).

shown in Figure 6.10 and connect one module at a time to the PC using the USB cable. Use 10 kΩ for the resistor when the thermistor is 10 kΩ nominal at 25°C; therefore, the voltage at A0 should be about half of 3.3 V at 25°C. Using a voltmeter, for each one of the modules verify that the voltage at pin A0 is a corresponding fraction 3.3 V for the ambient temperature you are working with.

Second, open the Arduino IDE, and under `File|Preferences` add the ESP8266 boards into the text box labeled "Additional Boards Manager URLs". Add the following link:

```
http://arduino.esp8266.com/stable/package_esp8266com_index.json
```

save and close preferences, then open `Tools|Board|Board Manager` and install the `esp8266` by `ESP8266 Community` package. Under ESP8266 boards, select `NodeMCU 1.0 (ESP-12 Module)` and port. Now, install libraries `ESP8266WiFi` (to handle the NodeMCU module) and `PubSubClient` (for the MQTT protocol).

Third, we write an Arduino script to program the nodes and publish data. You can find it in the RTEM repository download with the name `SensorNodeMQTT_Publish`. Now, we will explain how it is structured. Include the libraries

```
// libraries for ESP8266 and publishing
#include <ESP8266WiFi.h>
#include <PubSubClient.h>
```

Declare global variables for the sensor and circuit parameters

```
// thermistor parameter
float Rt0= 10.0; // in kohm
float B= 4116.0;
// sensor circuit parameter
float Vs = 3.3;
float Rf = 10.0; // in kohm
int pinSignal=A0;
```

as well as for the local Wi-Fi network name and password that the NodeMCU will connect to

```
// LAN WiFi name and password
const char* ssid = "your_network_name";
const char* pswd = "your password";
```

and now we declare variables related to MQTT publisher implementation. Provide the broker information loaded to the RPi. That is the IP address of the RPi, and the username and password you set in the previous step when loading the `mosquitto` broker.

```
// MQTT broker
const char* broker = "192.168.1.132";
const char* mqtt_username = "your_mqtt_username";
const char* mqtt_password = "your_mqtt_password";
```

Also declare variables for MQTT publishing, the topic (which two level in this case "Node1" and "Temp") separated by forward slash, the `clientID`, and a character string for the message `payload` (which will later be assigned the value from the sensor reading). Refer to the textbook companion to these lab guides and references therein for further details for a discussion of topics and payload.

```
// Topic to publish
const char* topic = "Node2/Temp";
// Node name
const char* clientID = "Node2";
// mqtt message payload
char payload[20];
```

ready to initialize the client as a publisher connecting to port 1883 which we defined in the previous step when setting up the broker.

```
// Initialize use port 1883
WiFiClient wifiClient;
PubSubClient client(broker, 1883, wifiClient);
```

The next section of the script is the sensor function, that applies the B equation of the thermistor and voltage divider equation, and return the temperature in °C.

```
float readSensor(){
  int din = analogRead(pinSignal);
  float Vout = din*Vs/1023;
  float Rt = Vs*Rf/Vout - Rf; // in kohm
  float TK = 1/(1/298.0+(1/B)*log(Rt/Rt0));
  float TC = TK - 273.0;
  //can remove comment next lines to verify calculations
  //Serial.print(din);Serial.print(',');
  //Serial.print(Vout);Serial.print(',');
  //Serial.print(Rt);Serial.print(',');
  //Serial.print(TK);Serial.print(',');
  //Serial.println(TC);
  return TC;
}
```

When we are ready to test you can remove the comments to verify that the sensor is functioning properly.

Now in the setup() function, start the serial stream at 115,200 baud (make sure to set this baud rate at your serial monitor later), the transducer pin mode, followed by connection to the local area network (LAN) by Wi-Fi, and several lines streamed to the serial monitor for verification.

```
void setup() {
  // Begin Serial on 115200
  Serial.begin(115200);
  // for sensor
  pinMode(pinSignal,INPUT);

  Serial.print("Connecting to ");
  Serial.println(ssid);

  // Connect to the WiFi
  WiFi.begin(ssid,pswd);

  // Wait until the connection has been confirmed
  while (WiFi.status() != WL_CONNECTED) {
    delay(500);
    Serial.print(".");
  }

  // Output the IP Address of the ESP8266
  Serial.println("WiFi connected");
  Serial.print("IP address: ");
  Serial.println(WiFi.localIP());
```

Later you should verify the IP address acquired by the node as displayed on the serial monitor. Now we connect the node to the broker

```
  // Connect to MQTT Broker
  if (client.connect(clientID, mqtt_username, mqtt_password)) {
    Serial.println("Connected to MQTT Broker!");
  }
  else {
    Serial.println("Connection to MQTT Broker failed...");
  }

}
```

The last curly brace was the end of the setup() function. In the loop() function, we read the sensor, and its returned value as a float is converted to string assigned to the message payload using the sprintf function

```
void loop() {
    float TC=readSensor();
    sprintf(payload, "%f", TC);
```

publish the message with topic and payload, and echo to the serial monitor for verification.

```
    // Publish to the MQTT Broker
    if (client.publish(topic,payload)) {
      Serial.print("Topic sent ");
      Serial.print(topic);
      Serial.print(" Payload sent ");
      Serial.println(payload);

    }
```

The remainder of the loop is used to establish what to do if the node fails to publish

```
    else {
      Serial.println("Failed to send, trying again");
      client.connect(clientID, mqtt_username, mqtt_password);
      delay(10); // avoid publish vs. connect
      client.publish(topic,payload);
    }
```

And then repeat every five seconds, and close the loop

```
    delay(5000);
}
```

Once compiled and uploaded for Node1, you can remove the USB cable from the PC and connect the node to a charging block for power. Repeat the process for Node 2 but of course edit the topic to "Node2/Temp" and clientID to "Node2".

SUBSCRIBER CLIENT

To complete the WSN, we write Python code for the subscriber client which will run in the RPi. Normally the subscribers will run on separate devices, we are using the same RPi just for learning in a simplified scenario. This Python script `RPiSubscribe.py` is available from the RTEM repository. We base this script on guidance to client subscribers using `paho mqtt client` available from Steve's Internet Guide (2022). Start with imports and note that after the one we have used before we import the `paho.mqtt.client` package

```
import os
from datetime import datetime,timezone
import serial
import time
from time import sleep
import paho.mqtt.client as mqtt
```

Next, we declare datalog file names for each sensor node but with the same header for each, consisting of the timestamp and temperature reading of the node which will be captured from the message payload of each

```
lognode1 = "data/node1-datalog.csv"
lognode2 = "data/node2-datalog.csv"
header = "TimeStamp,Temp\n"
```

now we open the files to append a new record or alternatively, write a header when it is the first time we have open the file.

```
if os.path.exists(lognode1):
        print("File exists, appending records\n")
        file1 = open(lognode1, "a")
else:
        print("File does not exists, writing header\n")
        print(header)
        file1 = open(lognode1, "w")
        file1.write(header)

if os.path.exists(lognode2):
        print("File exists, appending records\n")
        file2 = open(lognode2, "a")
else:
        print("File does not exists, writing header\n")
        print(header)
        file2 = open(lognode2, "w")
        file2.write(header)
```

now after the broker's IP address, and authentication conditions of the broker (username and password set previously)

```
# MQTT broker
mqtt_broker_ip = "192.168.1.132"
mqtt_username = "mqtt_username"
mqtt_password = "mqtt_password"
```

at this point we define the topics as an array of two elements, note that each array element has the two-level topic of a node and an additional required argument which is the quality of service (qos) which in this case we leave at the lowest 0 for simplicity.

```
# array for two topics the 0 is qos
mqtt_topic = [("Node1/Temp",0),("Node2/Temp",0)]
```

now we start the subscriber client

```
# subscriber client
client = mqtt.Client()
client.username_pw_set(mqtt_username, mqtt_password)
```

and define a function on _ connect for what to do when this client connects to the broker

```
# function to run when the MQTT client connects to the broker
def on_connect(client, userdata, flags, rc):
        print("Connected! Error code is ", str(rc))
        client.subscribe(mqtt_topic)
```

which is basically to subscribe to the topics. Importantly, now we write another function on _ message for what to do when the subscriber receives a message from the broker, which was in turn received by the broker and sent by one of the publisher nodes.

```
# function to run when a message is received
def on_message(client, userdata, msg):
        payload = str(msg.payload, 'utf-8')
        now = datetime.now(timezone.utc)
        timestamp = str(now.astimezone().strftime("%Y-%m-%d %H:%M:%S %z")+',')
        record = timestamp+payload
        print(msg.topic+" ",end="")
        print(record)
        if(msg.topic == 'Node1/Temp'):
                file1.write(record+"\n")
                file1.flush()
        if(msg.topic == 'Node2/Temp'):
                file2.write(record+"\n")
                file2.flush()
```

As we can see, this is a more complex function that encodes `msg.payload` as `utf-8` as the sensor temperature reading, composes a timestamp, concatenates it with the payload into a file record, echoes to the serial stream, and then according to the `msg.topic` writes the record to one of the two datalog files. We are almost done; the remainder of the script is to define how to apply these functions, connect to the broker using the previously defined port number, and loop "forever" to execute those functions every time messages are received.

```
# functions  to run
client.on_connect = on_connect
client.on_message = on_message

# connect to the broker on port 1883
client.connect(mqtt_broker_ip, 1883)

# let the client run
client.loop_forever()
client.disconnect()
```

Once we save this script under the working directory `labs` in the RPi, we execute it using Python from the working directory labs

```
python3 RPiSubscribe.py
```

At this point, we have all the systems running. At the PuTTY terminal, we should see the stream of messages by topic with created timestamp and file being created or appended if already exists as shown in Figure 6.11.

After compilation and upload of each node, at the Arduino IDE serial monitor we would see the streaming message sent, topic, and payload. For example, for node 2, see Figure 6.12. You would also see the IP address of the node, in the example shown in Figure 6.12, Node2 acquired address 192.168.1.31.

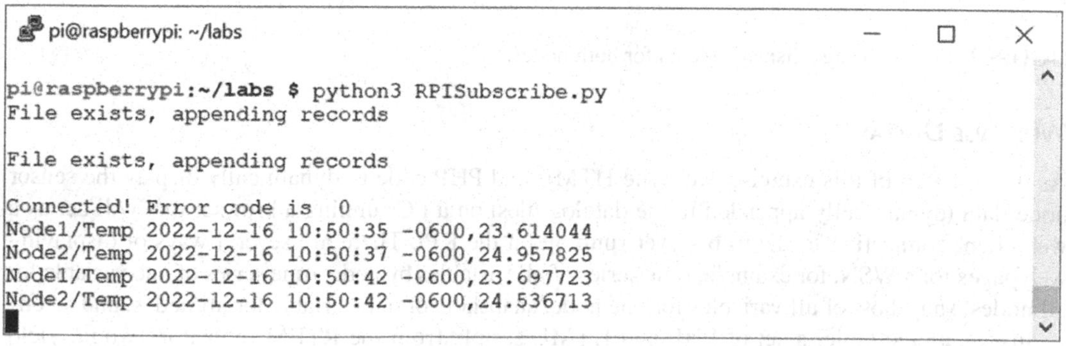

```
pi@raspberrypi: ~/labs                                    —    □    ×

pi@raspberrypi:~/labs $ python3 RPISubscribe.py
File exists, appending records

File exists, appending records

Connected! Error code is  0
Node1/Temp 2022-12-16 10:50:35 -0600,23.614044
Node2/Temp 2022-12-16 10:50:37 -0600,24.957825
Node1/Temp 2022-12-16 10:50:42 -0600,23.697723
Node2/Temp 2022-12-16 10:50:42 -0600,24.536713
```

FIGURE 6.11 PuTTY terminal showing the subscriber output.

Message (Enter to send message to 'NodeMCU 1.0 (ESP-12E Module)' on 'COM6')

Topic sent Node2/Temp Payload sent 23.781433
.... WiFi connected
IP address: 192.168.1.31
Connected to MQTT Broker!
Topic sent Node2/Temp Payload sent 23.614044
Topic sent Node2/Temp Payload sent 23.697723
Topic sent Node2/Temp Payload sent 23.697723

FIGURE 6.12   Serial monitor of Arduino IDE echoing the message published by the node.

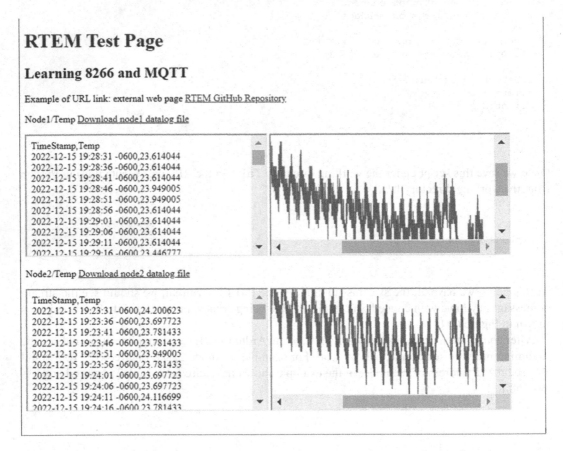

FIGURE 6.13   Web page displaying data for both nodes.

WEB PAGE DISPLAY

As the last step of this exercise, we write HTML and PHP code to dynamically display the sensor node data (dynamically appended to the datalog files) on a PC running a browser that will act as a web client connecting to the web server running on the RPi. There are several ways of displaying web pages for a WSN, for example, time series of all variables by node, time series of one variable for all nodes, snapshots of all variables for one node, snapshots of one variable for all nodes, and so on.

Just as an example, a set of PHP and HTML scripts from the RTEM repository would yield the display shown in Figure 6.13. This was produced using an element `frameset` to organize various iframes in cols or rows, and php scripts, as well as d3 scripts, that apply to each node file (Figure 6.13). Once you have all these PHP scripts and HTML documents loaded to the labs directory of the RPi, point your browser on the PC to the IP address of the RPi and verify that your display is similar to that shown in Figure 6.13.

WSN USING MOTEINO DEVICES BASED ON 915 MHZ RADIO

Figure 6.14 shows a Moteino USB, which is one model of Moteino devices; it has a micro-USB connector for serial communication and power, whereas other models use other types of serial connectors such as an FTDI adapter (LowPowerLab 2016, 2022).

The figure also illustrates the pinout configuration. One of the long sides has GND, Vin (input), 3v3 (output), together with digital pins 0 and 1 (these are also Tx and Rx), and analog pins A0–A7. A pin for the antenna is located near A1. The other long side has RST and digital pins 2–13. Analog pins A3 and A4 serve for I2C lines SDA and SCL, respectively, and digital pins 11, 12, and 13 serve for SPI lines, MOSI, MISO, and SCK, respectively. Moteinos can include onboard radio capabilities provided by a RFM69CW transceiver in the 433, 868, and 915 MHz range. The units used to write this lab guide are 868/915 MHz.

In Figure 6.15, we can see the setup for this WSN. Starting from the left-hand side of the diagram, a PC is connected by USB to a Moteino USB node to upload the node sketch. Keep in mind that in a real application there are other nodes, which are part of the WSN. Once a node is programmed, we will disconnect the node Moteino from the PC and power it via the USB cable, so that it becomes a stand-alone WSN node.

The Moteino gateway connects by USB to the RPi Zero (Figure 6.15). Data transmission occurs at a time interval set by the node sketch. Keep in mind that in a real application, the gateway will receive transmission from multiple nodes and therefore protocols to avoid collision must be implemented. The RPi, using a Python script for serial communication, will obtain all the data received by the gateway.

The RPi Zero connects to a LAN using Wi-Fi (2.4 GHz) and it is programmed using SSH from a PC connected to this LAN. The RPi will serve web pages over the LAN, and thus any device on the LAN (or on the internet if the private IP is routed to the internet) can then use a browser to visualize the activity on the WSN. In this lab, for simplicity, the browser could run on the same PC (used to program the Moteinos and program the RPi using SSH), or for a more realistic demonstration you could use a smartphone, or a tablet connected to the LAN.

We will write an Arduino script for the node and one for the gateway, as well as a Python program to build files from the gateway. To simplify programming, you may download the archive `Moteino _ Scripts.zip` from the RTEM repository and extract files from this archive to your working folder `labs`.

FIGURE 6.14 Moteino USB example illustrating layout, radio, and pin-out.

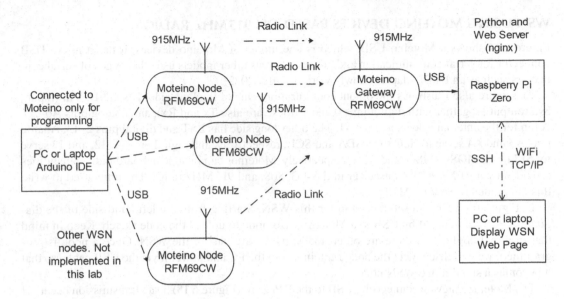

FIGURE 6.15 Diagram showing WSN Moteino setup.

WSN Node Using a Moteino

We will use a Moteino USB unit like the one shown in Figure 6.14, for each sensor node. A convenient setup to conduct the lab is to solder male pin headers to the Moteino node so that you can easily place it on a protoboard and facilitate wiring sensors. You can connect the Moteino USB directly to a PC using a USB cable with one end terminated as micro-USB and another terminated as regular USB. See the appendix available from eResources for connecting the Moteino using a FTDI adapter.

Wire your transducer to a Moteino node following Figure 6.16. You will use 3.3 V of the Moteino to power the voltage divider. Assume $R_f = 10$ kΩ and $R_0 = 10$ kΩ. Pin A0 is the input pin measuring the voltage output from the divider. Insert the Moteino header in the breadboard and connect it by USB to the PC (Figure 6.17).

The plan is to write an Arduino script to upload to the Moteino nodes using the Arduino IDE on a PC, but before doing this we need to setup the board and port, as well as install new libraries. Just to avoid confusion, in this exercise we are not using the Arduino UNO board, but the Moteino board; however, we are using the Arduino IDE to program the Moteino board. Connect a Moteino USB to the PC and under `File| Preferences`, in the box for Additional board manager URLs type the following (after a comma if you already have other additional URLs)

```
https://lowpowerlab.github.io/MoteinoCore/package_LowPowerLab_index.json
```

click ok and now go to `Tools|Board|Boards Manager`, select the Moteino AVR boards which include the Moteino USB and press install. Make sure you have selected the correct port. We will need additional libraries `LowPower` and `RFM69` to include in the script. Download and install these from the `LowPowerLab` GitHub repository (Low Power Lab 2023) or simply use the Arduino IDE Library manager.

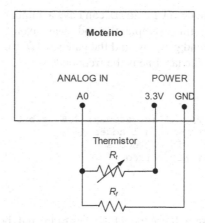

FIGURE 6.16 Sensor connected to Moteino pins.

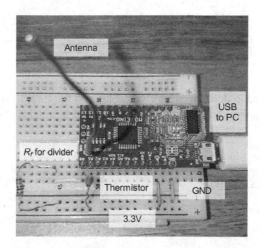

FIGURE 6.17 Sensor and USB connected to Moteino.

We will write a script `MoteinoSensorNode.ino` for the node (recall it is available from archive `WSN-Moteino _ Scripts.zip`). Start by including the libraries for the Moteino and radio RFM69

```
#include <LowPower.h>    //https://github.com/lowpowerlab/lowpower
#include <SPI.h>          //included with Arduino IDE install (www.arduino.cc)
#include <RFM69.h>        //https://www.github.com/lowpowerlab/rfm69
```

Next, we define variables. The node ID, NodeID, can have a number starting at 2, since we reserve ID number 1 for the gateway; in this example, the node was given ID 2. The Network ID, netID set to 100, is used for all nodes and gateway, and the gateway ID, gwayID of the Moteino gateway node is 1, as mentioned above. The last line is the frequency.

```
#define nodeID    2 // 2-254, 1 is gateway, 255 is broadcast
#define netID     100 // same on all nodes
#define gwayID    1 // same in all nodes
#define freq      RF69_915MHZ // freq
```

The next section declares that we will use the RFM69 as radio and the character array for the transmission packet

```
RFM69_ATC radio;
// max is 61
char packet[20];
```

followed by a sector devoted to defining sensor variables and the readSensor function which we have discussed in the previous exercise using NodeMCU.

```
// thermistor parameter
float Rt0= 10.0; // in kohm
float B= 4116.0;
// sensor circuit parameter
float Vs = 3.3;
float Rf = 10.0; // in kohm
int pinSignal=A0;

float readSensor(){
  int din = analogRead(pinSignal);
  float Vout = din*Vs/1023;
  float Rt = Vs*Rf/Vout - Rf; // in kohm
  float TK = 1/(1/298.0+(1/B)*log(Rt/Rt0));
  float TC = TK - 273.0;
  //can remove comment next lines to verify calculations
  //Serial.print(din);Serial.print(',');
  //Serial.print(Vout);Serial.print(',');
  //Serial.print(Rt);Serial.print(',');
  //Serial.print(TK);Serial.print(',');
  //Serial.println(TC);
  return TC;
}
```

Following that sector, we write a function to make the packet to be transmitted from the node

```
void makePacket(){
    char TCa[5];
    float TC = readSensor();
    // Converto float to string 2 decimal
    dtostrf(TC, 5, 2, TCa);
    sprintf(packet, "Temp,%s", TCa);
}
```

Note the use of function `dtostrf` to convert the sensor reading float `TC` to string `TCa`. A packet will consist of string `Temp` and the temperature value separated by a comma; for example, "Temp,21.56".

Now, follow with the `setup()` function, where we initialize the radio and print a message to the serial monitor to identify the node

```
void setup(){
  Serial.begin(9600);
  radio.initialize(freq, nodeID, netID);

  char titleMessage[25];
  Serial.print("\nNode "); Serial.print(nodeID);
  sprintf(titleMessage, " Tx at %d MHz", radio.getFrequency() / 1000000);
  Serial.println(titleMessage);
  Serial.print("Network ID: ");Serial.println(netID);

}
```

Finally, in the `loop()` function, we call the function to make the packet, echo the packet to the serial monitor, and transmit using `radio.sendWithRetry` with 5 retries and 100 wait interval. This is repeated every five seconds using `delay`.

```
void loop(){
  makePacket();   // make and check the packet
  Serial.print("packet "); Serial.print(packet);
  Serial.print(" Sending it to node "); Serial.print(gwayID);
  // retries 5 wait retry 100
  if (radio.sendWithRetry(gwayID, packet, strlen(packet),5,100))
    Serial.println(" ACK Recvd");
  else Serial.println(" ACK Not Recvd");
  delay(5000);
}
```

Once you compile and upload, look at the output using the Serial Monitor. Adjust the baud rate in the serial monitor (lower right-hand side) to match 9600 defined in the sketch (see Figure 6.18).

Each line consists of the packet a text message and an acknowledgment ACK received or not. The first time you do this, the ACK would not be received because you have not programmed the gateway yet. Once we have tested that the node transmits, we will disconnect it from the PC and power it independently via the USB cable. Typically, in the field we will deploy the node powered by a battery, likely recharged from a solar panel (to be discussed in Chapter 7), but for the sake of simplicity, we will power it now via the USB cable connected to a charging block.

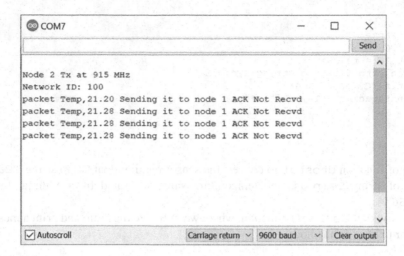

FIGURE 6.18   Serial monitor window with adjusted baud rate showing node transmission.

Now you can connect the other node to the PC, edit the `MoteinoSensorNode.ino` script to change only the node ID to 3

```
#define nodeID   3 // 2-254, 1 is gateway, 255 is broadcast
```

Compile, upload, and make sure it also transmits by inspecting the serial monitor.

WSN Gateway Using a Moteino and a RPi

Connect another Moteino USB to the RPi Zero using the USB cable and micro-USB adapter just as you connected the RPi Zero to the Arduino in previous labs (Figure 6.19).

Since the Moteino board requires additional packages, we will list the URLs to additional package indexes in the Arduino CLI configuration file. For this purpose, let us edit the `Arduino-cli.yaml` file

```
nano .arduino15/arduino-cli.yaml
```

and change the board_manager settings as follows:

```
board_manager:
    additional_urls:
        https://downloads.arduino.cc/packages/package_index.json
        https://lowpowerlab.github.io/MoteinoCore/package_LowPowerLab_index.json
```

See Figure 6.20.

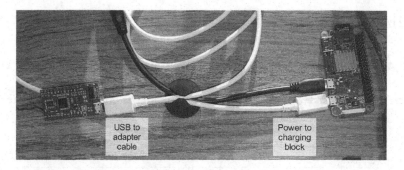

FIGURE 6.19   Connecting the RPi to the Gateway Moteino.

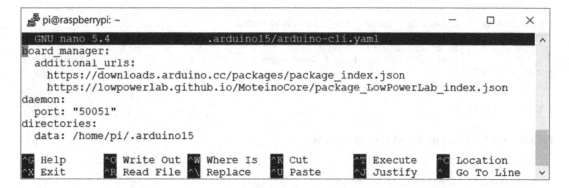

FIGURE 6.20   Configuration file.

```
pi@raspberrypi: ~                                              —    □    ✕

pi@raspberrypi:~ $ arduino-cli core update-index
Updating index: package_index.json downloaded
Updating index: package_index.json.sig downloaded
Updating index: package_index.json downloaded
Updating index: package_index.json.sig downloaded
Updating index: package_LowPowerLab_index.json downloaded
pi@raspberrypi:~ $ ▊
```

FIGURE 6.21   Results of update.

From now on, commands supporting custom cores will automatically use the additional URL from the configuration file. Let us update the RPi (Figure 6.21)

```
arduino-cli core update-index
```

Now, execute a search (Figure 6.22)

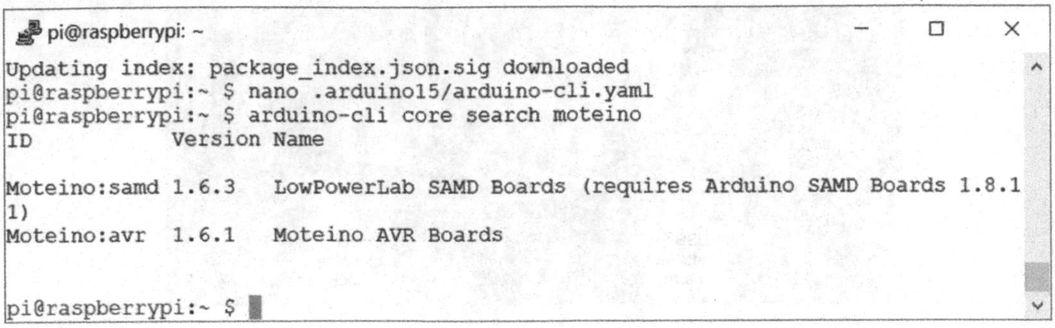

```
pi@raspberrypi: ~                                           —    □    ×
Updating index: package_index.json.sig downloaded
pi@raspberrypi:~ $ nano .arduino15/arduino-cli.yaml
pi@raspberrypi:~ $ arduino-cli core search moteino
ID          Version Name

Moteino:samd 1.6.3  LowPowerLab SAMD Boards (requires Arduino SAMD Boards 1.8.1
1)
Moteino:avr  1.6.1   Moteino AVR Boards


pi@raspberrypi:~ $ ▮
```

FIGURE 6.22   Arduino boards showing the Low Power Lab Moteino AVR Boards.

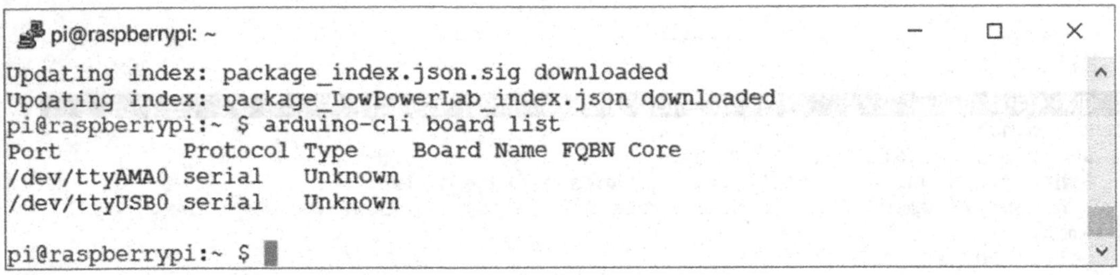

```
pi@raspberrypi: ~                                           —    □    ×
Updating index: package_index.json.sig downloaded
Updating index: package_LowPowerLab_index.json downloaded
pi@raspberrypi:~ $ arduino-cli board list
Port          Protocol Type     Board Name FQBN Core
/dev/ttyAMA0 serial    Unknown
/dev/ttyUSB0 serial    Unknown

pi@raspberrypi:~ $ ▮
```

FIGURE 6.23   Serial ports and Arduino boards.

```
arduino-cli core search moteino
```

List the boards (Figure 6.23)

```
arduino-cli board list
```

Note that now we are using the ttyUSB0 serial port, with an unknown device, which is the Moteino we can use the same fqbn as Arduino Uno.

Install the RFM69 _ LowPowerLab library in Arduino-cli (also install the SPI library if you have not yet)

```
arduino-cli lib install RFM69_LowPowerLab
```

We will write a script named MoteinoGateway.ino (also available in the archive downloaded from the RTEM repository). To facilitate editing the sfcript, you can use Geany on your PC and then copy it to the RPi.

Start with including the libraries

```
#include <RFM69.h>          //get it here: https://github.com/lowpowerlab/
rfm69
#include <SPI.h>            //included with Arduino IDE (www.arduino.cc)
```

Then, define node ID, network ID, and frequency, and radio as we did for the sensor node. Note in this case the `nodeID` is 1 because it is the gateway.

```
// settings
#define nodeID  1 //the ID of this node
#define netID   100 //the network ID of all nodes this node listens/talks to
#define freq    RF69_915MHZ

RFM69 radio;
```

Now we write the receiving function. The first part is to declare variables.

```
void listenReceiving(){
  int senderID; int targetID;  int rslRSSI;
  int dataLen; String dataRx; bool ackReq;
```

Pay attention to the following two logical if segments. The first one is executed upon receiving a packet and as soon as the packet is received by reading the buffer into `dataRx`, we send the acknowledgment.

```
  if (radio.receiveDone()){
    // save received info and message so that node can send ack right away
    senderID = radio.SENDERID; targetID= radio.TARGETID;  rslRSSI = radio.RSSI;
    dataLen= radio.DATALEN; ackReq= radio.ACKRequested();
    for (uint8_t i=0; i<radio.DATALEN; i++){
     dataRx += char(radio.DATA[i]);
    }
    // send ACK right away
    if (radio.ACKRequested()) radio.sendACK();
```

This is important so that the node receives an acknowledgment right away instead of after executing code to process the received data. We will take the time to do that after the ACK is sent as shown below.

```
    // now can take the time to print received info and message for confirmation
    Serial.print("Node");Serial.print(senderID, DEC); Serial.print(",");
    Serial.print(dataRx);Serial.print(",");
    Serial.print("RSL,");Serial.print(rslRSSI);Serial.print(",");
    if (ackReq) Serial.println("ACK sent");
    else Serial.println("No ACK Req");
  }
}
```

The next function is the `setup()` where we initialize the radio and print a message to identify the node.

```
void setup(){
  Serial.begin(9600);
  radio.initialize(freq, nodeID, netID);

  char titleMessage[25];
  sprintf(titleMessage, "Gateway node 1 at %d MHz", radio.getFrequency() /
1000000);
  Serial.println(titleMessage);
  Serial.print("Network ID: "); Serial.println(netID);
}
```

Finally, the `loop()` function is just a repetition of the receiving function.

```
void loop(){
    listenReceiving(); //listen for received packets
    delay(100); //wait before listening again
}
```

Once you have the script ready, copy it to your working directory `labs` in the RPi, either by opening a `cmd` windows and using `pscp` to secure-copy the file to the RPi or by using WinSCP. Verify that the file is in the folder.

Now we are going to compile and upload this script to the Moteino gateway using the RPi.

```
arduino-cli compile --fqbn arduino:avr:uno labs/MoteinoGatewayPi
arduino-cli upload -p /dev/ttyUSB0 --fqbn arduino:avr:uno labs/MoteinoGatewayPi
```

Once uploaded, for a quick check we can use the very simple Python code `test_serial.py` that we have used in previous labs to communicate serially with the Arduino but will change the port to ttyUSB0 (Figure 6.24). This is easily accomplished using nano. Let us rename the file as `test_serialUSB.py`

Run the Python code from the labs working directory

```
python3 test_serialUSB.py
```

You should see the transmission from node 2 and node 3 demonstrating that you have established radio communication between the nodes and the Gateway (Figure 6.25).

In this output, the received signal strength is shown in dBm (negative) as received by the gateway's radio. In this example, the Received Signal Level (RSL) of about −26 dBm for node 2 is very good and this result was obtained by having the sensor node and gateway node within several feet in the same room of a building. In contrast, the RSL of about −75 dBm for node 3 is marginal because the sensor and gateway nodes were in different parts of a building separated by about 50 ft. At this point, you can go back to look at the Serial Monitor on the Arduino IDE showing the sensor node and verify that ACK is received (Figure 6.26).

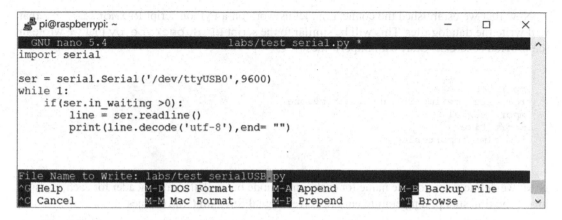

FIGURE 6.24 Python code for quick check of functioning nodes and gateway.

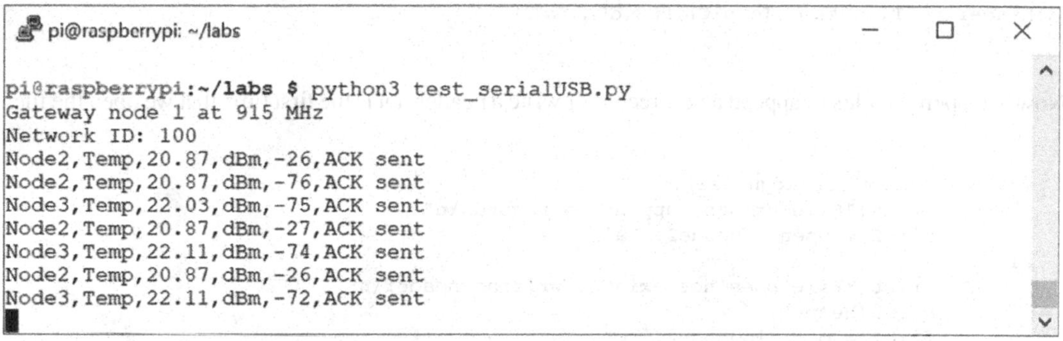

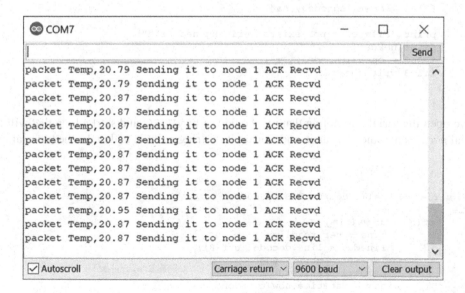

FIGURE 6.25 Moteino gateway serial stream output showing that the nodes are transmitting.

FIGURE 6.26 Serial monitor showing output from the sensor node.

Now that we established the connection, let us work on a Python script `RPiMoteino.py` that will write the datalog files. This will be similar to the script `RPiSubscribe.py` that we wrote in the previous exercise. Start with import lines

```
import os
from datetime import datetime,timezone
import serial
import time
from time import sleep
```

Next, we define a datalog file name for each sensor node but with the same header for each, consisting of the timestamp, temperature, and Received Signal Strength (RSS) values

```
lognode2 = "data/MoteinoNode2-datalog.csv"
lognode3 = "data/MoteinoNode3-datalog.csv"
header = "TimeStamp,Temp(C),RSS(dBm)\n"
```

Now, we open the files to append a new record or write a header if it is the first time that we open the file.

```
if os.path.exists(lognode2):
        print("File exists, appending records\n")
        file2 = open(lognode2, "a")
else:
        print("File does not exists, writing header\n")
        print(header)
        file2 = open(lognode2, "w")
        file2.write(header)

if os.path.exists(lognode3):
        print("File exists, appending records\n")
        file3 = open(lognode3, "a")
else:
        print("File does not exists, writing header\n")
        print(header)
        file3 = open(lognode3, "w")
        file3.write(header)
```

Now, we open the serial port noting that we should use `ttyUSB0`, and start a loop that will check the serial port, read a line, decode it, and convert it to string, print it, and make the timestamp

```
ser =serial.Serial("/dev/ttyUSB0",9600)
while 1:
        if(ser.in_waiting >0):
                line = ser.readline()
                linedec = line.decode('utf-8')
                linestr = str(linedec)
                print(linestr,end="")
                now = datetime.now(timezone.utc)
                timestamp = str(now.astimezone().strftime("%Y-%m-%d %H:%M:%S
%z")+',')
```

We must *parse* the incoming record, for example "Node2, Temp, 21.78, dBm, -90, ACK sent" into separate values so that we can filter which file to write to and what to write to that file.

It is a good opportunity to learn how to parse a string. We use `split` with comma separator that will produce an array `lineParsed` of six positions: ["Node2", "Temp", "21.78", "dBmV, "-90", "ACK sent"]. For example, we can address the node with `lineParsed[0]` which is "Node2" or the temperature value which is `lineParsed[2]`.

```
# parse the incoming line
lineParsed = linestr.split(sep=",")
```

Armed with these components, we can write a file for node 2 or node 3 with records with the time-stamp, followed by the temperature and dBm values.

```
if(lineParsed[0] == 'Node2'):
        values = lineParsed[2]+','+lineParsed[4]
        record = timestamp+values
        file2.write(record+"\n")
        file2.flush()
        print(record+"\n")
if(lineParsed[0] == 'Node3'):
        values = lineParsed[2]+','+lineParsed[4]
        record = timestamp+values
        file2.write(record+"\n")
        file2.flush()
        print(record+"\n")
```

We are done writing this Python script. We can copy to the `labs` working directory on the RPi and execute

```
python3 RPiMoteino.py
```

At the PuTTY terminal, we should see the stream of raw messages by node with timestamp and record being created by the parsing code. At the very beginning, we will see that the files are appended to or created (Figure 6.27).

Now, take note of the received signal strength and determine how the signal strength varies by changing the distance between your node and gateway, as well as conditions.

In the same manner as the WSN based on esp8266 and MQTT, we write HTML and PHP code to dynamically display the sensor node data (dynamically appended to the datalog files) on a PC running a browser that will act as a web client connecting to the web server running on the RPi. The PHP and HTML are available from the RTEM repository and follow closely what we already did in the previous exercise using ESP8266.

In this exercise, we have used a very simple deployment of Moteinos. More complicated experiments are described in the eResources online material, such as the use of encryption, the SPI flash memory, automatic transmission control. The latter reduces transmitting power to that just above a set threshold, thereby avoiding transmitting unnecessarily "loud" and wasting power. In the online material, we install LowPowerLab gateway and metrics in the RPI. This way we can examine WSN nodes, multigraph, download a datalog file, read it using R, and plot.

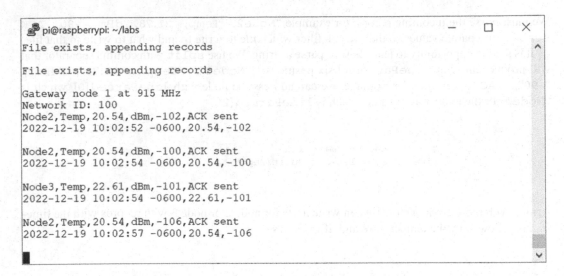

```
pi@raspberrypi: ~/labs                                          —    □    ×
File exists, appending records

File exists, appending records

Gateway node 1 at 915 MHz
Network ID: 100
Node2,Temp,20.54,dBm,-102,ACK sent
2022-12-19 10:02:52 -0600,20.54,-102

Node2,Temp,20.54,dBm,-100,ACK sent
2022-12-19 10:02:54 -0600,20.54,-100

Node3,Temp,22.61,dBm,-101,ACK sent
2022-12-19 10:02:54 -0600,22.61,-101

Node2,Temp,20.54,dBm,-106,ACK sent
2022-12-19 10:02:57 -0600,20.54,-106
```

FIGURE 6.27   Streaming messages and records on the RPi.

WSN USING LORA DEVICES

LoRa devices operating at 915 MHz have become popular for WSN and IoT applications since they provide a long range with low power. Commonly used LoRa radios are the RFM95W, which can be connected to an Arduino, or to an ESP device, or an RPi via an SPI interface. For this purpose, there are now breakout boards available to expand the 2 mm pitch pins of the radio to the 2.54 mm pitch compatible with common protoboards. Arduino libraries have also become available, for example Mistry (2022), as well as Python libraries (Python 2022). More information is available from the eResources.

EXERCISES

Exercise 6.1 Path loss using R.
    Use the radio link R functions to determine PL in dB and received power in dBm as a function of distance. Assuming antenna unity gain and transmit power of 1 W, analyze two scenarios. (1) Near field or short distance up to 10 km, for 915 MHz, 2.4 GHz, and 5 GHz. (2) Far field or long distance for *hthr* doubling from 10 to 160 m and distance doubling from 50 to 400 km.

Exercise 6.2 Fresnel zones using R.
    Use the radio link R functions to determine the first Fresnel zone clearance for 2.4 GHz, link distance 2 km, LoS 10 m, with an obstacle at 1 km from the transmitter. What is the obstacle height at which we start blocking more than 40% of the first zone?

Exercise 6.3 Using RadioMobile.
    Analyze the radio link between EESAT and DP following the steps given in the RadioMobile activity.

Exercise 6.4 WSN using NodeMCU, RPi, and MQTT.
    Implement a simple WSN consisting of two NodeMCUs with temperature transducer acting as MQTT publishers, and a RPi Zero W, acting as MQTT broker and subscriber client. Examine activity of each node, noting topic and payload transmitted and received.

Exercise 6.5 WSN using Moteinos and RPi.

Implement a simple WSN consisting of one Moteino node with temperature transducer, a Gateway Moteino, and the RPi Zero W. Examine activity of each node, noting packets transmitted and received, as well as transmit and receive power.

REFERENCES

Acevedo, M.F. 2024. *Real-Time Environmental Monitoring: Sensors and Systems - Textbook, Second Edition.* Boca Raton, FL: CRC Press, Taylor & Francis Group, 392 pp.

Adafruit. 2023. *RTClib*. Accessed January 2023. https://github.com/adafruit/RTClib.

Arduino. 2022a. *EthernetENC*. Accessed December 2022. https://reference.arduino.cc/reference/en/libraries/ethernetenc/.

Arduino. 2022b. *SD*. Accessed December 2022. https://www.arduino.cc/reference/en/libraries/sd/.

Arduino. 2022c. *SoftwareSerial Library*. Accessed December 2022. https://docs.arduino.cc/learn/built-in-libraries/software-serial.

Arduino. 2023a. *Arduino-CLI*. Accessed January 2023. https://github.com/arduino/arduino-cli.

Arduino. 2023b. *SPI*. Accessed January 2023. https://www.arduino.cc/en/Reference/SPI.

Arduino. 2016. *Download the Arduino Software*. Accessed August 2016. https://www.arduino.cc/en/Main/Software.

Components 101. 2022. *NodeMCU ESP8266*. https://components101.com/development-boards/nodemcu-esp8266-pinout-features-and-datasheet.

Coudé, R. 2022. *Radio Mobile*. Accessed December 2022. https://www.ve2dbe.com/rmonline_s.asp.

Espressif. 2022. *ESP8266EX Datasheet*. Accessed December 2022. https://www.espressif.com/sites/default/files/documentation/0a-esp8266ex_datasheet_en.pdf.

Geany. 2023. *Geany - The Flyweight IDE*. Accessed January 2023. https://www.geany.org/.

LowPowerLab. 2016. *Adventures in the Land of Low Power Embedded Systems*. Accessed August 2016. https://lowpowerlab.com/.

LowPowerLab. 2022. *Low PowerLab*. Accessed March 2022. https://lowpowerlab.com.

LowPowerLab. 2023. *LowPowerLab*. Accessed January 2023. https://github.com/lowpowerlab.

Posit. 2023. *RStudio IDE*. Accessed January 2023. https://posit.co/downloads/.

PuTTY. 2023. *Download PuTTY*. https://www.putty.org/.

Python. 2022. *pyLoRa 0.3.1*. https://pypi.org/project/pyLoRa/.

R Project. 2023. *The Comprehensive R Archive Network*. Accessed January 2023. http://cran.us.r-project.org/.

Sandeep Mistry. 2022. *Sandeepmistry/Arduino-LoRa*. Accessed December 2022. https://github.com/sandeepmistry/arduino-LoRa.

Steve's Internet Guide. 2022. *Paho Python MQTT Client Subscribe with Examples*. Accessed December 2022. http://www.steves-internet-guide.com/subscribing-topics-mqtt-client/.

Varnish, T. 2022. *How to Use MQTT with the Raspberry Pi and ESP8266*. Accessed December 2022. https://www.instructables.com/How-to-Use-MQTT-With-the-Raspberry-Pi-and-ESP8266/.

WinSCP. 2023. *Free Award-Winning File Manager*. Accessed January 2023. https://winscp.net/eng/index.php.

7 Environmental Monitoring and Electric Power

INTRODUCTION

In this session, we will conduct computer-based lab activities to learn about the relationships between environmental monitoring and electric power, focusing on two types of applications. One of these is powering the electronics of remote monitoring stations, along with wired and wireless sensor networks (WSNs), and the other is the application of monitoring to inform the design and operation of renewable power systems based on solar and wind. We conduct exercises using R to determine sun path and atmospheric effects at a given site, as well as functions of the R package `renpow` to analyze the I-V characteristics of photovoltaic (PV) cells. Then, we analyze solar radiation and wind speed data in the context of the resource available for electrical power production; for this purpose, we work with an example of data collected by a weather station and analyze the solar radiation and wind speed time series using package xts of R. Subsequently, we discuss the use of the solar radiation data to evaluate powering remote monitoring stations and energy storage using batteries.

MATERIALS

READINGS

For theoretical background, you can use Chapter 7 of Acevedo, M.F. 2024. *Real-Time Environmental Monitoring: Sensors and Systems - Textbook, Second Edition* which is a companion to these guides (Acevedo 2024). Other bibliographical references are cited throughout the guide. In addition, we will use the R package `renpow` developed for the textbook *Introduction to Renewable Power Systems and the Environment with R* (Acevedo 2018).

The following user manuals are recommended Datalogger CR3000 (Campbell Scientific Inc. 2018a), Pyranometer CS300 or CS301 (Campbell Scientific Inc. 2018b), Anemometer CS03101 (Campbell Scientific Inc. 2018c), and Anemometer CS05103 (Campbell Scientific Inc. 2020).

SOFTWARE (LINKS PROVIDED IN THE REFERENCES)

- R, for data analysis (R Project 2023)
- R package `renpow` available from the CRAN repository
- R package `tibble` available from the CRAN repository
- R package `xts` available from the CRAN repository

DATA FILES (AVAILABLE FROM THE GITHUB RTEM)

RTEM GitHub repository https://github.com/mfacevedol/rtem.

- Archive CR3000Tower.zip

Supplementary support material including additional screenshots, images, and procedures are available from the publisher eResources web page provided for this book.

SUN PATH AT A LOCATION: ELEVATION AND AZIMUTH

One important aspect of the solar resource is calculating the position of the sun in the sky for any time during the day and all days of the year. This is important for collecting energy by PV panels or by concentrated solar power (CSP), for aiming total atmospheric gas monitors (Lab guide 11), as well as for the design of buildings.

First, using R, install and load package renpow, which supports the renewable electric power system textbook by Acevedo (2018). Using the sun.path function of package renpow, we can calculate sun elevation and azimuth at a given latitude and day of the year. For example, use the latitude of Dallas, Texas, for day 20 of the year (January 20) and print results to a pdf in the output folder

```
wd=7; ht=7; outfile <- "output/sunpath.pdf"
pdf(outfile,wd,ht)
panels(wd,ht,1,1,pty="m")
# latitude dallas (DFW)
sun.path(lat=32.90,nday=20)
dev.off()
```

and obtain results for sun angle in Figure 7.1 (upper graph) and azimuth (lower graph) as a function of hours before noon; negative are morning hours and positive are afternoon hours.

Azimuth is given with respect to the south, negative are angles to the east, while positive are angles to the west. These results are combined in an azimuth-elevation plane, in the second page of the pdf, where the hour is implicit and given by the graph marker (Figure 7.2).

Now, for selected days of the year (January 21, March 21, and June 21), corresponding to the winter solstice, vernal equinox, and summer solstice, respectively (for the northern hemisphere), we obtain a sun path diagram (Figure 7.3) with the following function call to sun.diagram function of renpow

```
# latitude Dallas (DFW)
sun.diagram(lat=32.90)
```

The autumn equinox is not plotted but is the same as the vernal equinox. This chart summarizes the position of the sun in the sky for given days of the year and has many applications; for example, it can be used to analyze the potential shading of a solar panel given the distance, height, and azimuth of potentially shading obstacles such as trees and buildings. For examples, you can refer to Acevedo (2018).

SOLAR RADIATION: ATMOSPHERIC EFFECTS

Together with the sun's position, we must consider the effect of the atmosphere in modifying solar radiation received at the Earth's surface. This is also relevant to study total column concentration of atmospheric gases as we will discuss in Lab guide 11. As an example, we will assume a hypothetical collector located at Golden, Colorado, USA and compare energy collected on a clear day at either equinox, with the energy collected on clear days at the winter and summer solstices.

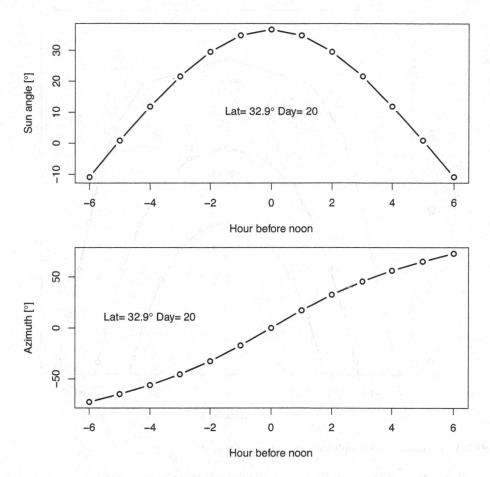

FIGURE 7.1 Hourly elevation and azimuth during a day.

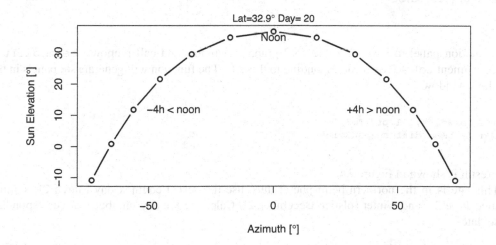

FIGURE 7.2 Elevation vs. Azimuth. Time is implicit and marked by circles every hour on the curve.

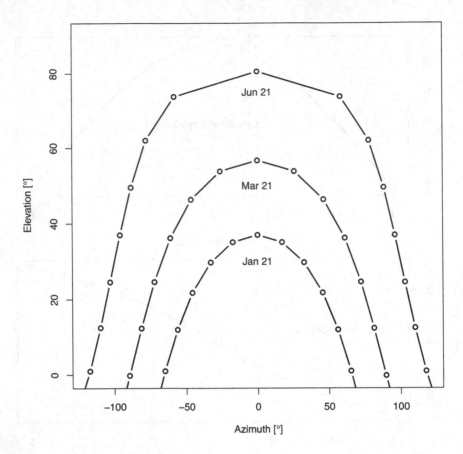

FIGURE 7.3   Elevation vs. Azimuth for sample days of the year.

We will start by opening a pdf file to store all the graphics to be produced

```
wd=7; ht=7; outfile <- paste(path,"Golden-three-days-collector.pdf",sep="")
pdf(outfile,wd,ht)
```

Use function panel to have only one 7×7 graphics window and call renpow's beam.diffuse with argument tauGolden corresponding to this site. The function will generate six panels in that graphics window

```
panels(wd,ht,1,1,pty="m")
Ibd <- beam.diffuse(tauGolden)
```

The result is shown in Figure 7.4.

This site is in the northern hemisphere; thus, use the vernal equinox day March 21, summer solstice June 21, and winter solstice December 21. Calculate the day number that corresponds to these dates

```
days <- days.mo(21)$day.mo[c(3,6,12)]
> days
[1]   80 172 355
>
```

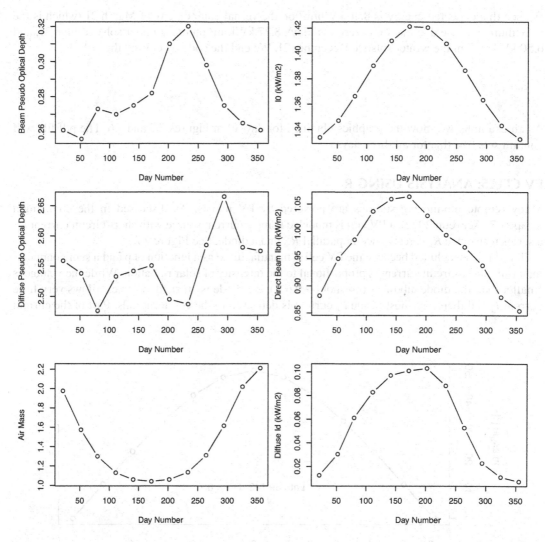

FIGURE 7.4 Golden, CO, optical depth (beam and diffuse), air mass, I0, Direct beam Ibn, and diffuse radiation ID.

These are days 80, 172, 355 of the year. Now we will generate panels of half the height, apply sun. path and collector functions for each one of these days, using a loop, and query the daily energy collected (total) in position 4 of the I.h array

```
panels(wd,ht/2,1,1,pty="m")
E <- array()
for(i in 1:3){
  sunpath <- sun.path(lat=Ibd$lat,nday=days[i])
  Ic <- collector(Ibd, sunpath, tilt=Ibd$lat,azi.c=0,fr=0.2)
  E[i] <- Ic$I.h[4]
}
> E
[1] 8.462 8.269 6.497
>
```

We see that expected energy is 8.46 kWh/m² on the vernal equinox day of March 21 (which is the maximum because declination is zero that day), 8.27 kWh/m² on the summer solstice June 21, and 6.50 kWh/m² on the winter solstice December 21. We end the code by closing the pdf file.

```
dev.off()
```

As an example, we show the graphics obtained for day 80 in Figures 7.5 and 7.6. The pdf file will contain sets like this for all three days.

PV CELLS: ANALYSIS USING R

Many remote monitoring stations are powered by PV panels. As discussed in the companion Chapter 7 (Acevedo 2022), a PV cell is modeled using a current source with short-circuit current I_{sc}, a series resistance R_s, a resistance in parallel R_p, and a diode. See Figure 7.7.

The diode is included because the PV cell is manufactured as a junction of p and n semiconductor material. Short-circuit current is proportional to the intensity of solar radiation. While the voltage is smaller than the diode elbow or threshold (~0.6 V), the diode is "turned off", i.e., allows very little current I_d and therefore most of the I_{sc} current is delivered to the cell terminals. Part of the current

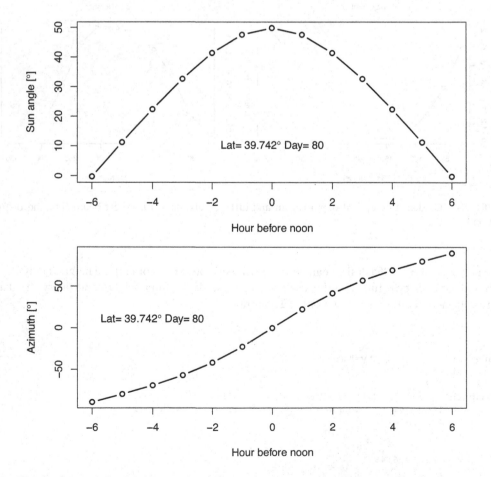

FIGURE 7.5   Sun path for day 80.

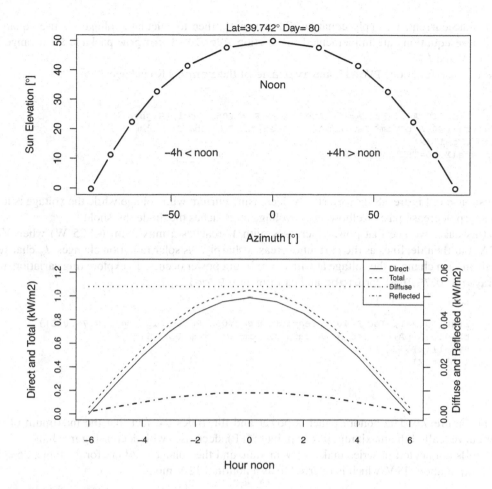

FIGURE 7.6 Sun diagram and radiation received at the collector: direct, diffuse, reflected, and total.

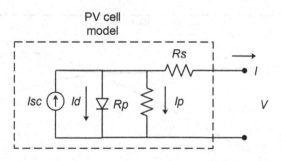

FIGURE 7.7 PV cell model.

is leaked via the parallel resistor R_p, the remaining is limited by the series resistor R_s. As the voltage increases and the diode "turns on", more current is diverted through the diode, and the output current decreases rapidly.

$$I = I_{sc} - I_0 \left[\exp(38.9 \times V_d) - 1 \right] - \frac{V_d}{R_p} \qquad (7.1)$$

$$V = V_d - I \times R_s \qquad (7.2)$$

To calculate current, we apply equation (7.1) given V_d, then to calculate voltage we use equation (7.2). These equations are implemented in function `PVcell` of `renpow` package for a range of values of V and I.

We will use functions `PVcell` and `ivplane` of the `renpow` R package.

```
x <- list(I0.A=1, Isc.A=40, Area=100, Rs=0.05, Rp=1, Light=1)
# units: I0.A pA/cm² Isc.A mA/cm² Area cm² Rs  ohm Rp  ohm
X <- PVcell(x)
ivplane(X,x0=TRUE,y0=TRUE)
```

The response in Figure 7.8 shows a slowly decreasing current with voltage while the voltage is low, and a sharp decrease past an elbow as the voltage approaches the diode threshold.

In this case, we see that power increases with V, reaches a maximum (~1.25 W) when V is ~0.42 V, but then declines as the current decreases sharply. As solar radiation changes, I_{sc} changes, and this in turn changes the voltage at which maximum power occurs. To explore this variation, use an array $c(1,0.75,0.5)$ for `Light` and call function `PVcell.plot`

```
x <- list(I0.A=1, Isc.A=40, Area=100, Rs=0.05, Rp=1, Light=c(1,0.75,0.5))
# units: I0.A pA/cm² Isc.A mA/cm² Area cm² Rs  ohm Rp  ohm
X <- PVcell(x)
PVcell.plot(X)
```

yielding Figure 7.9. The bottom panel is power and illustrates the fact that the maximum of the power curve, called the maximum power point (MPP), decreases with decreasing irradiance.

PV cells connected in series make a PV module and the voltages add up, for example, 36 cells would output about 18 V, which is referred to as a nominal 12-V module.

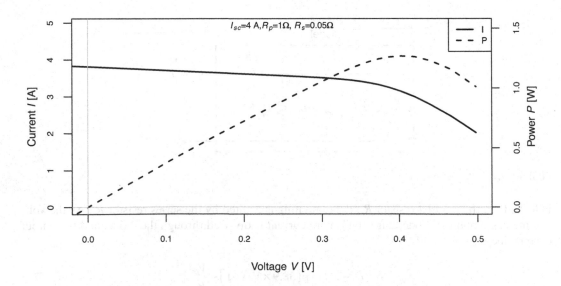

FIGURE 7.8 I-V characteristic of PV cell.

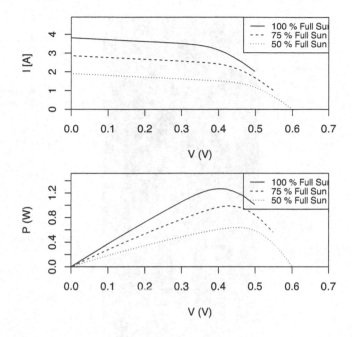

FIGURE 7.9   Current and power as a function of light as a percent of full sun.

MONITORING THE SOLAR RESOURCE: TIME SERIES

In this section, we will look at data for solar radiation received at a site and its dynamics. This will allow us to understand the solar resource available at a site for electricity production, both for supporting the design of a PV or CSP facility and to power a remote monitoring station or WSN using PV panels. In addition, it is an example of atmospheric modeling, which will continue to discuss in Lab 11.

As an example, we will use data from a weather station at the UNT Discovery Park campus, consisting of a tower and Campbell Scientific meteorological sensors (Figure 7.10) above and below ground, read by a Campbell Scientific datalogger CR3000 (Campbell Scientific Inc. 2018a).

Above ground, the sensors include:

- CS107 temperature probes to measure air temperature at two heights
- CS301 pyranometer
- CS03101 wind speed sensor
- CS700 rain gage
- CS106 barometric pressure sensor
- CS05103 wind speed and direction monitor
- HMP45C air temperature and relative humidity
- LWS Dielectric leaf wetness sensor.

Below ground, there are four CS616 water content reflectometers and four CS107 temperature probes.

A pyranometer measures solar radiation with a silicon PV detector mounted on a cosine-corrected head, its output is a current, which is converted to voltage by a potentiometer in the sensor head. The CS301 pyranometer (Figure 7.11) connects to a single ended (SE) analog channel of the datalogger and is read using instruction VoltSE( ). The sensitivity of the pyranometer is 0.2 mV/Wm^2.

The CS301 manual provides information on wiring, installation, and specifications (Campbell Scientific Inc. 2018b). A brief extract of the specifications given in the manual is

- Power requirements: none, self-powered
- Sensitivity: 5 W/m^{2}mV (0.2 mV/W m^2)

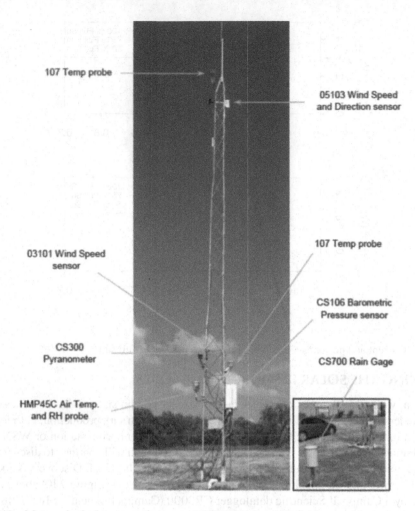

FIGURE 7.10   Meteorological tower and sensors at UNT Discovery Park campus.

- Absolute accuracy: ±5% for daily total radiation
- Cosine correction error: ±5% at 75° zenith angle ±2% at 45° zenith angle
- Response time: <1 ms
- Temperature response: 0.04% ±0.04% per C
- Long-term stability: <2% per year
- Operating temperature: −40°C to 70°C
- Relative humidity: 0%–100%
- Output: 0.2 mV/W m²
- Measurement range: 0–2000 W/m² (full sunlight ≈ 1000 W/m²)
- Light spectrum: waveband: 360–1120 nm (wavelengths where the response is 10% of maximum)

We can look at online data for this pyranometer using the web server on the CR3000 datalogger (Figure 7.12). With the work you have done over the previous lab sessions, you now understand how a web page like this is developed and presented.

The solar radiation data is `SlrW` and includes max, min, avg, sd, and total. See the example in Figure 7.13.

FIGURE 7.11 CS301 pyranometer installed on an arm anchored on the meteorological tower (side view and top view).

Download archive `CR3000Tower.zip` from the RTEM repository. Extract file `CR3000_March28-2019.csv`, save it in your `labs/data` folder, and open it using Geany or Notepad++. This file has data from all station sensors since 10/27/2018 10:00 am and ending on 03/28/2019 8:00 am. General information is given in the first line, followed by three lines of header containing names, units, and information of variables. The records start in line 5

```
"2018-10-27
10:00:00",0,12.9,4.625,0,2.77,1.009,233.1,18.45,67.15,0,381.2,357.6,369.3,6.9
85,188.3432,9.06,1.909,0,0.112,0.37,18.6,18.67,260.2,259.9,260.1,0.069,8.5,0,
0,15.01,37.82,62.49,17.47,0.01,0.65,0.008,0.595,0.01,0.667,0.01,0.654

"2018-10-27
10:10:00",1,12.92,5.085,0.767,2.742,1.128,231.5,19.28,58.65,0,407.1,381.5,394
.2,7.505,236.4967,462,4.928,0,0.884,1.32,19.54,19.96,260.1,259.4,259.7,0.165,
10,0,0,15.02,37.79,64.55,17.46,0.494,33.14,0.387,30.34,0.529,34.02,0.503,33.
36

"2018-10-27
10:20:00",2,13,7.716,1.863,4.649,1.118,198,19.64,60.17,0,433.5,407.4,420.9,7.
699,252.5234,462,6.94,0,3.184,1.346,19.52,19.52,259.7,259.4,259.6,0.069,10,0,
0,15.04,37.86,67.08,17.45,0.494,33.14,0.387,30.34,0.529,34.02,0.503,33.36
```

and these long lines are shown here wrapped and separated to facilitate seeing the contents. Note that data are collected every ten minutes during the first few records.

As discussed throughout the prior lab guides, an important type of monitoring data set is organized by a time stamp. We have seen several examples related to datalogging in Lab sessions 3 through 6, and we will expand working with this type of data in this lab session as well as in the next several lab sessions.

CR3000 Datalogger Home Page

- Newest Record from Status

- Newest Record from CR3000

- Display Last 24 Records from DataTable CR3000

- Newest Record from Public

FIGURE 7.12   CR3000 web server.

Current Record: 11757
Record Date: 2019-03-26 18:45:00.0

Batt_Volt	13.07
WS_mph_Max	10.04
WS_mph_Min	4.494
WS_mph_Avg	6.368
WS_mph_Std	1.553
WindDir	82.5
AirTC_Avg	19.2
RH	40.17
Rain_in_Tot	0
SlrW_Max	160.4
SlrW_Min	134.8
SlrW_Avg	149.2
SlrW_Std	8.82
SlrkJ_Tot	44.7578
BP_mmHg_Avg	457.2

FIGURE 7.13   Example of data available from the CR3000 web server.

To explore the data, we scroll down to the end of the file and look at the last few lines

```
"2019-03-28
07:50:00",12202,13.23,13.55,9.69,12.01,0.926,146.2,14.61,70.08,0,44.04,39.89,
42.1,1.332,12.62976,466.3,12.64,7.779,9.98,1.297,14.66,14.82,264.5,264.5,264.
5,0,5,0,0,15.49,40.78,"NAN",15.96,0.451,32.06,0.198,24.45,0.484,32.9,0.456,32
.18

"2019-03-28
07:55:00",12203,13.24,15.02,8.22,11.84,1.735,153.4,14.61,70.22,0,43.55,42.05,
42.99,0.389,12.89739,466.3,13.31,6.102,9.85,1.827,14.64,14.77,264.5,264.3,264
.5,0.03,5,0,0,15.48,41.04,"NAN",15.96,0.451,32.06,0.198,24.45,0.484,32.9,0.45
6,32.18

"2019-03-28
08:00:00",12204,13.23,14.95,7.65,10.92,1.909,131.5,14.59,70.62,0,52.85,43.05,
47.58,2.963,14.27522,466.3,12.81,5.934,8.52,1.922,14.6,14.74,264.5,264.5,264.
5,0,5,0,0,15.46,40.81,"NAN",15.96,0.451,32.06,0.198,24.45,0.484,32.9,0.456,32
.18
```

Note that the last measurements have an interval of five minutes and therefore the measurement interval changed during the data collection months, and we must take this into account when analyzing the time series.

Let us read the data into R and convert to a time series. First, set the path to the file

```
folder <- "data/"
# select the filename
file <- "CR3000_March28-2019.csv"
folder.file <- paste(folder,file,sep="")
```

Then, read the headers line by line; recall that the first line is general information, and the next lines are header for names, units, and calculations performed on variables.

```
# read file headers
header.meta <- scan(folder.file, sep=",", what="", nlines=1)
header.vars <- scan(folder.file, sep=",", what="", nlines=1,skip=1)
header.units <- scan(folder.file, sep=",", what="", nlines=1,skip=2)
header.calc <- scan(folder.file, sep=",", what="", nlines=1,skip=3)
```

We can read all records after the headers

```
# read all after header
tt.xx <- read.table(folder.file, sep=",", skip=4)
```

and assign names based on the header element containing variable labels

```
names(tt.xx) <- paste(header.vars,"(",header.units,")",sep="")
```

Thus, the names (tt.xx) array has labels for the variables. The first 16 are

```
 [1] "TIMESTAMP(TS)"              "RECORD(RN)"
 [3] "Batt_Volt(Volts)"          "WS_mph_Max(miles/hour)"
 [5] "WS_mph_Min(miles/hour)"    "WS_mph_Avg(miles/hour)"
 [7] "WS_mph_Std(miles/hour)"    "WindDir(Degrees)"
 [9] "AirTC_Avg(Deg C)"          "RH(%)"
[11] "Rain_in_Tot(inch)"         "SlrW_Max(W/mÂ²)"
[13] "SlrW_Min(W/mÂ²)"           "SlrW_Avg(W/mÂ²)"
[15] "SlrW_Std(W/mÂ²)"           "SlrkJ_Tot(kJ/mÂ²)"
```

TIBBLES

A related concept to the R data.frame is a tibble, or tbl_df, based on the ability to do less (i.e., do not change variable names) and warn more (e.g., when a variable does not exist) (Tibble 2023; R for Data Science 2023). Printing tibbles to the console make it easier to explore large datasets containing complex objects. For this purpose, install and loading package tibble.

```
install.packages("tibble")
library(tibble)
```

make a tibble out of tt.xx

```
tt.xx.tb <- as_tibble(tt.xx)
```

and check it

```
> tt.xx.tb
# A tibble: 27,518 × 42
    TIMESTAMP(T...¹ RECOR...² Batt_...³ WS_mp...⁴ WS_mp...⁵ WS_mp...⁶ WS_mp...⁷ WindD...⁸ AirTC...⁹
    <chr>               <int>    <dbl>    <dbl>    <dbl>    <dbl>    <dbl>    <dbl>    <dbl>
  1 2018-10-27 1...         0     12.9     4.62    0         2.77     1.01     233.     18.4
  2 2018-10-27 1...         1     12.9     5.08    0.767     2.74     1.13     232.     19.3
  3 2018-10-27 1...         2     13       7.72    1.86      4.65     1.12     198      19.6
  4 2018-10-27 1...         3     13.0     8.9     3.33      6.47     1.30     173.     19.6
  5 2018-10-27 1...         4     13.0     9.07    1.58      5.09     1.85     206.     19.9
  6 2018-10-27 1...         5     13.0     8.66    3.05      5.56     1.35     232      20.3
  7 2018-10-27 1...         6     13       7.94    2.39      5.24     1.39     251.     20.6
  8 2018-10-27 1...         7     12.9     8.68    2.52      4.79     1.22     244.     20.9
  9 2018-10-27 1...         8     12.9     9.75    1.95      4.78     1.61     252.     21.4
 10 2018-10-27 1...         9     12.9     9.73    2.13      4.99     1.64     266      21.7
# ... with 27,508 more rows, 33 more variables: 'RH(%)' <dbl>,
#    'Rain_in_Tot(inch)' <dbl>, 'SlrW_Max(W/m²)' <dbl>, 'SlrW_Min(W/m²)' <dbl>,
#    'SlrW_Avg(W/m²)' <dbl>, 'SlrW_Std(W/m²)' <dbl>, 'SlrkJ_Tot(kJ/m²)' <dbl>,
#    'BP_mmHg_Avg(mmHg)' <dbl>, 'WS_mph_2m_Max(miles/hour)' <dbl>,
#    'WS_mph_2m_Min(miles/hour)' <dbl>, 'WS_mph_2m_Avg(miles/hour)' <dbl>,
#    'WS_mph_2m_Std(miles/hour)' <dbl>, 'AirTC_2m_Avg(Deg C)' <dbl>,
#    'AirTC_10m_Avg(Deg C)' <dbl>, 'LWmV_Max(mV)' <dbl>, 'LWmV_Min(mV)' <dbl>,
...
# i Use 'print(n = ...)' to see more rows, and 'colnames()' to see all
variable names
>
```

We can use print to view more rows, for example, 20 rows

```
print(n=20,tt.xx.tb)
```

or colnames to view the column names

```
colnames(tt.xx.tb)
```

SOLAR RADIATION

From this dataset, we want the data for timestamp (variable 1) and variables 12–16 for solar radiation SlrW, which includes Max, Min, Avg, Std, and total energy (SlrJ_tot). These are maximum, minimum, average, standard deviation, and integrated solar radiation over the interval of measurement.

As mentioned above, these measurements started in October 2018 with an interval of ten minutes and changed later to five minutes. For example, `SlrW _ Avg` is the average of solar radiation over a ten-minute interval during the first measurements.

When we look at the first record using

```
> tt.xx[1,1]
[1] "2018-10-27 10:00:00"
```

we see the timestamp "2018-10-27 10:00:00" has the format "%Y-%m-%d %H:%M:%S"

```
"2018-10-27 10:00:00",0,12.9,4.625,0,2.77,1.009,233.1,18.45,67.15,0,381.2,357
.6,369.3,6.985,188.3432,9.06,1.909,0,0.112,0.37,18.6,18.67,260.2,259.9,260.1,
0.069,8.5,0,0,15.01,37.82,62.49,17.47,0.01,0.65,0.008,0.595,0.01,0.667,0.01,0
.654
```

Use `strptime`, removing all records with missing timestamp

```
# read time sequence from file
tt.raw <- strptime(tt.xx[,1], format="%Y-%m-%d %H:%M:%S",tz="")
# remove timestamps = NA
tt <- tt.raw[-which(is.na(tt.raw))]

# all data but time
xx <- tt.xx[-which(is.na(tt.raw)),-1] # remove those with time = NA
names(xx) <- names(tt.xx[-1])
```

Now, we make sure the package `xts` is loaded and create a time series `x.t` object

```
# make sure package xts is loaded
require(xts)
# create time series using array with the data and time base tt
x.t <- xts(xx,tt)
```

For a quick check of the time series created, ask for the first two records and ten variables

```
> x.t[1:2,1:10]
                    RECORD(RN) Batt_Volt(Volts) WS_mph_Max(miles/hour)
2018-10-27 10:00:00          0            12.90                  4.625
2018-10-27 10:10:00          1            12.92                  5.085
                    WS_mph_Min(miles/hour) WS_mph_Avg(miles/hour)
2018-10-27 10:00:00                  0.000                  2.770
2018-10-27 10:10:00                  0.767                  2.742
                    WS_mph_Std(miles/hour) WindDir(Degrees) AirTC_Avg(Deg C)
2018-10-27 10:00:00                  1.009            233.1            18.45
2018-10-27 10:10:00                  1.128            231.5            19.28
                    RH(%) Rain_in_Tot(inch)
2018-10-27 10:00:00 67.15                 0
2018-10-27 10:10:00 58.65                 0
>
```

For such large datasets, it will be useful to have the dimensions of the series, which can be obtained by

```
nvars <- dim(x.t)[2]
ntt <- dim(x.t)[1]
nvars;ntt
```

Or use tibble

```
> as_tibble(x.t)
# A tibble: 27,506 × 41
   'RECORD(RN)' Batt_V…¹ WS_mp…² WS_mp…³ WS_mp…⁴ WS_mp…⁵ WindD…⁶ AirTC…⁷ 'RH(%)'
          <dbl>    <dbl>   <dbl>   <dbl>   <dbl>   <dbl>   <dbl>   <dbl>   <dbl>
 1            0     12.9    4.62       0    2.77    1.01    233.    18.4    67.2
 2            1     12.9    5.08   0.767    2.74    1.13    232.    19.3    58.6
 3            2     13      7.72    1.86    4.65    1.12     198    19.6    60.2
 4            3     13.0     8.9    3.33    6.47    1.30    173.    19.6    61.3
 5            4     13.0    9.07    1.58    5.09    1.85     206    19.9    59.2
 6            5     13.0    8.66    3.05    5.56    1.35     232    20.3    56.4
 7            6     13      7.94    2.39    5.24    1.39     251.    20.6    56.8
 8            7     12.9    8.68    2.52    4.79    1.22     244.    20.9    56.3
 9            8     12.9    9.75    1.95    4.78    1.61     252.    21.4    54.2
10            9     12.9    9.73    2.13    4.99    1.64     266    21.7    52.4
# … with 27,496 more rows, 32 more variables: 'Rain_in_Tot(inch)' <dbl>,
#    'SlrW_Max(W/m²)' <dbl>, 'SlrW_Min(W/m²)' <dbl>, 'SlrW_Avg(W/m²)' <dbl>,
#    'SlrW_Std(W/m²)' <dbl>, 'SlrkJ_Tot(kJ/m²)' <dbl>,
#    'BP_mmHg_Avg(mmHg)' <dbl>, 'WS_mph_2m_Max(miles/hour)' <dbl>,
#    'WS_mph_2m_Min(miles/hour)' <dbl>, 'WS_mph_2m_Avg(miles/hour)' <dbl>,
#    'WS_mph_2m_Std(miles/hour)' <dbl>, 'AirTC_2m_Avg(Deg C)' <dbl>,
#    'AirTC_10m_Avg(Deg C)' <dbl>, 'LWmV_Max(mV)' <dbl>, 'LWmV_Min(mV)' <dbl>, …
# i Use 'print(n = ...)' to see more rows, and 'colnames()' to see all
variable names
>
```

Inspecting the dataset, we see that solar radiation is in columns 11–15 giving variables Max, Min, Avg, Std, and kJtot. Let us select the first three; that is max, min, and avg, into a new variable `slr` and change names for readability.

```
# solar radiation max, min, avg
slr <- x.t[,c(11:13)]
names(slr) <- c("Slr_Max(W/m2)", "Slr_Min(W/m2)","Slr_Avg(W/m2)")
```

We can query the first few records and last few records.

```
> slr[1:5,];slr[(ntt-5):ntt,]
                     Slr_Max(W/m2) Slr_Min(W/m2) Slr_Avg(W/m2)
2018-10-27 10:00:00       381.2         357.6         369.3
2018-10-27 10:10:00       407.1         381.5         394.2
2018-10-27 10:20:00       433.5         407.4         420.9
2018-10-27 10:30:00       459.3         434.2         446.7
2018-10-27 10:40:00       481.9         460.4         471.4
                     Slr_Max(W/m2) Slr_Min(W/m2) Slr_Avg(W/m2)
2019-03-28 07:35:00       20.28         14.79         18.45
2019-03-28 07:40:00       26.92         19.61         22.55
2019-03-28 07:45:00       39.56         27.26         33.74
2019-03-28 07:50:00       44.04         39.89         42.10
2019-03-28 07:55:00       43.55         42.05         42.99
2019-03-28 08:00:00       52.85         43.05         47.58
>
```

Note how the interval between measurements changed from ten to five minutes.

TIME-SERIES: WEEKLY, DAILY, HOURLY

We will now learn how to use package xts to look at the time series in a variety of periods, for example, weekly, daily, and monthly. We will focus on the average values; the reader interested in other values, i.e., minimum and maximum can consult the appendix to this lab guide available from the eResources. Divide the average value series by weeks

```
# divide average series by weeks
slr.wk <- split(slr[,3], f="weeks")
length(slr.wk)
```

the resulting object is a list of length 23. The elements of a list can be queried using double square braces, for example,

```
slr.wk[[1]]
```

is the first element and the elements of one element of the list can be obtained by square braces. For example,

```
slr.wk[[1]][1:10]
```

are the first ten entries of the first element of the list.

```
> slr.wk[[1]][1:10]
                    Slr_Avg(W/m2)
2018-10-27 10:00:00         369.3
2018-10-27 10:10:00         394.2
2018-10-27 10:20:00         420.9
2018-10-27 10:30:00         446.7
2018-10-27 10:40:00         471.4
2018-10-27 10:50:00         493.4
2018-10-27 11:00:00         514.0
2018-10-27 11:10:00         534.7
2018-10-27 11:20:00         553.5
2018-10-27 11:30:00         570.0
>
```

Once you explore this list of values by week, select element 7 which is a week in December and plot (Figure 7.14)

```
plot(slr.wk[[7]], main=file)
mtext("Solar radiation (W/m2)",3,-1)
```

Note how December 4 and 5 were mostly clear days with irradiance reaching 550 W/m2 whereas December 6, 7, and 8 were likely cloudy with irradiance not exceeding 100 W/m^2, except for a spike occurring on December 7. Three consecutive days like these can lead to low power production by PV panels, and this is significant for the battery bank design of off-grid remote stations.

Now for contrast, plot one week in March, for example, week 20 (Figure 7.15)

```
plot(slr.wk[[20]], main=file)
mtext("Solar radiation (W/m2)",3,-1)
```

We see that several days reached 800 W/m^2, which is higher than the week in December.

We can change scale and look at hourly values; for example, to obtain hourly averages, using xts find endpoints by hours and then use period.apply based on these endpoints

```
# hourly average
ends.h <- endpoints(slr,"hours")
slr.h <- period.apply(slr, ends.h , mean)
```

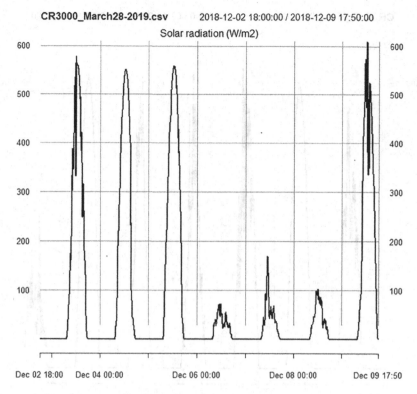

FIGURE 7.14 Solar radiation for one week in December 2018.

and for a quick check look at the first few records

```
> ends.h[1:10]
 [1]  0  6 12 18 24 30 36 42 48 54
> slr.h[1:10]
                     Slr_Max(W/m2)  Slr_Min(W/m2)  Slr_Avg(W/m2)
2018-10-27 10:50:00     444.41667    420.58333333      432.65000
2018-10-27 11:50:00     568.00000    550.51666667      559.63333
2018-10-27 12:50:00     640.28333    631.48333333      635.96667
2018-10-27 13:50:00     658.71667    654.26666667      656.41667
2018-10-27 14:50:00     624.08333    601.38333333      616.36667
2018-10-27 15:50:00     535.41667    515.18333333      525.43333
2018-10-27 16:50:00     392.35000    365.30000000      378.91667
2018-10-27 17:50:00     219.85000    185.35000000      205.48333
2018-10-27 18:50:00      50.78333     31.71800000       39.18983
2018-10-27 19:50:00       0.36000      0.05533333        0.16250
>
```

You can see that the endpoints are in steps of 6, that is 0, 6, 12, … which makes sense because the interval is ten minutes, and indeed the new reported values are hourly.

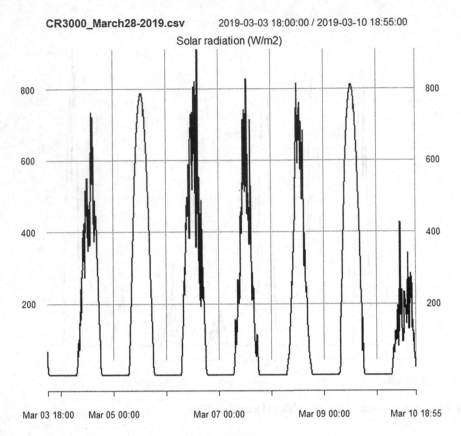

FIGURE 7.15   Solar radiation for one week in March 2019.

We can plot these hourly values for the entire period and assert the logical `multi.panel` argument so that we can see all three variables

```
# plot multipanel
plot(slr.h,main=file, multi.panel=TRUE)
mtext("Hourly Average Solar radiation (W/m2)",3,0)
```

with resulting plots in Figure 7.16.

ENERGY: HOURS OF FULL SUN

When calculating totals for a time interval, it is important to consider the time difference between measurements. We can get these by subtracting the time stamps of adjacent records

```
dtt<-array()
for(i in 2:length(tt)) dtt[i] <- tt[i]-tt[i-1];dtt[1]<- dtt[2]
```

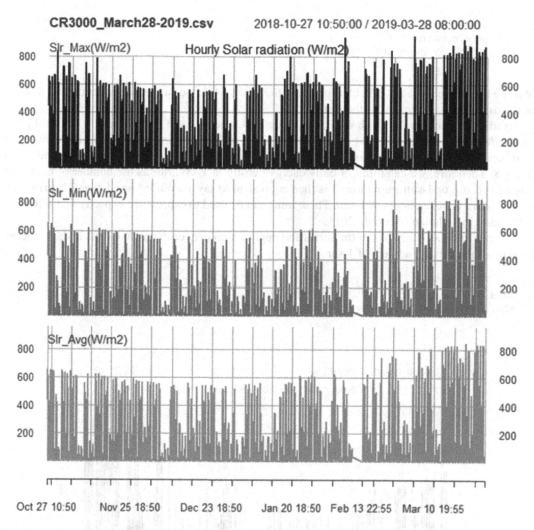

FIGURE 7.16   Solar radiation hourly average.

it is prudent to query the results to confirm that the interval changed from ten to five minutes.

```
> dtt[1:10]; dtt[(ntt-10):ntt]
 [1] 10 10 10 10 10 10 10 10 10 10
 [1] 5 5 5 5 5 5 5 5 5 5 5
>
```

then we can use the `dtt` interval values when calculating energy by integrating power.

```
# energy kWh/m2
sle <- slr*dtt/(60*1000)
```

Here, we divide by 1000 to give the energy in kWh, a commonly used unit for electrical energy. The energy values can be plotted using the following code yielding the results of Figure 7.17.

```
plot(sle.h,main=file)
mtext("Hourly solar radiation energy (kWh/m2)",3,-1)
```

We can see the seasonality effect by following the peaks while we can assess variability by looking at the differences in the size of the peaks.

Irradiance reaching the Earth surface in a clear day, when the sun is at the zenith, can be slightly above 70% of the solar constant or 1 kW/m² ($\approx 0.7 \times 1.377 \approx 1\,\mathrm{kW/m^2}$). This is such a convenient number that has been defined as *1-sun* equivalent of irradiance, or 1-sun insolation denoted with letter S. Therefore, we can think of daily energy density in 1 kWh/m² as the number of hours per day of full sun or 1-sun insolation. It is then equivalent to say 7 kWh/m² per day as 7 hours of full sun, or seven hours of 1-sun. These units can be used to express the capacity factor (CF) of a PV solar installation (Acevedo 2018).

We can find day endpoints in the series and apply a sum over each day to calculate daily energy to express the results in hours of 1-sun.

```
ends.d <- endpoints(sle[,3],"days")
sle.d <- period.apply(sle[,3], ends.d, sum)
```

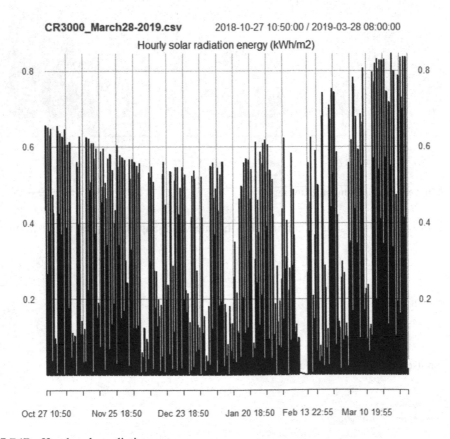

FIGURE 7.17   Hourly solar radiation energy.

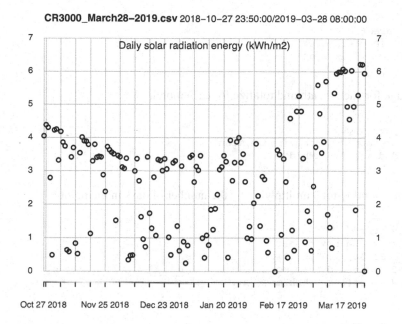

FIGURE 7.18   Solar insolation in hours of 1-sun equivalent.

and plot using type points "p" instead of lines for visualization of the value for the day (Figure 7.18)

```
plot(sle.d,type="p", ylim=c(0,7), main=file)
mtext("Daily solar radiation energy (kWh/m2)",3,-1)
```

The highest values (approximately six hours) would correspond to the higher insolation days in March, many of the lowest values would correspond to lower solar radiation days in December, and many of the values around the mean would correspond to higher insolation days from November through February. This result illustrates how measuring the number of hours of 1-sun for a site provides guidance on PV panel installed capacity design.

PRODUCTION

As just explained, for a given site, insolation S can be given in kWh/m^2 per day, or equivalently as hours per day of 1-sun, or hours of peak sun. We can then use area A of solar panel (in m^2) and efficiency η to evaluate energy produced by the system at a given site $E = S \times A \times \eta$ in kWh (Acevedo 2018). For example, assume we have 1 m^2 of panel with an efficiency of 15% at a site that has 6.0 kWh/m^2 per day on a given day. Energy production is $E = S \times A \times \eta = 6 \times 1 \times 0.15\,\text{kWh} = 0.9\,\text{kWh}$ for that day.

We will use this concept to calculate potential production given the solar radiation data we analyzed previously. Use efficiency 15% and area 1 m^2

```
eff=.15; area=1
prd.d <- sle.d*eff*area
names(prd.d) <- "kWh/d"
```

we can average for each month and totalize for each month

```
# monthly average
prd.d.mo <- apply.monthly(prd.d,mean)
names(prd.d.mo) <- "kWh/d"
```

we can get the monthly total multiplying by the number of days in the month

```
# monthly total
# days per month nov to mar
d.mo <- c(31,30,31,31,28,31)
prd.mo.eff <- prd.d.mo*d.mo
names(prd.mo.eff) <- "kWh/mo"
```

and querying the results

```
prd.mo.eff
```

An alternative approach is the concept of CF can be evaluated by dividing insolation S in hours/day of peak sun by 24 hours/day as a multiplier of installed capacity P in kW which is the rated power of a panel at 1-sun, and a time period H in hours; that is to say, $E = P \times CF \times H$. For instance, monthly energy E in kWh/month can be estimated from installed capacity P (in kW), and CF using the number of hours in a month $h_{mo} = 30$ days $\times$ 24 hours/day = 720 hours for 30-day months or use $h_{mo} = 31$ days $\times$ 24 hours/day = 744 hours for those months with 31 days.

$$E = P \times CF \times h_{mo} \tag{7.3}$$

Calculate CF for each month

```
cf.d <- 100*sle.d/24
names(cf.d) <- "%"
```

we can average for each month

```
# CF monthly average
cf.mo <- apply.monthly(cf.d,mean)
```

and querying the results

```
> cf.mo
                           %
2018-10-31 23:50:00 13.351232
2018-11-30 23:50:00 12.791279
2018-12-31 23:50:00  9.378433
```

```
2019-01-31 23:50:00  9.834456
2019-02-28 23:55:00  9.767300
2019-03-28 08:00:00 17.406926
>
```

Interpreting these results, we can see a much higher CF for March than for December. Now use an installed capacity of 150 W or rated power at 1-sun or 0.15 kW and days of the month

```
# days per month nov to mar
d.mo <- c(31,30,31,31,28,31)
# installed power capacity
pow <- 0.150
prd.mo.cf <- pow*(cf.mo/100)*d.mo*24
names(prd.mo.cf) <- "kWh/mo"
```

which we can query

```
> prd.mo.cf
                     kWh/mo
2018-10-31 23:50:00 14.899975
2018-11-30 23:50:00 13.814581
2018-12-31 23:50:00 10.466331
2019-01-31 23:50:00 10.975253
2019-02-28 23:55:00  9.845438
2019-03-28 08:00:00 19.426130
>
```

and as expected yields the same results as shown previously.

IMPLICATIONS FOR POWERING REMOTE STATIONS AND WSN

As we see in Figure 7.14, occasionally one can have a succession of several days with low solar radiation leading to no or low solar electricity production. Suppose we have an off-grid monitoring system powered by one 12V module of 150 W rated power. Suppose the system daily demand is 10 W at 12 V DC and assume 100% efficiency of other components (e.g., charge controller).

What is the DC daily energy demand? Multiply 10 W by 24 hours to get 0.24 kWh/day. What would be the minimum daily hours of 1-sun to cover this demand? The expected daily energy from the PV panel is the rated power times the hours of 1-sun. Therefore, the minimum

$$\frac{0.24 \text{ kWh/day}}{0.15 \text{ kW}} = 1.6 \text{ hours/day} \quad \text{or } 1.6 \text{ hours of 1-sun per day. Looking at Figure 7.18, we can}$$

see that there are many instances of energy <1.6 hours, particularly in December 2018. In fact, we can find them using function which and plot using R (Figure 7.19)

```
sle.d.low <- sle.d[which(sle.d<= 1.6)]
plot(sle.d.low,main=file)
mtext("Daily solar energy less than 1.6 kWh/m2",3,-1)
```

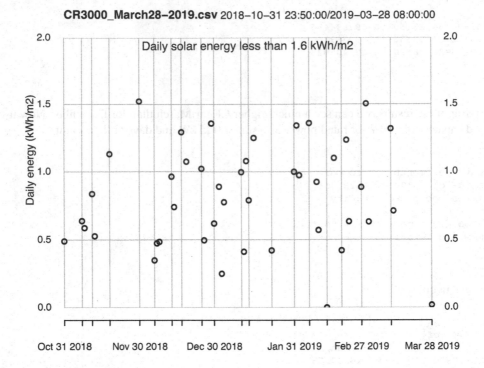

FIGURE 7.19   Days with <1.6 hours of 1-sun.

Of interest as well is how often do these events occur in proportion to all days in the data

```
> dim(sle.d.low)[1]/dim(sle.d)[1]
[1] 0.2913907
>
```

which means 29% of days were <1.6 hours/day. The next question would be how many ampere-hours (Ah) are required of the battery to provide three days of power of consecutive low insolation of only 1.0 hours/day of full sun? Assume the battery should not discharge more than 80% (this is called Depth of Discharge, DOD). For days of only 1.0 hour/day of 1 sun, the PV produces $E_{\text{low}} = 0.150\,\text{W} \times 1\,\text{hours/day} = 0.15\,\text{kWh/day}$. Thus, the battery must have a capacity of

$$C = \frac{E_{\text{DC}} - E_{\text{low}}}{V} \times \frac{\text{days}}{\text{DOD}_{\text{max}}} = \frac{(240 - 150)\,\text{Wh/day}}{12\,\text{V}} \times \frac{3\,\text{day}}{0.8} \simeq 28\,\text{Ah} \qquad (7.4)$$

WIND SPEED STATISTICS

As we discussed in Chapter 7 of the companion textbook, a good fit to wind speed is often found for a Weibull distribution with shape $k = 2$, or a Rayleigh distribution and that the mean or expected value of a Rayleigh pdf is related directly to the scale parameter c by $\mu_v = E(v) = \sqrt{\pi}c\,/\,2$. We can use function weibull.plot of package renpow, which makes use of dweibull and pweibull of R. Load the third column of ws.10 m, that is if you have not yet or if you started a new R session

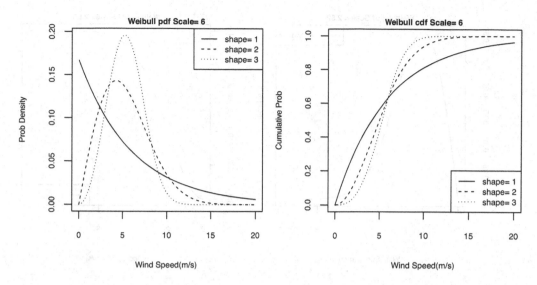

FIGURE 7.20 Weibull distribution for three values of the shape parameter. Shape=2 corresponds to the Rayleigh distribution.

```
require(renpow)
```

and now plot (Figure 7.20)

```
weibull.plot(xmax=20,scale=6,shape=c(1,2,3))
```

A first step to find a fit of wind speed data to a Rayleigh distribution is to assume shape $k = 2$, and determine the scale from the average wind speed; that is to say $c \simeq 2\bar{v} / \sqrt{\pi}$ where $\bar{v}$ is the sample mean or arithmetic average of all wind speed values. For example, take a site with an average wind of 2.5 m/s, use $c = 2\bar{v} / \sqrt{\pi} = 2 \times 2.5 / 1.77 = 2.82$. Suppose we want to determine the probability of two scenarios: (1) that the wind speed is above 2 m/s and (2) that the wind speed is above 3 m/s. We can plot the pdf and cdf using `weibull.plot` and add vertical lines at the desired values.

```
weibull.plot(xmax=10,scale=2*2.5/sqrt(pi),shape=2)
abline(v=2,lty=2);abline(v=3,lty=2)
```

The plot is shown in Figure 7.21, including the two vertical lines for $v=2$ m/s and $v=3$ m/s on the cdf to emphasize the points at which we seek the probability. We can see then that approximately wind speed would above 2 m/s $\Pr[v > 2] = 1 - \Pr[v \leq 2] \approx 1 - 0.4 = 0.6$ and above 3 m/s $\Pr[v > 3] = 1 - \Pr[v \leq 3] = 1 - 0.65 = 0.35$. We can find the exact values

```
> 1 - pweibull(c(2,3),scale=2*2.5/sqrt(pi),shape=2)
[1] 0.6049226 0.3227190
```

These are $\Pr[v > 2] = 1 - \Pr[v \leq 2] = 0.605$ and $\Pr[v > 3] = 1 - \Pr[v \leq 3] = 0.323$. We conclude that the wind will exceed 2 m/s 60% of the time and 3 m/s 32.3% of the time.

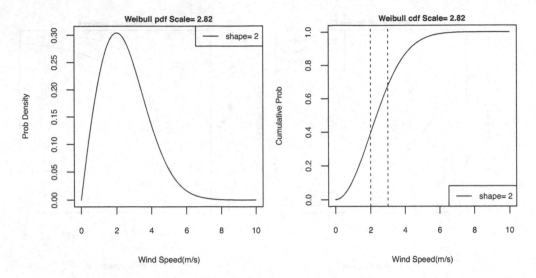

FIGURE 7.21   Estimate of a Rayleigh for an average wind speed of 2.5 m/s.

MONITORING THE WIND RESOURCE: WIND SPEED STATISTICS

In this section, we analyze wind resource available at a site for wind powered electricity production, both for supporting the design of a wind power facility and to power a remote monitoring station using a wind generator. In addition, it is an example of atmospheric modeling, which will continue to discuss in Lab guide 11.

A classical `anemometer` or wind monitor measures wind speed and direction. It consists of two major parts: the rotating cups or propeller for wind speed and the vane for wind direction. Speed is proportional to the number of rotations of the cups or propeller detected as electronic pulses produced by Hall Effect sensors. Direction is proportional to output voltage from a potentiometer acting as a voltage divider transducer.

As described previously in this chapter, at the UNT Discovery Park campus, we have a weather station consisting of a tower and Campbell Scientific meteorological sensors read by a Campbell Scientific datalogger CR3000 (Campbell Scientific Inc. 2018a). As shown in Figure 7.10, there are two anemometers, one of them near the top (10 m above the ground) and another just above the datalogger box (2 m above the ground). The anemometer at 2 m of height is a Campbell Scientific 03101 anemometer (Campbell Scientific Inc. 2018c) at 2 m, providing wind speed only, and the one at 10 m is a Campbell Scientific 05103 anemometer (Campbell Scientific Inc. 2020), providing wind speed and direction. Both are wired to the CR3000 datalogger, to measure and store wind speed and direction at both heights.

The 03101 can measure the wind speed within the range of 0–112 mph and operates over a −50° to +50°C temperature range. The 05301 can measure the wind speed within the range of 0–224 mph and wind direction 360° mechanical and 355° electrical. The temperature range within which it can operate is −50° to +50°C.

Using data from these instruments, we will focus on calculating wind statistics at 10 m, and the relation between wind at 2 m to that at 10 m. As we noted previously, the datalog file from the CR3000 shows a header containing the variable names. For wind, the variables corresponding to 10 m are `WS _ mph _ Max`, `WS _ mph _ Min`, `WS _ mph _ Avg`, `WS _ mph _ Std`, `WindDir`. The ones corresponding to 2m are `WS _ mph _ 2m _ Max`, `WS _ mph _ 2m _ Min`, `WS _ mph _ _ 2m _ Avg`, `WS _ mph _ 2m _ Std`.

When you examine the names of the object `tt.xx` created in the solar radiation section you will see that these variables are in positions 3–7 for the 10-m wind speed, and positions 17–20 for the 2-m wind speed.

```
names(x.t)[1:20]
 [1] "RECORD(RN)"                    "Batt_Volt(Volts)"
 [3] "WS_mph_Max(miles/hour)"        "WS_mph_Min(miles/hour)"
 [5] "WS_mph_Avg(miles/hour)"        "WS_mph_Std(miles/hour)"
 [7] "WindDir(Degrees)"              "AirTC_Avg(Deg C)"
 [9] "RH(%)"                         "Rain_in_Tot(inch)"
[11] "SlrW_Max(W/mÂ²)"               "SlrW_Min(W/mÂ²)"
[13] "SlrW_Avg(W/mÂ²)"               "SlrW_Std(W/mÂ²)"
[15] "SlrkJ_Tot(kJ/mÂ²)"             "BP_mmHg_Avg(mmHg)"
[17] "WS_mph_2m_Max(miles/hour)"     "WS_mph_2m_Min(miles/hour)"
[19] "WS_mph_2m_Avg(miles/hour)"     "WS_mph_2m_Std(miles/hour)"
>
```

We use the array `tt.xx` created previously in monitoring the solar resource section, and extract all the information minus the timestamp, since we will not need it to calculate statistics.

```
xx <- tt.xx[,-1]
```

Create new variables for wind speed at 10 m (positions 3–5) and at 2 m (positions 17–19) while multiplying by 0.44704 to convert from mph to m/s.

```
ws.10m <- xx[,3:5]*0.44704
ws.2m <- xx[,17:19]*0.44704
```

We will focus on average wind speed at 10 m which is the third column of ws.10m, that is ws.10m[,3] and apply function `fit.wind` of `renpow`.

```
fit.wind(ws.10m[,3])
```

which consists of calculating average, assuming shape parameter equal to 2, i.e., a Rayleigh distribution, using the average to calculate the scale parameter of the Rayleigh distribution, and to visualize the fit of these data to the Rayleigh plots a histogram, a density approximation, a cumulative density function (CDF), and an empirical CDF (ECDF) (Figure 7.22). You may want to review these concepts from Lab guide 1. We can see that at least by visual inspection, we have a reasonable fit to the PDF and CDF.

Our next analysis is the relationship between wind speed at 2 m and that at 10 m following the models given in Chapter 7 of the companion textbook $\frac{v}{v_0} = \left(\frac{H}{H_0}\right)^\alpha$ and $\frac{v}{v_0} = \frac{\ln(H/l)}{\ln(H_0/l)}$, where H_0 is a reference height, v is the wind speed at height H, v_0 is the wind speed at the reference height, and α and l are parameters characterizing the terrain conditions. Here, we assume the reference height is 2 m for the purposes of finding the values of α and l by regression. For this purpose, we bind the average values at each height and make a list using the respective heights,

```
v1.v2 <-cbind(ws.2m[,3],ws.10m[,3])
x <- list(v1.v2=v1.v2,H1=2,H2=10)
```

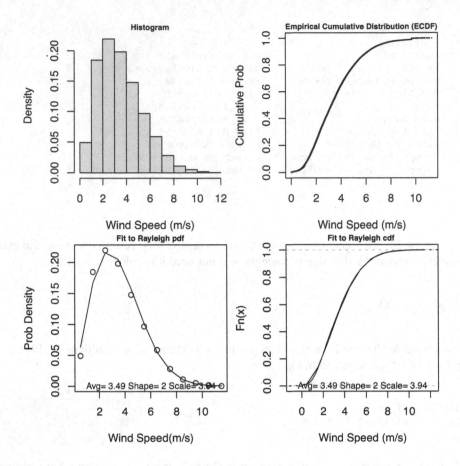

FIGURE 7.22   Fit of wind speed at 10 m to a Rayleigh distribution.

and then apply function `cal.vH` of `renpow`

```
>  cal.vH(x)
$alpha
[1] 0.141

$rough
[1] 0.004

>
```

The graphical result is shown in Figure 7.23; the values of $\alpha = 0.141$ and $l = 0.004$ indicate a relatively low terrain friction. Armed with the values of the coefficients, we can use the models to predict the wind speed at the design height of the tower for a wind generator.

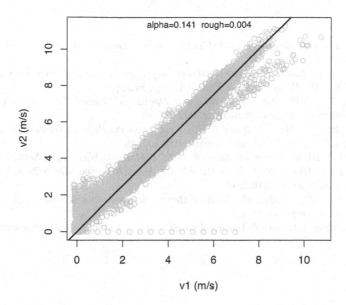

FIGURE 7.23   Regression between wind speed of the two anemometers to determine model coefficients.

EXERCISES

Exercise 7.1 Sun diagram and optical path.

Plot a sun diagram for days January 21, March 21, and June 21 at the Dallas, Texas USA, latitude. Discuss the values at the solstices and equinoxes. Produce sun diagram using the file with an optical path for Golden, Colorado, USA, and calculate radiation on a collector tilted at the latitude, facing S ($\phi c = 0°$) and 10% contribution of reflected radiation from surrounding surfaces.

Exercise 7.2 Solar radiation time series.

Use the information and method given in the activity related to monitoring the solar resource. (1) Read the datalogger file and extract solar radiation data. (2) Divide by week and plot irradiance for two different weeks. (3) Calculate and plot hourly averages of irradiance. (4) Calculate and plot hourly and daily energy.

Exercise 7.3 PV energy production.

Use the information and method given in the activity related to production and powering remote stations. (1) Calculate expected production from insolation and efficiency; interpret your results and compare December and March. (2) Calculate expected production from CF and installed power; interpret your results and compare December and March. (3) Design battery capacity for the data provided.

Exercise 7.4 Wind speed statistics.

Assume wind speed at a site follows a Rayleigh distribution and the average wind speed is 3 m/s. Estimate and plot the distribution and estimate the percent of time that wind speed exceeds 4 m/s.

Exercise 7.5 Wind speed and surface friction.

Use the information and method given in the activity related to analysis of data from monitoring wind speed. (1) Estimate the Rayleigh distribution and (2) estimate friction coefficient and roughness length.

REFERENCES

Acevedo, M.F. 2018. *Introduction to Renewable Electric Power Systems and the Environment with R*. Boca Raton, FL: CRC Press, 439 pp.

Acevedo, M.F. 2024. *Real-Time Environmental Monitoring: Sensors and Systems - Textbook, Second Edition*. Boca Raton, FL: CRC Press, Taylor & Francis Group, 392 pp.

Campbell Scientific Inc. 2018a. *CR300 Micrologger: Operators Manual*. Accessed March 2021. https://s.campbellsci.com/documents/us/manuals/cr3000.pdf.

Campbell Scientific Inc. 2018b. *CS300 and CS301 Pyranometers: Product Manual*. Accessed March 2021. https://s.campbellsci.com/documents/us/manuals/cs300.pdf.

Campbell Scientific Inc. 2018c. *Instruction Manual: 03002 R.M. Young Wind Sentry Set 03101 R.M. Young Wind Sentry Anemometer 0330 1 R.M. Young Wind Sentry Vane*. Accessed March 2021. https://s.campbellsci.com/documents/us/manuals/03002.pdf.

Campbell Scientific Inc. 2020. *Product Manual Wind Monitor Series*. Accessed March 2021. https://s.campbellsci.com/documents/us/manuals/05103.pdf.

R Project. 2023. *The Comprehensive R Archive Network*. Accessed January 2023. http://cran.us.r-project.org/.

8 Remote Monitoring of the Environment

INTRODUCTION

In this session, we will conduct computer-based lab activities to learn about remote sensing applications to environmental monitoring. We will focus on using the image processing capabilities of the R system using several geospatial packages to read a Landsat 8 raster file, display images for the various bands, and obtain descriptive statistics for each. Then, we look at relationships among bands and calculation of normalized difference indices such as the ones for vegetation and water; using these results, we learn how to perform reclassification by selecting various intervals. Subsequently, we look at multivariate analysis application to remote sensing, in particular, principal component analysis (PCA) and cluster analysis; the latter we expand to k-means clustering to perform unsupervised classification.

MATERIALS

READINGS

For theoretical background, you can use Chapter 8 of Acevedo, M.F. 2024. *Real-Time Environmental Monitoring: Sensors and Systems - Textbook, Second Edition* which is a companion to these guides (Acevedo 2024). This guide follows the Remote Sensing Image Analysis with R by Ghosh and Hijmans (2019). Other bibliographical references are cited throughout the guide.

SOFTWARE

- R, for data analysis (R Project 2023)
- R packages `raster`, `terra`, `rgdal`, `sp`, and `rasterVis` available from the CRAN repository

DATA FILES

RTEM GitHub repository https://github.com/mfacevedol/rtem.

- Archive `nt20210618.zip`

SUPPLEMENTARY SUPPORT MATERIAL

Supplementary support material including additional screenshots, images, and procedures are available from the publisher eResources web page provided for this book. In particular, the images in this guide are printed in gray scale, but full color images are available from the online eResources and better illustrate the results of image processing.

DOI: 10.1201/9781003184362-8

RASTER OBJECTS FROM LANDSAT FILES

As discussed in the companion Chapter 8 (Acevedo 2023), a raster file consists of a grid of pixels. We will work with images for bands 1–7 of Landsat 8 for June 18, 2021 of an area in the vicinity of Denton, Texas, USA. These files are in archive nt20210618.zip of lab8 of the RTEM repository, where x is 1,…,7 containing files nt20210618 _ 1.tif through nt20210618 _ 7.tif, where 1, 2, …, 7 refer to band 1, band 2, …, band 7. The nt in these files stands for North Texas.

The original images correspond to a scene downloaded from Earth Explorer (USGS 2022) and correspond to Path 27, Row 37 of the Landsat World Reference System 2 (WRS-2), and are LC08 _ L2SP _ 027037 _ 20210618 _ 20210628 _ 02 _ T1 _ SR _ B1.TIF, etc. Note that the filename contains information about the image. More information about WRS-2 is given in the textbook companion to these lab guides.

Download the archive and extract as a folder nt in your labs/data where labs is your working directory used for R, so that you have a folder labs/data/nt containing seven tif files with names nt20210618 _ 1.tif through nt20210618 _ 7.tif. Install and load R package raster which has dependencies on packages terra and sp.

```
install.packages("raster")
```

You could also install terra and sp and then install raster.

Once installed successfully, load package raster

```
library(raster)
```

Create filenames for the files we want to use, apply the stack function of package raster to create a RasterStack object for these Landsat bands, and give names to the layers of the stack

```
files <- paste0('data/nt/nt20210618_', 1:7, ".tif")
nt7 <- stack(files)
names(nt7) <- c('CoastalAerosol', 'Blue', 'Green', 'Red', 'NIR', 'SWIR1',
'SWIR2')
```

check the properties of the images by querying the contents of nt

```
> nt7
class       : RasterStack
dimensions  : 1167, 1167, 1361889, 7  (nrow, ncol, ncell, nlayers)
resolution  : 30, 30  (x, y)
extent      : 664995, 700005, 3671805, 3706815  (xmin, xmax, ymin, ymax)
crs         : +proj=utm +zone=14 +datum=WGS84 +units=m +no_defs
names       : CoastalAerosol,  Blue,  Green,   Red,   NIR, SWIR1,  SWIR2
min values  :             41,    93,   4316,  4997,  6656,  7086,   7296
max values  :          35878, 39515,  44720, 49025, 52342, 56235,  58644

>
```

Note the dimensions are 1167 rows and 1167 columns, or a total of 1,361,889 cells of $30 \times 30 \, m^2$, and seven layers. The extent is given by a previously applied image crop operation of the original image and was selected to contain water, vegetation, urban, and rural areas around the City of Denton, Texas. The coordinate reference system (CRS) is UTM Zone 14, with datum WGS84. The names were assigned according to the names function previously applied, and the minimum and maximum values for each band, are within the 0 to $2^{16} - 1 = 65{,}535$, range of unsigned integers for the radiometric resolution of Landsat 8.

These properties can also be checked using functions of the raster package

```
# coordinate reference system
crs(nt7)
# Number of cells
ncell(nt7)
# rows, columns, layers
dim(nt7)
# spatial resolution
res(nt7)
# Number of bands
nlayers(nt7)
```

Each band can be addressed as a component of nt7 by $name; for example, the Green band, nt7$Green

```
> nt7$Green
class       : RasterLayer
dimensions  : 1167, 1167, 1361889  (nrow, ncol, ncell)
resolution  : 30, 30  (x, y)
extent      : 664995, 700005, 3671805, 3706815  (xmin, xmax, ymin, ymax)
crs         : +proj=utm +zone=14 +datum=WGS84 +units=m +no_defs
source      : nt20210618_3.tif
names       : Green
values      : 4316, 44720  (min, max)

>
```

Note that in this case the object is of class RasterLayer. Various layers can be compared to check that bands have the same extent, number of rows and columns, projection, resolution, and origin

```
> compareRaster(nt7$Blue,nt7$Red)
[1] TRUE
>
```

IMAGE DISPLAY

We can display a map or image of each band using the plot function of the raster package (Figure 8.1).

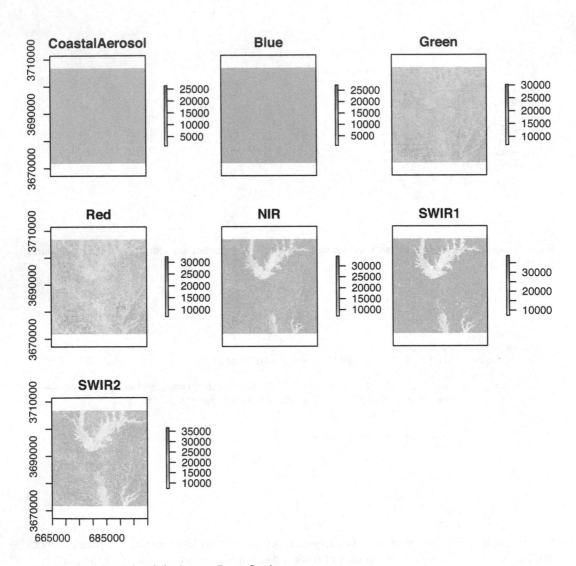

FIGURE 8.1   Seven bands in the nt7 RasterStack.

For simplicity of comparison across bands, we will conduct the subsequent work using values scaled from 0 to 1, which are obtained by dividing the original values into $2^{16} - 1 = 65,535$.

```
nt <- nt7
for(i in 1:7) nt[[i]] <- nt7[[i]]/(2^16-1)
```

which yields

```
> nt
class        : RasterStack
dimensions   : 1167, 1167, 1361889, 7  (nrow, ncol, ncell, nlayers)
resolution   : 30, 30  (x, y)
extent       : 664995, 700005, 3671805, 3706815  (xmin, xmax, ymin, ymax)
crs          : +proj=utm +zone=14 +datum=WGS84 +units=m +no_defs
names        : CoastalAerosol,    Blue,     Green,      Red,     NIR,      SWIR1,      SWIR2
min values   : 0.0006256199, 0.0014190890, 0.0658579385, 0.0762493324,
0.1015640497, 0.1081254292, 0.1113298238
max values   : 0.5474632,    0.6029603,    0.6823835,    0.7480735,
0.7986877,    0.8580911,    0.8948501

>
```

We will use this new scaled `RasterStack` to plot four of the seven bands. In this case, we will use function panels of package `renpow`, which is a function we have used in previous labs

```
require(renpow)
panels(8,8,2,2,pty='m')
for(i in 3:6)
plot(nt[[i]],main=names(nt)[i],cex.main=0.8)
```

which produces Figure 8.2.

As we can see, displaying the contents of only one band (particularly, green and red) does not show many features of the terrain; interestingly, the near-infrared (NIR) and short-wave infrared 1 (SWIR1) bands do show differences between water (lakes) and land. However, we can compose a *true or natural color composite*, meaning that we assign the red, green, and blue bands to the red, green, and blue channels of the monitor (Figure 8.3). This image is better appreciated in color and recall that full color images are available from the eResources online.

```
nt.RGB <- stack(nt$Red,nt$Green,nt$Blue)
panels(6,6,1,1,int='i')
plotRGB(nt.RGB, axes = TRUE, stretch = "lin", main = "NT True Color Composite")
```

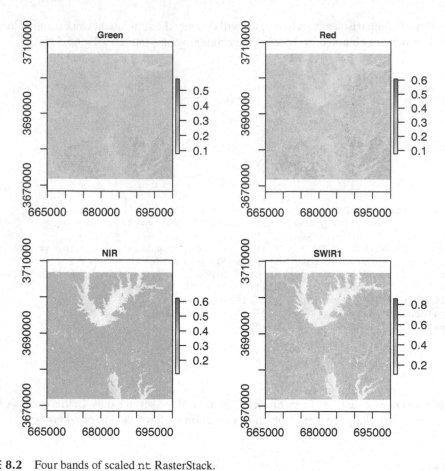

FIGURE 8.2 Four bands of scaled nt RasterStack.

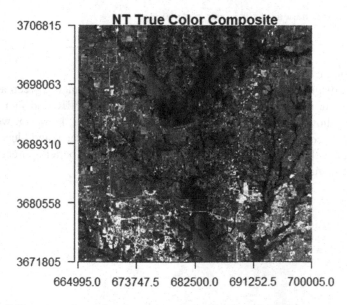

FIGURE 8.3 North Texas area image bands as true or natural color composite.

An alternative to making a stack with three layers of a larger stack is to use function subset

```
nt.RGB <- subset(nt, c(4,3,2))
```

To help the reader be acquainted with the area of the image, Figure 8.4 composed using Goggle Earth shows the cities, towns, creeks, lakes, roads, and rural landscapes. These features can be identified in the natural color image of Figure 8.3 where we can visualize the lakes (Ray Roberts to the North and Lewisville to the south), including differences in water quality, roads and highways, urban areas in the southwest part of the image (corresponding to the City of Denton).

Often differences can be better visualized in a *false color composite* that assigns the NIR, red, and green bands to the red, green, and blue channels of the monitor (Figure 8.5)

FIGURE 8.4 North Texas area image from Google Earth.

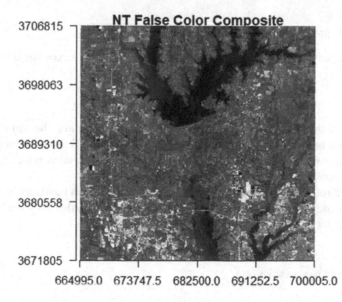

FIGURE 8.5 False color composite.

```
nt.FCC <- stack(nt$NIR, nt$Red, nt$Green)
panels(6,6,1,1,int='i')
plotRGB(nt.FCC, axes=TRUE, stretch="lin", main="NT False Color Composite")
```

The contrast between the lakes and the surrounding land areas is enhanced, as well as the contrast between the urban areas and surrounding rural areas. This image is better appreciated in color and recall it is available online.

DESCRIPTIVE STATISTICS

We gain insight regarding the contents of each band by looking at descriptive statistics and the distribution of the values in the band. For example, using summary we can see that the green band has a tight distribution around its median of ~0.15

```
> summary(nt$Green)
             Green
Min.    0.06585794
1st Qu. 0.14547951
Median  0.15249866
3rd Qu. 0.16008240
Max.    0.68238346
NA's    0.00000000
```

Or more visually we can apply function hist that we learn in previous lab sessions to a set of bands, say 3:6 or the 3 (Green), 4 (Red), 5 (NIR), and 6 (SWIR1) (Figure 8.6) and arranged in four panels

```
panels(8,8,2,2,pty='m')
for(i in 3:6)
hist(nt[[i]],xlab='',xlim=c(0,0.6),prob=T,ylim=c(0,15),main=names(nt)[i],cex.
main=0.8)
```

Indeed, the values of the green band do not show much variability; and the same seems to be the case for the red band but with opposite skewness. Contrastingly, NIR shows bimodality due to the differences between water and land, with higher values due to terrestrial vegetation. That difference is attenuated in the SWIR1 band.

Let us look at the relationship between pixels of the blue and green bands using function pairs applied to a stack made of blue and green bands (Figure 8.7), to note that they are highly correlated as shown by scatterplot and the high correlation coefficient 0.97.

```
pairs(stack(nt$Blue,nt$Green))
```

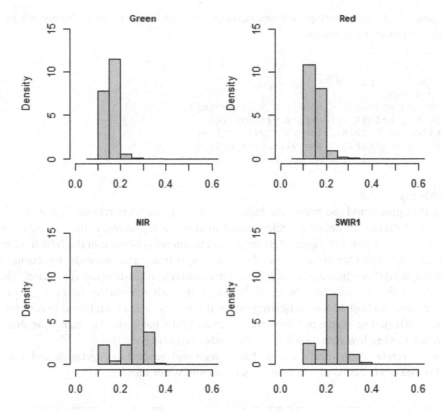

FIGURE 8.6   Histogram of four bands illustrating differences in distribution.

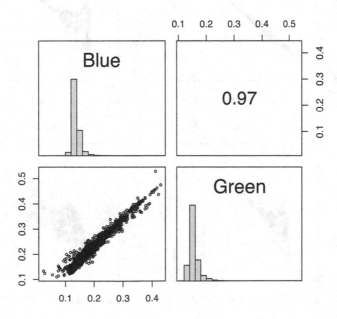

FIGURE 8.7   Relationship between blue (horizontal axis in scatterplot) and green (vertical axis in scatterplot) bands.

We can compare the relationship between pairs of the four bands given in Figure 8.6 by plotting scatterplots on comparable scales

```
panels(8,8,2,2,pty='m')
xylim=c(0,0.7)
plot(nt$Red,nt$Green,xlim=xylim,ylim=xylim)
plot(nt$Red,nt$NIR,xlim=xylim,ylim=xylim)
plot(nt$Green,nt$NIR,xlim=xylim,ylim=xylim)
plot(nt$SWIR1,nt$NIR,xlim=xylim,ylim=xylim)
```

which yields Figure 8.8.

We see that green and red bands are highly correlated, but their relationship with NIR shows an interesting pattern. An important relationship to analyze vegetation is the one between the red and NIR (upper right panel of Figure 8.8), since vegetation reflects more in the NIR than in the red. Note that the scatterplot for these two bands has a unique triangular shape. Its top corner is due to pixels with high NIR (vertical axis) and low red (horizontal axis), indicating vegetation. The bottom corner has low reflectance for both bands, indicating water, whereas the furthest corner corresponds to high reflectance in both bands, indicating exposed surfaces such as bright soil or concrete (Ghosh and Hijmans 2019). The scatterplot between green and NIR has a similar triangular shape, with a bottom corner having low reflectance for both bands, indicating water.

The unique relationship of NIR with red and green with respect to vegetation and water will be employed in the next section to construct vegetation and water indices.

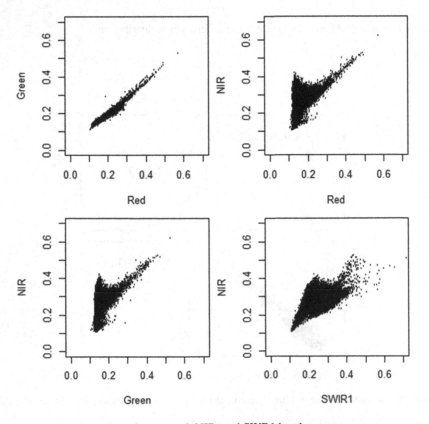

FIGURE 8.8   Relationship of pairs of green, red, NIR, and SWR1 bands.

ANALYSIS USING INDICES

Several indices, such as those for vegetation and water, use a *normalized difference* that consists of taking the difference between values of two bands and dividing by the sum of these two bands

$$nd = \frac{b_k - b_i}{b_k + b_i} \tag{8.1}$$

These indices vary between −1 and +1. Let us build a function that will be used to calculate various indices. Write and load this function.

```
vi <- function(img, k, i) {
  bk <- img[[k]]
  bi <- img[[i]]
  vi <- (bk - bi) / (bk + bi)
  return(vi)
}
```

We can use it, for instance, to calculate the *Normalized Difference Vegetation Index* (NDVI), which uses band 5 (or NIR) for b_k and band 4 (or red) for b_i.

$$NDVI = \frac{b_5 - b_4}{b_5 + b_4} \tag{8.2}$$

Then, we can display as an image using plot (Figure 8.9).

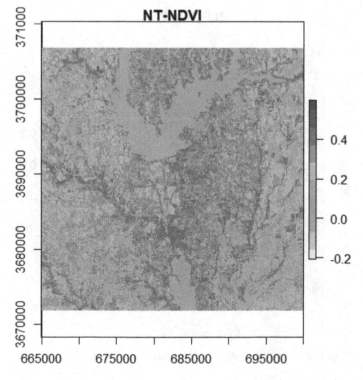

FIGURE 8.9   North Texas NDVI.

```
ndvi <- vi(nt, 5, 4)
panels(6,6,1,1,int='i')
plot(ndvi, col = rev(terrain.colors(10)), main = "NT NDVI")
```

Higher values of NDVI, darker (or green in the color version) areas (above ~0.3) indicate vegetation. In this case, we see, for example, higher values along Clear Creek that runs Southeast toward Lake Lewisville, and particularly an area named the Greenbelt Corridor, just to the north of the Lake Lewisville headwaters (see Figure 8.4 for locations).

Now, we can calculate the *Normalized Difference Water Index* (NDWI), which uses band 3 (or green) for b_k and band 5 (or NIR) for b_i.

$$NDWI = \frac{b_3 - b_5}{b_3 + b_5} \qquad (8.3)$$

Then, we can display as an image using plot (Figure 8.9).

```
ndwi <- vi(nt, 3, 5)
panels(6,6,1,1,int='i')
plot(ndwi, col = rainbow(10), main = "NT NDWI")
```

Higher values of NDWI, darker (or blue in the color version) areas (above ~0.2) indicate water, and these pixels show clearly for the lakes (Figure 8.10).

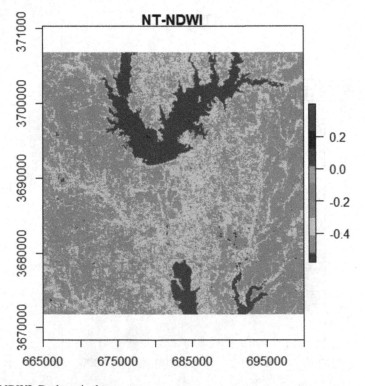

FIGURE 8.10   NDWI. Darker pixels are water.

It is helpful to look at the statistics of NDVI and NDWI

```
> summary(ndvi)
          layer
Min.    -0.2065928
1st Qu.  0.1867683
Median   0.2776227
3rd Qu.  0.3385055
Max.     0.5953191
NA's     0.0000000
>
```

which has a median of ~0.28 close to 0.3, and most values are positive. Whereas for NDWI,

```
> summary(ndwi)
          layer
Min.    -0.6069396
1st Qu. -0.3184302
Median  -0.2718604
3rd Qu. -0.2041498
Max.     0.3991775
NA's     0.0000000
>
```

which has a median of ~−0.27 and most values are negative. If we look at the histograms of NDVI and NDWI (Figure 8.11)

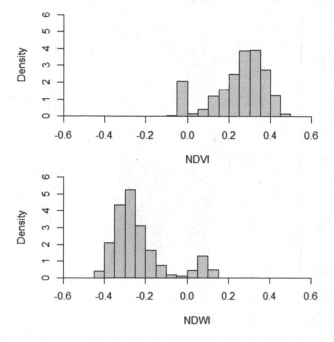

FIGURE 8.11   North Texas histogram of NDVI and NDWI.

```
panels(6,6,2,1,int='i')
hist(ndvi,main = "", xlab = "NDVI", prob=T,xlim=c(-0.6,0.7),ylim=c(0,6))
hist(ndwi,main = "", xlab = "NDWI", prob=T,xlim=c(-0.6,0.7),ylim=c(0,6))
```

We see how these two indices give us contrasting information, the pixels with low negative values of NDVI centered around −0.05, correspond to water, that coincides with the positive values of NDWI ~0.1. Contrastingly, high positive values of NDVI for vegetation correspond to negative values of NDWI. We can use this information to extract areas in vegetation or water using *reclassification*.

IMAGE RECLASSIFICATION

A useful function is `reclassify`, which can select pixels above a threshold or in an interval, to have a better idea of what they correspond to in the image. We can use it to mask out with NA the pixels with NDVI lower than 0.3, which will show as white when we plot the image (Figure 8.12)

```
veg <- reclassify(ndvi, c (-Inf, 0.3, NA))
panels(6,6,1,1,int='i')
plot(veg, main='Vegetation')
```

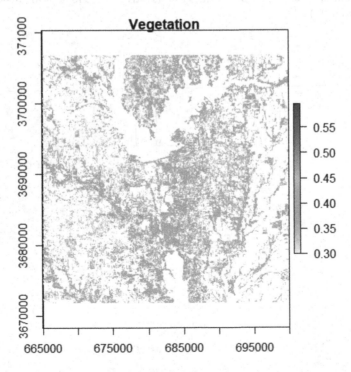

FIGURE 8.12 NT NDVI reclassified to values above 0.3.

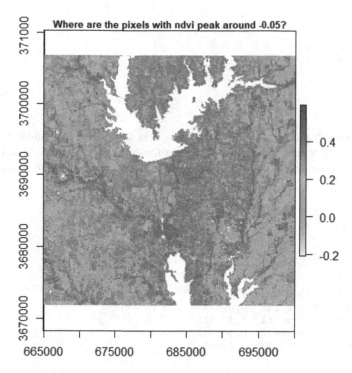

FIGURE 8.13   Finding out where are the pixels with NDVI ~−0.05.

We confirm the vegetation pattern we had determined previously.

As another example let us find out where are those pixels with NDVI around -0.05 (Figure 8.13) by masking them with NA

```
peak0.05 <- reclassify(ndvi, c(-0.1, 0, NA))
panels(6,6,1,1,int='i')
plot(peak0.05, main = 'Where are the pixels with ndvi peak around -0.05?',
cex.main=0.8)
```

We confirm that those pixels (white in the image) correspond to the lakes and some smaller bodies of water. The reader can use similar ideas to reclassify the NDWI to confirm those pixels.

PCA: A SIMPLE EXAMPLE

Before we use available PCA functions in R for remote sensing analysis and for better understanding of the fundamentals of PCA, we will develop an example from scratch based on eigendecomposition of a three-variable data matrix, which is based on a simpler two-variable example analyzed with R in (Acevedo 2013).

```
> X <- cbind(c(3,4,5,6,7), c(14,13,6,5,0), c(5,7,10,11,15))
> X
     [,1] [,2] [,3]
[1,]    3   14    5
[2,]    4   13    7
```

```
[3,]    5    6    10
[4,]    6    5    11
[5,]    7    0    15
>
```

When variables have values in disparate ranges, it is important to standardize the variables before performing PCA or to perform PCA using the correlation matrix instead of the covariance matrix. First, standardize each variable by subtracting the mean and dividing by the standard deviation

```
> x <- X
> for(i in 1:3) x[,i] <- (X[,i]-mean(X[,i]))/sd(X[,i])
> x
             [,1]         [,2]         [,3]
[1,] -1.2649111   1.0927804  -1.1957131
[2,] -0.6324555   0.9220335  -0.6758378
[3,]  0.0000000  -0.2731951   0.1039750
[4,]  0.6324555  -0.4439421   0.3639127
[5,]  1.2649111  -1.2976768   1.4036632
>
```

calculate the covariance matrix of the standardized variables and the correlation matrix of the original values

```
# covariance matrix of standardized values
Cov.x <-  var(x)
# same as cor(X) or correlation matrix of original values
Cor.X <- cor(X)
```

to realize that these two matrices are the same

```
> Cov.x
             [,1]         [,2]         [,3]
[1,]  1.0000000  -0.9719086   0.9863939
[2,] -0.9719086   1.0000000  -0.9853149
[3,]  0.9863939  -0.9853149   1.0000000
> Cor.X
             [,1]         [,2]         [,3]
[1,]  1.0000000  -0.9719086   0.9863939
[2,] -0.9719086   1.0000000  -0.9853149
[3,]  0.9863939  -0.9853149   1.0000000
>
```

Next, we calculate eigenvalues, and the proportion of variance explained by each one

```
# eigen decomp
ec <-eigen(Cor.X)
# total correlation
tot<- sum(ec$values)
# proportion of total variance
> round(ec$values,3)
[1] 2.962 0.028 0.009
> round(ec$values/tot,3)
[1] 0.987 0.009 0.003
>
```

The first eigenvalue 2.96 is a very large fraction of total correlation (3.00), then the variance explained by the first eigenvalue is $100 \times (2.96/3) = 98.7\%$. Therefore, only one component explains almost all the variance.

The *loadings* are the eigenvectors, and together form a *rotation* matrix. Then, project new coordinates or *scores*

```
> # loadings
> A <- ec$vectors
> # scores
> z <- x%*%A
> z
               [,1]            [,2]           [,3]
[1,]  -2.0516030  -0.12164531  -0.005148426
[2,]  -1.2874614   0.20770077   0.078778099
[3,]   0.2176704  -0.19400828  -0.020658719
[4,]   0.8312647   0.12828964  -0.148518486
[5,]   2.2901293  -0.02033682   0.095547532
>
```

Compare the original data vs. rotated data (scores) (Figure 8.14).

```
#plot data and transformed data
panels(6,6,2,2,pty="s")
# characters to display values
pchn <- as.character(seq(1:dim(x)[1]))

plot(x[,1],x[,2], xlab="x1", ylab="x2", pch=pchn)
abline(h=0,v=0,lty=2)
plot(z[,1],z[,2], xlab="z1", ylab="z2", pch=pchn)
abline(h=0,v=0,lty=2)

plot(x[,1],x[,3], xlab="x1", ylab="x3", pch=pchn)
abline(h=0,v=0,lty=2)
plot(z[,1],z[,3], xlab="z1", ylab="z3", pch=pchn)
abline(h=0,v=0,lty=2)
```

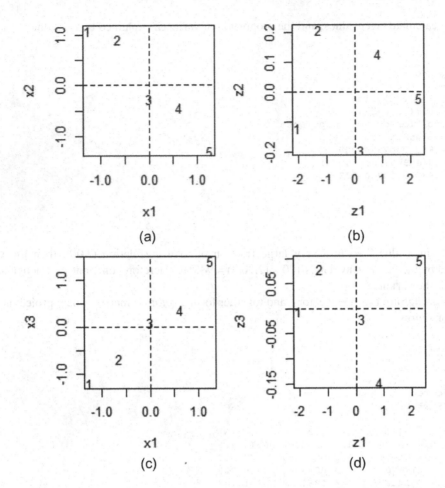

FIGURE 8.14 Original data x (a and c) and transformed data (scores) z (b and d) using the observation number for labels.

There are two major functions in R to perform PCA: `princomp`, which is based on the eigen function, and `prcomp`, which is based on singular value decomposition. We will work out the same simple example using these functions. We can apply the `princomp` function to matrix x of standardized variables using `cor=FALSE` by default or to matrix X of original values with argument `cor=TRUE`

```
> pca3 <- princomp(X,cor=T)
> summary(pca3)
Importance of components:
                          Comp.1      Comp.2       Comp.3
Standard deviation      1.7211701 0.167666498 0.097270691
Proportion of Variance  0.9874755 0.009370685 0.003153862
Cumulative Proportion   0.9874755 0.996846138 1.000000000
>
```

As expected, we obtain that one component explains most (98.7%) of the variance and the dataset can be reduced to a univariate one.

We can obtain the loadings and scores as components of `princomp` result

```
> pca3$loadings

Loadings:
      Comp.1 Comp.2 Comp.3
[1,]   0.577  0.693  0.433
[2,] -0.576  0.720 -0.386
[3,]   0.579        -0.815

                Comp.1 Comp.2 Comp.3
SS loadings      1.000  1.000  1.000
Proportion Var   0.333  0.333  0.333
Cumulative Var   0.333  0.667  1.000
> pca3$scores
          Comp.1        Comp.2        Comp.3
[1,] -2.2937618 -0.13600359  0.005756115
[2,] -1.4394256  0.23221652 -0.088076593
[3,]  0.2433629 -0.21690785  0.023097150
[4,]  0.9293822  0.14343218  0.166048716
[5,]  2.5604424 -0.02273726 -0.106825388
>
```

We can produce several useful plots: first, the loadings can be visualized as a `barplot`, second, we can produce a plot of variances for each component, and finally we can make biplots (Figure 8.15).

```
panels(7,6,2,2,pty="m")
barplot(loadings(pca3), beside=T, main="Loadings", cex.main=0.8)
plot(pca3, type="l", main="Screeplot",cex.main=0.8)
biplot(pca3,choices=1:2,cex.axis=0.8,)
biplot(pca3,choices=c(1,3),cex.axis=0.8)
```

Now, we can apply the `prcomp` function to matrix x of standardized variables using `scale.=FALSE` by default or to matrix X of original values with argument `scale.=TRUE`

```
> pca3 <- prcomp(X,scale.=TRUE,retx=TRUE)
> summary(pca3)
Importance of components:
                        PC1     PC2     PC3
Standard deviation    1.7212 0.16767 0.09727
Proportion of Variance 0.9875 0.00937 0.00315
Cumulative Proportion  0.9875 0.99685 1.00000
>
```

which are the same results obtained with `princomp`. Here, we have used argument `retx=TRUE` to obtain the rotated values.

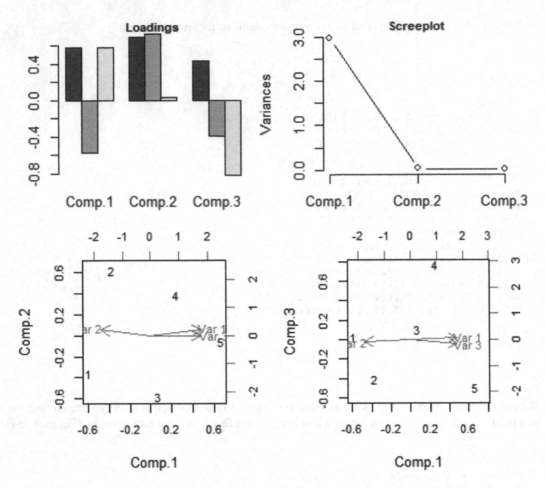

FIGURE 8.15 Loadings, screeplot, and biplots for pca3 using `princomp`.

We can obtain the loadings (named rotation in `prcomp`) and scores returned as component `x` because we used argument `retx=TRUE`

```
> pca3$rotation
              PC1          PC2          PC3
[1,]   0.5765464   0.69296092    0.4328965
[2,]  -0.5763335   0.72046509   -0.3857069
[3,]   0.5791666   0.02711481   -0.8147581
> pca3$x
              PC1          PC2          PC3
[1,]  -2.0516030  -0.12164531    0.005148426
[2,]  -1.2874614   0.20770077   -0.078778099
[3,]   0.2176704  -0.19400828    0.020658719
[4,]   0.8312647   0.12828964    0.148518486
[5,]   2.2901293  -0.02033682   -0.095547532
>
```

We can produce the same plots as in Figure 8.15, except using "rotation" instead of "loadings".

```
panels(7,6,2,2,pty="m")
barplot(pca3$rotation, beside=T, main="Rotation", cex.main=0.8)
plot(pca3, type="l", main="Screeplot",cex.main=0.8)
biplot(pca3,choices=1:2,cex.axis=0.8,)
biplot(pca3,choices=c(1,3),cex.axis=0.8)
```

PCA APPLIED TO REMOTE SENSING IMAGES

We will demonstrate the application of prcomp to the nt raster set. Since the number of pixels is very large (1,361,889), we will take a random sample of size 10,000

```
set.seed(1)
nt.rs <- sampleRandom(nt, 10000)
```

and now perform PCA

```
nt.pca <- prcomp(nt.rs, scale. = TRUE, retx=TRUE)
> summary(nt.pca)
Importance of components:
                          PC1     PC2     PC3      PC4      PC5      PC6      PC7
Standard deviation      2.3160  1.0655  0.59640  0.33085  0.12900  0.12200  0.06440
Proportion of Variance  0.7663  0.1622  0.05081  0.01564  0.00238  0.00213  0.00059
Cumulative Proportion   0.7663  0.9284  0.97927  0.99490  0.99728  0.99941  1.00000
>
```

This means that the first three components explain 98% of the variance. The plots can also provide insight. For instance, the loadings or rotation indicate that the first component loadings are of the same sign and nearly similar magnitude for the first four bands (visible), whereas the second component loadings are different signs for the last three bands (NIR and SWIR) (Figure 8.16).

```
barplot(nt.pca$rotation, beside=T, main="Rotation")
```

A biplot of the first two components

```
biplot(nt.pca,choices=c(1,2),xlabs=rep(".",10000), ylabs=1:7)
```

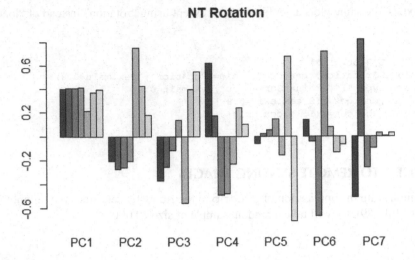

FIGURE 8.16   PCA rotation plot of nt image.

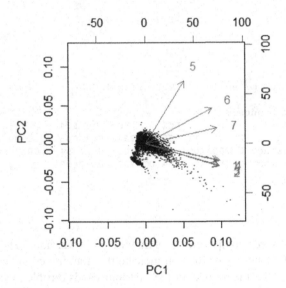

FIGURE 8.17   PCA biplot for PC1 and PC2 of nt image.

shows that all bands 1–7 follow the PC1 axis but bands 5, 6, 7 (NIR and SWIR bands) separate from bands 1–4 (visible) for the PC2 axis (Figure 8.17)

Let us predict the nt raster stack using just the first three components, PC1, PC2, and PC3, and create an image assigning the first three components to the RGB channels (Figure 8.18)

```
nt.pca.pr <- predict(nt, nt.pca, index = 1:3)
panels(6,6,1,1,int='i')
plotRGB(nt.pca.pr, axes=TRUE, stretch="lin", main="NT PCA Color Composite")
```

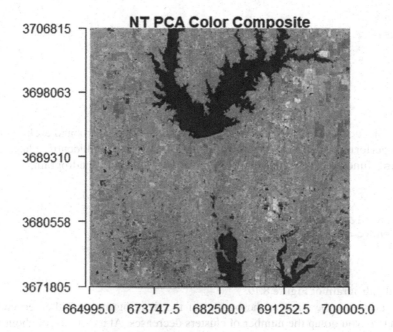

FIGURE 8.18   NT RGB image of first three PCA components.

We see a clear distinction of pixels by types of land use, highlighting the lakes, vegetated areas, rural areas, and urban areas. The full color version of this image available online shows these differences more clearly.

CLUSTER ANALYSIS: A SIMPLE EXAMPLE

Before presenting cluster methods applied to remote sensing images for unsupervised classification, we will gain a basic understanding of cluster analysis using the same simple example presented when explaining PCA, that is, three variables and five observations of each

```
> X
     [,1] [,2] [,3]
[1,]    3   14    5
[2,]    4   13    7
[3,]    5    6   10
[4,]    6    5   11
[5,]    7    0   15
```

You can perform hierarchical agglomerative cluster analysis using functions dist and hclust; function dist is used to compute a distance matrix, then the result is used in hclust. Apply dist with default Euclidean

```
dx <- dist(X)
> dx
           1          2          3          4
2  2.449490
```

```
3  9.643651  7.681146
4 11.224972  9.165151  1.732051
5 17.663522 15.556349  8.062258  6.480741
>
```

Note that the object dx is the lower half of the symmetrical 5 × 5 matrix and excludes the diagonal. Now, we will perform hierarchical clustering, plot a dendrogram, and identify clusters. To do this, run the hclust function using dx with default method and plot as dendrogram,

```
hx <- hclust(dx)
plot(as.dendrogram(hx))
```

to obtain the dendrogram of Figure 8.19.

The observation numbers are at the bottom of the dendrogram, and the vertical axis is distance; note that as you go up the number of clusters decreases. At a distance of about five, you can distinguish three clusters, but as distance increases to about eight we have two clusters; one cluster includes observations 1 and 2 and the other includes observations 3–5. These clusters can be obtained by applying function cutree to the hx object

```
> cutree(hx,k=2)
[1] 1 1 2 2 2
>
```

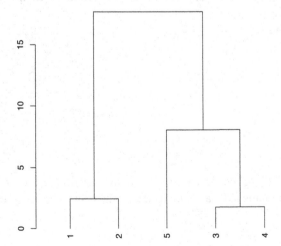

FIGURE 8.19 Dendrogram of X.

A useful tool to visualize the clusters at a given level is the `rect.hclust` function. For instance, to draw rectangles around two clusters use argument k=2

```
rect.hclust(hx, k=2)
```

to obtain the two boxes shown in Figure 8.20. Alternatively, we can use various arguments of `rect.clust` to select the clusters. One can go back to nature of the dataset to reach conclusions; however, many times, interpreting a dendrogram is difficult.

Let us employ the same dataset X to perform K-means clustering aiming to find two clusters

```
kmeans(X,centers=2)
```

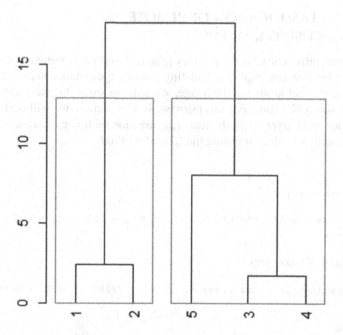

FIGURE 8.20   Dendrogram plus cluster selection.

looking at part of the results

```
> kmeans(X,centers=2)
K-means clustering with 2 clusters of sizes 2, 3

Cluster means:
  [,1]      [,2] [,3]
1 3.5 13.500000    6
2 6.0  3.666667   12

Clustering vector:
[1] 1 1 2 2 2

Within cluster sum of squares by cluster:
[1]  3.00000 36.66667
 (between_SS / total_SS =  80.8 %)
```

We can see from the clustering vector that we obtain one cluster consisting of observations 1 and 2 and a second cluster with observations 3–5. This result is consistent with the hierarchical method applied previously to this dataset.

UNSUPERVISED CLASSIFICATION OF REMOTE SENSING IMAGES USING K-MEANS

Hierarchical agglomerative clustering is not very practical to classify remote sensing images due to a large number of pixels, which implies calculating a very large distance matrix. However, it can be applied to a relatively small sample of the image. We will not study this method but rather focus on divisive clustering using K-means. For this purpose, in this section, we will perform unsupervised classification of the ndvi layer using the kmeans function with five clusters, a maximum of 100 iterations, starting with 1 random set using the Lloyd method

```
nr <- getValues(ndvi)
set.seed(1)
kmncluster <- kmeans(nr, centers = 5, iter.max = 100, nstart = 1,
algorithm="Lloyd")
```

Let us look at part of the results

```
K-means clustering with 5 clusters of sizes 318828, 280010, 181036, 424718,
157297

Cluster means:
       [,1]
1  0.2354232
2  0.3916355
3  0.1363333
4  0.3109618
5 -0.0250777
```

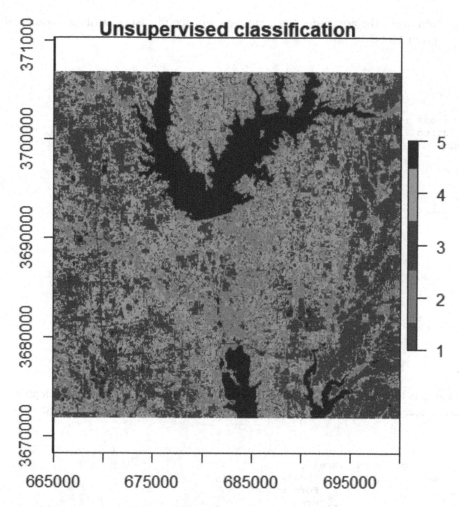

FIGURE 8.21 Unsupervised classification of NDVI.

From our knowledge of the NDVI, cluster five which is centered around −0.025 must correspond to the lakes, cluster 2 must be dense vegetation, cluster 4 must relate to vegetation, and clusters 1 and 3 must relate to open areas or urban areas. Next, we assign cluster number to the pixels, create colors to correspond to these classes, and display the image (Figure 8.21).

```
knr <- raster(ndvi)
values(knr) <- kmncluster$cluster
class.color <- c("brown", "green", "red","cyan","blue")
panels(6,6,1,1,int='i')
plot(knr, main = 'Unsupervised classification', col=class.color)
```

We will now learn how to assign class labels to the unsupervised classification of the image. For this purpose, we use `ratify` to create the *Raster Attribute Table* (*RAT*) of the image

```
knr.rat <- ratify(knr)
```

now when you query the new raster `knr.rat` you will see the list of attributes printed after the values and given ID 1-5.

```
> knr.rat
class       : RasterLayer
dimensions  : 1167, 1167, 1361889  (nrow, ncol, ncell)
resolution  : 30, 30  (x, y)
extent      : 664995, 700005, 3671805, 3706815  (xmin, xmax, ymin, ymax)
crs         : +proj=utm +zone=14 +datum=WGS84 +units=m +no_defs
source      : memory
names       : layer
values      : 1, 5  (min, max)
attributes  :
 ID
 1
 2
 3
 4
 5

>
```

Our next step is to create a data frame of `ID` and `class` labels based on our assumption of what these classes are. At this point, these types are hypothetical since this is unsupervised classification.

```
# build df class and label
knr.class <- matrix(c(1,"Open",
                      2,"Forest",
                      3,"Urban",
                      4,"Grass",
                      5,"Water"),
             ncol=2,byrow=T)
knr.class <- data.frame(knr.class)
names(knr.class) <- c("ID","Label")
```

we are using `ID` for the first column because that is the name in the RAT as we discovered above. The result is

```
> knr.class
  ID  Label
1  1   Open
2  2 Forest
3  3  Urban
4  4  Grass
5  5  Water
>
```

Now, we will assign the labels to the raster

```
# attach Label to raster attributes
levels(knr.rat) <- list(knr.class)
```

and repeat the query of `knr.rat` to verify that `Label` is added to the attributes

```
> knr.rat
class      : RasterLayer
dimensions : 1167, 1167, 1361889  (nrow, ncol, ncell)
resolution : 30, 30  (x, y)
extent     : 664995, 700005, 3671805, 3706815  (xmin, xmax, ymin, ymax)
crs        : +proj=utm +zone=14 +datum=WGS84 +units=m +no_defs
source     : memory
names      : layer
values     : 1, 5  (min, max)
attributes :
 ID  Label
  1   Open
  2 Forest
  3  Urban
  4  Grass
  5  Water

>
```

To visualize an image based on classes rather than values, we can use the function `levelplot` of package `rasterVis`. Install and load this package

```
install.packages("rasterVis")
library(rasterVis)
```

and now use `levelplot` to obtain the image shown in Figure 8.22

```
levelplot(knr.rat, main="Unsupervised Classes")
```

The labels shown in the image legend are based on our assumption of what these classes are. At this point, these types are hypothetical since this is unsupervised classification. In the next chapter, we learn supervised classification and how to evaluate the accuracy of the classes.

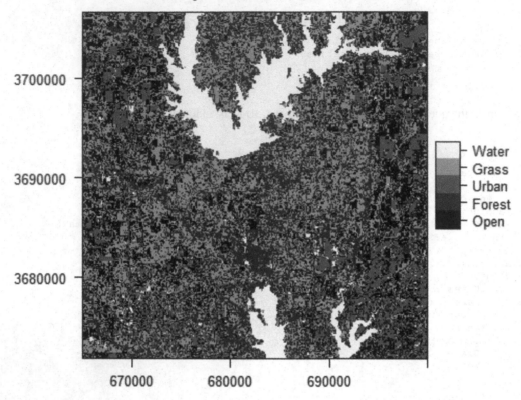

FIGURE 8.22   Image based on classes rather than values.

WRITING GEOTIFF FILES FROM RASTER

The resultant images we have created so far in this guide are stored in the R workspace .Rdata in your working directory. However, often we need to export these to files so that they can be used by other applications; for example, remote sensing analysis and geographic information system (GIS) software such as IDRISI, GRASS, and QGIS. As an example, we will see in Chapter 10, how to import some of the images developed in this chapter to the QGIS geographic information system software for analysis.

Let us see how to write files for ndvi and knr. We will assume all files will be written to folder data/nt/. We will use the function writeRaster and first let us see some of the arguments and its corresponding options. Argument format includes some of the following options

```
File type        Long name         default extension        Multiband support
raster  'Native' raster package format           .grd   Yes
ascii   ESRI Ascii    .asc    No
IDRISI  IDRISI .rst    No
GTiff   GeoTiff .tif    Yes
ENVI    ENVI .hdr Labelled        .envi   Yes
HFA     Erdas Imagine Images (.img)     .img    Yes
```

We will select `GTiff` being a convenient format to work across platforms.

Some of the options for argument `dataType` relevant to our discussion are

```
Datatype definition    minimum possible value maximum possible value
INT1S   -127    127
INT1U   0       255
INT2S   -32,767 32,767
INT2U   0       65,534
INT4S   -2,147,483,647 2,147,483,647
FLT4S   -3.4e+38        3.4e+38
```

the file is written in FLT4S when this argument is not specified.

For `ndvi`, we only need to specify the format since the values are floating point

```
writeRaster(ndvi, filename="data/nt/nt20210618_ndvi.tif", format= 'GTiff',
overwrite=TRUE)
```

we can use `overwrite=TRUE` just in case we need to rewrite the file. For `knr`, that has only five values 1–5 we may use a `datatype` for unsigned integer

```
writeRaster(knr, filename="data/nt/nt20210618_knr.tif", format= 'GTiff',
datatype='INT2U', overwrite=TRUE)
```

verify that the files are in the `data/nt` folder. You would note that the `ndvi` file is nearly 6 MB whereas the `knr` file is only ~700 kB. Hence, the importance of the data type when exporting files.

EXERCISES

Exercise 8.1 Composites.

Create true color and false color composites of the North Texas image.

Exercise 8.2 Band correlations.

Explore the relationship between the NIR and Red bands of the North Texas image.

Exercise 8.3 NDVI and reclasification.

Reclassify the NDVI image to mask those pixels not related to vegetation and reveal the pixels with low NDVI values. Display the resulting images.

Exercise 8.4 NDWI and reclassification.

Reclassify the NDWI image to mask those pixels not related to water and reveal the pixels with positive NDWI values. Display the resulting images.

Exercise 8.5 PCA and unsupervised classification using a simple example.
   Perform PCA, hierachical cluster analysis, and k-means using R and the simple example of a 5×3 data matrix as given in the guide.

Exercise 8.6 PCA of remote sensing image.
   Perform PCA of the North Texas image to explain land use pixels using three components and display the resulting image.

Exercise 8.7 Unsupervised classification of remote sensing image.
   Perform unsupervised classification of the North Texas image to cluster the NDVI pixels in five classes and display the resulting image.

Exercise 8.8 Exporting image files.
   Write files for ndvi and knr to your file system, verify completion and size of the files.

REFERENCES

Acevedo, M.F. 2013. *Data Analysis and Statistics for Geography, Environmental Science & Engineering. Applications to Sustainability*. Boca Raton, FL: CRC Press, Taylor & Francis Group, 535 pp.

Acevedo, M.F. 2024. *Real-Time Environmental Monitoring: Sensors and Systems - Textbook*, Second Edition. Boca Raton, FL: CRC Press, Taylor & Francis Group, 392 pp.

Ghosh, A., and R.J. Hijmans. 2019. *Remote Sensing Image Analysis with R*. Berlin/Heidelberg, Germany: Springer, 48 pp.

R Project. 2023. *The Comprehensive R Archive Network*. Accessed January 2023. http://cran.us.r-project.org/.

USGS. 2022. *EarthExplorer*. Accessed December 2022. https://earthexplorer.usgs.gov/.

9 Probability, Statistics, and Machine Learning

INTRODUCTION

In this lab session, we cover data analysis and machine learning (ML) of environmental monitoring data focusing on probability, statistics, and multiple regression analysis. We use probability to formulate risk and decision calculations employing Bayes' theorem together with false negative and false positive errors. Discrete random variables are introduced to support the analysis of counts and proportions. We write an R program to automate analysis of the error or confusion matrix since these calculations will be done frequently, including repetitive calculations in a loop when performing cross-validation. Then, we learn multiple linear regression, including approaches to select the explanatory variables, emphasizing collinearity issues and stepwise regression. This is followed by computing classification and regression trees (CART) and its application, together with cross-validation, to supervised classification of remote sensing images.

MATERIALS

READINGS

For theoretical background, you can use Chapter 9 of Acevedo, M.F. 2024. *Real-Time Environmental Monitoring: Sensors and Systems - Textbook, Second Edition* which is a companion to these guides (Acevedo 2024). Other bibliographical references are cited throughout the guide.

SOFTWARE

- R, for data analysis (R Project 2023)
- R packages `caTools`, `naivebayes`, `caret`, `rpart`, `rpart.plot`, `rasterVis`, `dismo`

DATA FILES (AVAILABLE FROM THE RTEM GITHUB REPOSITORY)

- Archive `classdata.zip` for classification exercises.
- Archive `nt-ml.zip` for remote sensing supervised classification.

SUPPLEMENTARY SUPPORT MATERIAL

Supplementary support material including additional screenshots, images, and procedures are available from the publisher's eResources web page provided for this book.

PROBABILITY-BASED CALCULATION: APPLICATION OF BAYES' RULE

Consider Bayes' rule for two events A (contaminated water) and B (uncontaminated water), and a water quality test with C (test negative) and D (test positive). Can we back-calculate? e.g., what is the probability that we have contaminated water given a positive test result? In other words, what is the probability $P[A|D]$ of A given D? Use Bayes' rule to calculate

DOI: 10.1201/9781003184362-9

$$P[A \mid D] = \frac{P[AD]}{P[D]} = \frac{P[D \mid A]P[A]}{P[D \mid A]P[A] + P[D \mid B]P[B]}$$

Similarly, e.g., what is the probability that we have uncontaminated water given a negative test result? In other words, what is the probability $P[B|C]$ of B given C Use Bayes' theorem to calculate

$$P[B \mid C] = \frac{P[BC]}{P[C]} = \frac{P[C \mid B]P[B]}{P[C \mid A]P[A] + P[D \mid B]P[B]}$$

First, let us build a function in R to calculate probabilities $P[A|D]$ and $P[B|C]$, assuming we know the base rate $P[A]$, and conditional probabilities $P[C|A]$ or false negative, $P[D|B]$ false positive.

```
bayes2 <- function(pA,pC.A,pD.B){
  # pA is the base rate
  # pC.A is prob of C given A false negative
  # pD.B is prob of D given B false positive
  # calculate probs not given
  pB <- 1- pA; pD.A <- 1-pC.A; pC.B <- 1-pD.B
  # calculate joint prob
  pAC <- pC.A*pA; pAD <- pD.A*pA
  pBC <- pC.B*pB; pBD <- pD.B*pB
  # calc total from joint
  pC<- pAC+pBC; pD <- pAD+pBD
  # Bayes
  pA.C <- pAC/pC; pA.D <- pAD/pD
  pB.D <- pBD/pD; pB.C <- pBC/pC
  return(list(pA.C=round(pA.C,3),pA.D=round(pA.D,3),pB.C=round(pB.C,3),pB.
D=round(pB.D,3)))
}
```

Suppose we know the base rate $P[A] = 0.2$ of contaminated water and want to calculate $P[A|D]$ given that we know the probability $P[C|A] = 0.03$ of false negative and $P[D|B] = 0.07$ of false positive for a water quality test.

```
# application to contamination
p.cont=0.2 # well contaminated event A
# false positive P(D|B) and false negative P(C|A) of water quality test
Fpos <- 0.07; Fneg <- 0.03
# apply Bayes
bayes2(pA=p.cont,pC.A=Fneg,pD.B=Fpos)
$pA.C
[1] 0.008

$pA.D
[1] 0.776

$pB.C
[1] 0.992
```

```
$pB.D
[1] 0.224

>
```

We see that the probability of a correct prediction that the well water is contaminated given a positive test is 77.6% and that the well water is not contaminated given a negative test is 99.2%. This means the test is a good predictor of noncontaminated water but not a good predictor of contaminated water. Similarly, an incorrect prediction of contaminated water given a negative test is 0.8% and of noncontaminated water given a positive test is 22.4%. We conclude that the risk of incorrectly predicting water contamination from positive results is high. This occurs because the test's false positive error has a relatively high probability of 0.07. If we reduce it say to 0.01 the probability that the well water is uncontaminated given a positive test has been reduced to 4%.

```
> Fpos <- 0.01; Fneg <- 0.03
> # apply Bayes
> bayes2(pA=p.cont,pC.A=Fneg,pD.B=Fpos)
$pA.C
[1] 0.008

$pA.D
[1] 0.96

$pB.C
[1] 0.992

$pB.D
[1] 0.04

>
>
```

We can explore how $P[A|C]$ varies as we change the probabilities of false negative $P[C|A]$ and false positive $P[D|B]$. The following script automates the calculation and produces graphical output (Figure 9.1).

```
# pA =contamination p[A]
# Fneg = false negative p[C|A]
# Fpos = false positive p[D|B]
# fix pA and explore changes of p[A|C]
# as we vary Fpos and Fneg
# fix pA
p.cont=0.2
# sequence of values
Fpos <- seq(0,1,0.05); Fneg <- seq(0,1,0.2)
# false neg pC.A false pos pD.B
```

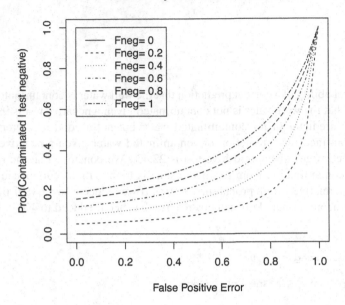

FIGURE 9.1 Bayes' rule: effect of false positive and false negative errors.

```
cont.neg <- sapply(Fneg, function(Fneg) bayes2(pA=p.cont,pC.A=Fneg,pD.
B=Fpos)$pA.C)
matplot(Fpos,cont.neg, type="l",lty=1:length(Fneg), col=1,
        xlab="False Positive Error", ylab="Prob(Contaminated | test negative)")
legend(0,1, paste("Fneg=",as.character(Fneg)), lty=1:length(Fneg), col=1)
title(paste("Probability Contaminated =",pA))
```

We can see how the false positive and false negative errors have an impact on the risk of predicting a contaminated well if the test is negative.

CONFUSION MATRIX OF BINARY CLASSIFICATION AND BAYES' RULE

In this exercise, we will use file wq-test-wells.csv contained in the classdata.zip archive available from the RTEM repository. Place it in your labs/data folder. And read it using read. table using comma as a separator

```
wells <- read.table("data/wq-test-wells.csv", sep=",", header=T)
```

check the dimension of the resulting data frame

```
> dim(wells)
[1] 600   2
>
```

which means we have 600 wells and two variables. Verify the first few records

```
> wells[1:10,]
   Test Contam
1   Neg    Non
2   Neg    Non
3   Neg    Non
4   Neg    Non
5   Neg    Non
6   Neg    Non
7   Pos   Cont
8   Neg    Non
9   Neg    Non
10  Neg    Non
>
```

Or use tibble

```
> as_tibble(wells)
# A tibble: 600 × 2
   Test   Contam
   <chr>  <chr>
 1 Neg    Non
 2 Neg    Non
 3 Neg    Non
 4 Neg    Non
 5 Neg    Non
 6 Neg    Non
 7 Pos    Cont
 8 Neg    Non
 9 Neg    Non
10 Neg    Non
# … with 590 more rows
# i Use 'print(n = ...)' to see more rows
>
```

The first column is a categorical variable with values Neg and Pos for testing positive and negative, respectively, and the second column indicates whether the well is contaminated (value Cont) or not (value Non).

The base rate of contaminated wells can be calculated by finding how many records have value "Cont" and dividing by the total number of wells

```
#base rate
pA <- length(which(wells$Contam=="Cont"))/length(wells$Contam)
```

We are using the same notation pA as in the previous section; the result is

```
> round(pA,3)
[1] 0.207
>
```

or a base rate of ~21%.

The plan is to create a confusion matrix; for this purpose, first we reorder the values of each factor so that Pos and Cont are in the same position, as well as Neg and Non

```
Test <- factor(wells$Test, levels=c("Pos","Neg"))
Cont <- factor(wells$Contam, levels=c("Cont","Non"))
```

Now, we can create the matrix using table

```
tb <- table(Test,Cont)
```

and check the result

```
> tb
      Cont
Test  Cont Non
  Pos  121  25
  Neg    3 451
>
```

We note that we have 121 true positive cases, 451 true negative cases, 3 false negative cases, and 25 false positive cases. The sum of all entries must match the number 600 of wells

```
> sum(tb)
[1] 600
>
```

The confusion matrix can be used to calculate accuracy as the sum of correct predictions (true negative and true positive) divided by the total

```
acc <- sum(diag(tb)) / sum(tb)
> acc
[1] 0.9533333
>
```

Therefore, the accuracy is 95.3% which means we have $1 - 0.953 = 0.047\%$ or 4.7% misclassification rate. Later, in the guide, we will build a function to calculate more complete statistics of a confusion matrix.

To map these results to the Bayes' rule, we can calculate conditional probabilities from these entries; using the same notation as in the previous section, the probabilities for false negative and false positive can be calculated by dividing the corresponding column total

```
pC.A <- tb[2,1]/sum(tb[,1]) # false negative
pD.B <- tb[1,2]/sum(tb[,2]) # false positive
```

round to three decimals

```
> round(pC.A,3)
[1] 0.024
> round(pD.B,3)
[1] 0.053
```

We can calculate the probabilities of correct predictions

```
> pA.D <- tb[1,1]/sum(tb[1,]) # correct positive
> pB.C <- tb[2,2]/sum(tb[2,]) # correct negative
> round(pA.D,3)
[1] 0.829
> round(pB.C,3)
[1] 0.993
>
```

Indicating that the test does well (0.99) when predicting noncontaminated water but not so well (0.83) in predicting contaminated water. As explained in the previous section, this is due to the high value for a false positive probability of 0.053.

NAÏVE BAYES' CLASSIFIER

In this section, we use the same dataset wells built in the previous section to learn how to apply a naïve Bayes' classifier, which is a ML technique. We typically use a naïve classifier when we have multiple variables, but we use it in this case of only one variable for didactic purposes since you will be able to connect the ideas to the previous section.

When using ML, divide the dataset into a subset for *training* the classifier and a subset to *test* or *evaluate* the classifier. For this purpose, we will use package caTools of R, thus install and load this package

```
install.packages("caTools")
library(caTools)
```

We will use function sample.split to divide the dataset into train and eval

```
split <- sample.split(wells)
train <- subset(wells, split == "TRUE")
eval <- subset(wells, split == "FALSE")
```

let us see what we got

```
> dim(train)
[1] 300    2
> dim(eval)
[1] 300    2
>
>
```

which means the two datasets are the same size of 300 records.

Now, we install and load package `naivebayes`

```
install.packages("naivebayes")
library(naivebayes)
```

once loaded we will apply function `naive_bayes` to the entire dataset

```
set.seed(120)
wq.test <- naive_bayes(Contam ~ ., data = wells)
```

We can inspect the results and interpret them based on the previous exercise. Note that the prior probability for contaminated water is the base rate for contamination pA ~ 0.2 and the table calculated under the "Test (Bernoulli)" corresponds to the posterior or conditional probabilities, where you can see the false negative 0.024 and false positive 0.053.

```
> wq.test
================================ Naive Bayes ================================
Call:
naive_bayes.formula(formula = Contam ~ ., data = wells)
----------------------------------------------------------------------------
Laplace smoothing: 0
----------------------------------------------------------------------------
A priori probabilities:
      Cont        Non
0.2066667 0.7933333
----------------------------------------------------------------------------
Tables:
----------------------------------------------------------------------------
  ::: Test (Bernoulli)
----------------------------------------------------------------------------

Test       Cont        Non
  Neg 0.02419355 0.94747899
  Pos 0.97580645 0.05252101
----------------------------------------------------------------------------
>
```

We can plot the result for visualization to obtain Figure 9.2 which in this case is very simple illustrating the effect of only one variable. The bars are the elements of the matrix.

```
plot(wq.test)
```

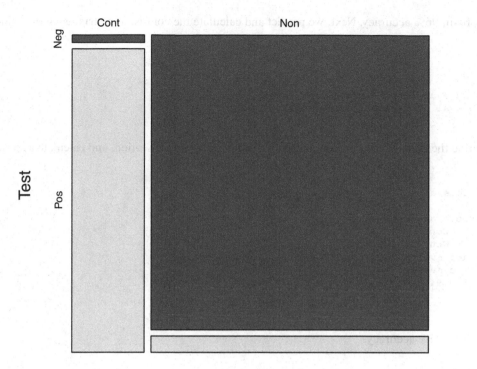

FIGURE 9.2   Naïve Bayes' classifier plot.

Now, we predict and calculate the confusion matrix using the training set

```
pred.train <- predict(wq.test, train)
tb.train <- table(pred.train, train$Contam)
```

examine the table in the same manner as we did in the previous section and calculate accuracy

```
> tb.train

pred.train Cont Non
      Cont   58   12
      Non     0  230
> acc.train <- sum(diag(tb.train)) / sum(tb.train)
> acc.train
[1] 0.96
>
```

We obtain 96% accuracy. Next, we predict and calculate the confusion matrix using the evaluation set

```
pred.eval <- predict(wq.test, eval)
tb.eval <- table(pred.eval, eval$Contam)
```

examine the table in the same manner as we did in the previous section, and calculate accuracy

```
> tb.eval

pred.eval Cont Non
     Cont   63  13
     Non     3 221
> acc.eval <- sum(diag(tb.eval)) / sum(tb.eval)
> acc.eval
[1] 0.9466667
>
```

yielding ~95% accuracy.

COUNT TESTS AND THE BINOMIAL

Suppose we hypothesize that 20% of the pixels in an area of a remote sensing image corresponds to grassland cover. We sample coordinates at random to select 20 locations, then survey these and count the number of times we find grass. Suppose we find grass in five sites; should we reject the hypothetical 20% of grass pixels? To answer this question, we apply the binomial test using the function binom.test

```
> binom.test(5,20,p=0.2)

        Exact binomial test

data:  5 and 20
number of successes = 5, number of trials = 20, p-value = 0.5764
alternative hypothesis: true probability of success is not equal to 0.2
95 percent confidence interval:
 0.08657147 0.49104587
sample estimates:
probability of success
              0.25

>
```

We get a high p-value of ~0.58, indicating that we should not reject a 20% grass cover. In addition, the 95% confidence interval indicates that 20% is within the interval for 25% estimate in cover. In this case, we conclude that the cover could be 20%.

For comparison, suppose we count eight sample sites covered with grass. Now re-run the binomial test

```
> binom.test(8,20,p=0.2)

        Exact binomial test

data:  8 and 20
number of successes = 8, number of trials = 20, p-value = 0.04367
alternative hypothesis: true probability of success is not equal to 0.2
95 percent confidence interval:
 0.1911901 0.6394574
sample estimates:
probability of success
                  0.4
>
```

With this p-value ~0.04, we could reject the null of 20% grass cover at α slightly <5%. The lower bound of the 95% confidence interval for an estimated cover of 0.4 barely includes the hypothetical 20%. In this case, we would conclude that the cover is not 20%.

As we have practiced many times throughout the lab manual, you may want to address only a component of the R object; in this case, the object is the list resulting from the test, and you can do this using the $ dollar sign and the name of the component. For example,

```
> binom.test(8,20,p=0.2)$p.value
[1] 0.04367188
```

gives you only the p-value. The online help tells you about the components produced for each type of function. For example, for the `binom.test`, you find this is in the help

```
statistic      the number of successes.
parameter      the number of trials.
p.value             the p-value of the test.
conf.int       a confidence interval for the probability of success.
estimate       the estimated probability of success.
null.value     the probability of success under the null, p.
alternative    a character string describing the alternative hypothesis.
method the character string "Exact binomial test".
data.name      a character string giving the names of the data.
```

COMPARING PROPORTIONS

As an example, suppose we hypothesize that the proportion of grass pixels in a remote sensing image is the same regardless of the location or area of the image. We sample pixels around three locations and obtain 10, 12, 13 grass pixels out of 80, 90, 85 examined for each area. We can use the function prop.test with the number of grass pixels and the sample counts

```
> xo <- c(10,12,13); xt <- c(80,90,85)
> prop.test(xo,xt)

        3-sample test for equality of proportions without continuity correction

data:  xo out of xt
X-squared = 0.2898, df = 2, p-value = 0.8651
alternative hypothesis: two.sided
sample estimates:
   prop 1     prop 2     prop 3
0.1250000 0.1333333 0.1529412

>
```

Given the high p-value ~0.87, we should not reject H0 and conclude that the proportions may be the same regardless of location.

Alternatively, we obtain the same result using chisq.test on the contingency table of positive (a pixel is grass) and negative (a pixel is not grass) counts

```
> table <- rbind(xo,xc=(xt-xo))
> table
   [,1] [,2] [,3]
xo   10   12   13
xc   70   78   72
> chisq.test(table)

        Pearson's Chi-squared test

data:  table
X-squared = 0.2898, df = 2, p-value = 0.8651

>
```

Part of the results given in the output window is reproduced here. The high p-value indicates that we should not reject H_0, and the proportions may be the same regardless of location.

```
> .Table  # Counts
    1  2  3
1 10 12 13
2 70 78 72
> .Test <- chisq.test(.Table, correct=FALSE)
> .Test
        Pearson's Chi-squared test
data:  .Table
X-squared = 0.2898, df = 2, p-value = 0.8651
> .Test$expected # Expected Counts
```

```
          1         2         3
1 10.98039 12.35294 11.66667
2 69.01961 77.64706 73.33333
> round(.Test$residuals^2, 2) # Chi-square Components
     1    2    3
1 0.09 0.01 0.15
2 0.01 0.00 0.02
```

As another example, let us analyze the example in Table 9.1 already discussed in Chapter 9 of the textbook. Enter values as a matrix object and apply the function chisq.test
to get the same results as we did by calculations in Chapter 9 and conclude that there are differences in vegetation cover with elevation.

SUPERVISED CLASSIFICATION: ERROR MATRIX ANALYSIS

Let us build a function to calculate all metrics of an error or confusion matrix

```
err.mat <- function(x) {
  N <- sum(x)
  po <- sum(diag(x))/N
  rm <- rowSums(x)/N
  cm <- colSums(x)/N
  pe <- sum(rm*cm)
  kappa = (po-pe)/(1-pe)
  prd.acc <- diag(x)/ colSums(x)
  err.om <- 1- prd.acc
  #err.co <- (rowSums(x)-diag(x))/rowSums(x)
  usr.acc <- diag(x)/ rowSums(x)
  err.co <- 1- usr.acc
  #err.om <- (colSums(x)-diag(x))/colSums(x)

  return(list(Po=round(po,3),Pe=round(pe,3), kappa=round(kappa,3),
            prd.acc=round(prd.acc,3),usr.acc=round(usr.acc,3),
            err.co=round(err.co,3),err.om=round(err.om,3) ))
}
```

TABLE 9.1

Cross-Tabulation of Topographic Class and Vegetation Cover

	Crop Field	Forest	Grassland
Lowlands	12	20	8
Uplands	8	10	22

TABLE 9.2

Confusion Matrix: Correct Class in Diagonal and Incorrect Classification in Off-Diagonal Entries

		Ground			
		Grass	Forest	Urban	Row Total
	Grass	75	10	5	90
Image	Forest	10	80	5	95
	Urban	8	4	85	97
	Col Total	93	94	95	282

And apply it to the confusion matrix discussed in Chapter 9 reproduced here as Table 9.2 for easy reference.

```
> x <- matrix(c(75,10,5, 10,80,5, 8,4,85),ncol=3,byrow=T)
> err.mat(x)
$Po
[1] 0.851

$Pe
[1] 0.333

$kappa
[1] 0.777

$prd.acc
[1] 0.806 0.851 0.895

$usr.acc
[1] 0.833 0.842 0.876

$err.co
[1] 0.167 0.158 0.124

$err.om
[1] 0.194 0.149 0.105

>
```

We can see the overall accuracy or Po is 85.1% and the Pe is 33.3%, which yields a kappa coefficient value of 77.7%. The producer and user accuracy for all classes is above 80%, being the highest for class 3 or urban. Consequently, errors of commission and omission are all below 20%, being the lowest for class 3.

CONTINGENCY TABLES APPLIED TO CONTINUOUS DATA

We can cross-tabulate in R using function `table` and dividing continuous variables in bins. As an example, let us use the R dataset `airquality` which we employed in Lab session 1 when we studied simple linear regression. Recall that ozone is an important urban air pollution problem:

excessive ozone in the lower troposphere, related to emissions and photochemistry, and meteorological variables.

Check the contents of the `airquality` dataset by printing a few lines or using tibble

```
> as_tibble(airquality)
# A tibble: 153 × 6
   Ozone Solar.R  Wind  Temp Month   Day
   <int>   <int> <dbl> <int> <int> <int>
 1    41     190   7.4    67     5     1
 2    36     118   8      72     5     2
 3    12     149  12.6    74     5     3
 4    18     313  11.5    62     5     4
 5    NA      NA  14.3    56     5     5
 6    28      NA  14.9    66     5     6
 7    23     299   8.6    65     5     7
 8    19      99  13.8    59     5     8
 9     8      19  20.1    61     5     9
10    NA     194   8.6    69     5    10
# ... with 143 more rows
# i Use 'print(n = ...)' to see more rows
```

which has 153 records and 6 variables. You should notice that there are NA values in the dataset. We start by building a new set with no NAs using `na.omit`

```
airqual <- na.omit(airquality)
```

check the new dimensions and note that we have 111 valid observations

```
> dim(airqual)
[1] 111   6
>
```

Let us attach this dataset so that we can use the names directly

```
attach(airqual)
```

Suppose we want to build a two-way table relating `Temp` and `Month`. First, let us see how many different values these factors have

```
> unique(Month)
[1] 5 6 7 8 9
> unique(Temp)
 [1] 67 72 74 62 65 59 61 69 66 68 58 64 57 73 81 79 76 82 90 87 77 84 85 83 88
[26] 92 89 80 86 78 97 94 96 91 93 75 71 63 70
>
```

We can see that `Temp` has too many unique values for a practical cross-tabulation and thus we want to select fewer intervals. We do this with the `cut` function using `breaks` based on the quantiles

```
Tcut <- cut(Temp, quantile(Temp))
Levels: (57,71] (71,79] (79,84.5] (84.5,97]
```

Now, we have four bins: from $57 < \text{Temp} \leq 71$ and so on. Then, we apply `table` with these four levels for `Temp` and the five months 5, ..., 9.

```
T.Mo <- table(Tcut,Month)
> T.Mo
           Month
Tcut        5  6  7  8  9
  (57,71]  17  1  0  0 10
  (71,79]   5  4  2  8  9
  (79,84.5] 1  2 12  6  5
  (84.5,97] 0  2 12  9  5
>
```

We can note the higher frequency of high level of `Temp` as we increase the `Month`, which we can verify with `chisq.test`

```
> chisq.test(T.Mo)

        Pearson's Chi-squared test

data:  T.Mo
X-squared = 63.602, df = 12, p-value = 4.938e-09

Warning message:
In chisq.test(T.Mo) : Chi-squared approximation may be incorrect
>
```

Indeed, the low p-value means that we can reject the null H0 of no association and conclude that `Temp` is associated with `Month`. The warning message indicates that the conclusion may not be correct.

Let us try `Temp` and `Ozone`. First cut `Ozone` by quantiles then table, and then apply `chisq.test`

```
> Ocut <- cut(Ozone, quantile(Ozone))
> T.O <- table(Tcut,Ocut)
> T.O
            Ocut
Tcut        (1,18] (18,31] (31,62] (62,168]
  (57,71]       15       9       3        0
  (71,79]       10      10       7        1
  (79,84.5]      4       6      11        5
  (84.5,97]      0       0       6       22
```

We note that Ozone has been cut into four levels as well. Larger values of ozone occur for the bins with higher temperature and low values of ozone occur for the bins with lower temperature. Let us verify with chisq.test

```
> chisq.test(T.O)

        Pearson's Chi-squared test

data:  T.O
X-squared = 76.309, df = 9, p-value = 8.713e-13
```

Indeed, this very low p-value means we can reject the null of no association and conclude that high values of Ozone are associated with high values of Temp. We will not detach the airqual dataset because it will be used in the next section.

MULTIPLE LINEAR REGRESSION (MLR)

We will use the same airqual dataset of the previous section, which is the same as airquality with no-data values (NA) omitted. In Lab session 1, we did a simple regression of ozone vs. temperature and found that temperature only explained ~48% of variation in ozone and therefore concluded that other variables should be included. First, visually explore by looking at pairwise scatter plots including radiation and wind. Simply, use function pairs

```
>pairs(airqual)
```

Figure 9.3 shows a matrix of scatter plots. From the top row of panels, we see that there seems to be an increasing trend of ozone with increased radiation and temperature and a decreasing trend with wind. Month and Day in this case are not useful. Thus, let us try to include radiation and wind in a multiple regression model.

There are several ways of proceeding. (1) Include all variables at once, (2) Drop and add variables from a previously built model with fewer variables, and (3) An automatic process using the stepwise procedure. Let us pursue (1) first. Recall that we attached airqual in the previous section

```
ozone.mlm <- lm(Ozone ~ Solar.R + Temp + Wind)
> summary(ozone.mlm)

Call:
lm(formula = Ozone ~ Solar.R + Temp + Wind)

Residuals:
    Min      1Q  Median      3Q     Max
-40.485 -14.219  -3.551  10.097  95.619

Coefficients:
             Estimate Std. Error t value Pr(>|t|)
(Intercept) -64.34208   23.05472  -2.791  0.00623 **
Solar.R       0.05982    0.02319   2.580  0.01124 *
```

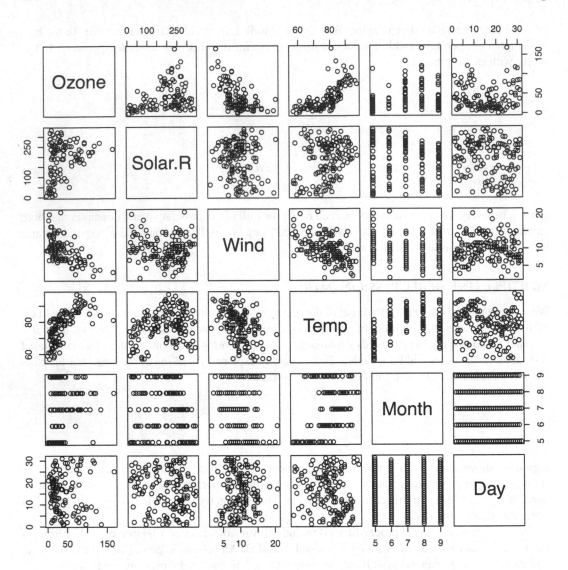

FIGURE 9.3   Scatterplot pairs.

```
Temp            1.65209     0.25353    6.516 2.42e-09 ***
Wind           -3.33359     0.65441   -5.094 1.52e-06 ***
---
Signif. codes:  0 '***' 0.001 '**' 0.01 '*' 0.05 '.' 0.1 ' ' 1

Residual standard error: 21.18 on 107 degrees of freedom
Multiple R-squared:  0.6059,    Adjusted R-squared:  0.5948
F-statistic: 54.83 on 3 and 107 DF,  p-value: < 2.2e-16


>
```

The t tests for the coefficients indicate significance for all of them. Note that we increased the explained variation to ~60% and that the p-value for the variance test F is very low (highly significant). The resulting model is

$$\text{Ozone} = -64.34 + 0.05982 \times \text{Solar.R} + 1.65 \times \text{Temp} - 3.33 \times \text{Wind} \qquad (9.1)$$

as expected, ozone increases with air temperature and radiation and decreases with wind speed. The negative value of ozone forced at zero values for `Solar.R`, `Temp`, and `Wind` due to the negative intercept is an invalid extrapolation considering the range of values for which the relationship is derived.

We can obtain evaluation plots as we did for single regression using

```
panels(6,6,2,2, pty="m")
plot(ozone.mlm)
```

to yield Figure 9.4. Here, we see a relatively random behavior of the residuals except for the low end of the ozone values. There is not much improvement in the relative spread or influence of outliers. To produce a plot of ozone vs. fitted values of ozone (Figure 9.5), we extract the `$fitted.values` from the `lm` object, plot the values as circles, and overlay the 1:1 line

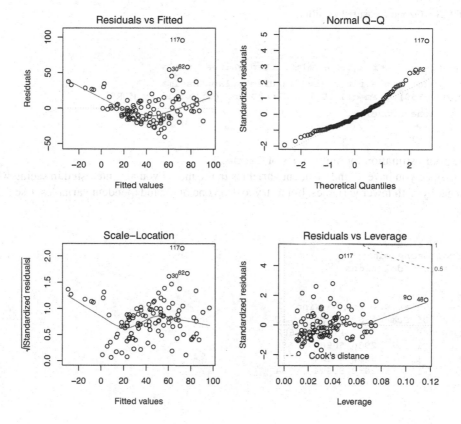

FIGURE 9.4 Diagnostic plots for residuals.

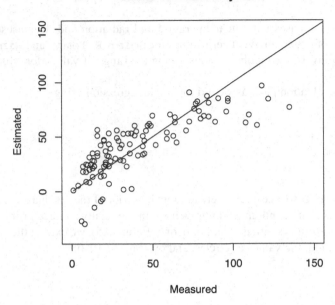

FIGURE 9.5   Ozone measured vs. fitted.

```
ozone.fit <- matrix(ozone.mlm$fitted.values, ncol=1)
plot(Ozone, ozone.fit,xlab="Measured",ylab="Estimated",
xlim=c(0,150),ylim=c(0,150), main="Ozone estimated by MLR")
lines(Ozone,Ozone)
```

We note poor estimation for high values of Ozone.

Often, once you have all independent variables in the model you are interested in seeing whether you can get by with fewer variables. Let us try to drop one of the independent variables. Use function drop1

```
> drop1(ozone.mlm)
Single term deletions

Model:
Ozone ~ Solar.R + Temp + Wind
        Df Sum of Sq   RSS    AIC
<none>                48003    682
Solar.R  1     2986  50989    686
Temp     1    19050  67053    717
Wind     1    11642  59644    704
>
```

In this case, the Akaike Information Criterion (AIC) is lowest for the current model (line labeled <none>) 682 and therefore you do not want to drop any of the three terms.

Now to illustrate approach (2), we create an `ozone.lm` object from `airqual` that relates ozone to temperature

```
> ozone.lm <- lm(Ozone ~ Temp)
```

Use function `add1`

```
> ozone.add <- add1(ozone.lm, ~ Temp + Solar.R + Wind)
> ozone.add
Single term additions

Model:
Ozone ~ Temp
        Df Sum of Sq    RSS    AIC
<none>                 62367 706.77
Solar.R  1    2723.1  59644 703.82
Wind     1   11378.5  50989 686.41
>
```

```
> ozone.tw <- lm(Ozone ~ Temp + Wind)
> summary(ozone.tw)

Call:
lm(formula = Ozone ~ Temp + Wind)

Residuals:
    Min      1Q  Median      3Q     Max
-42.156 -13.216  -3.123  10.598  98.492

Coefficients:
            Estimate Std. Error t value Pr(>|t|)
(Intercept) -67.3220    23.6210  -2.850  0.00524 **
Temp          1.8276     0.2506   7.294 5.29e-11 ***
Wind         -3.2948     0.6711  -4.909 3.26e-06 ***
---
Signif. codes:  0 '***' 0.001 '**' 0.01 '*' 0.05 '.' 0.1 ' ' 1

Residual standard error: 21.73 on 108 degrees of freedom
Multiple R-squared:  0.5814,  Adjusted R-squared:  0.5736
F-statistic: 74.99 on 2 and 108 DF,  p-value: < 2.2e-16

>
```

Based on the lowest AIC for wind (686), we decide to add it as an independent variable. You can repeat this by running a new lm with just temperature and wind as object `ozone.tw`
The new R^2 is 0.57, which is only slightly lower than the one for the three-variable model (0.59).

When you have many independent variables, the drop/add can be tedious. Now let us try approach (3) and perform *stepwise regression*. This is an automatic procedure that only requires fewer commands. The procedure tries to minimize the AIC and stops when the AIC for the current model (the "`none`" row) is the lowest among all the other terms.

```
> ozone0.lm <- lm(Ozone ~ 1,)
> step(ozone0.lm, ~ Solar.R + Temp + Wind)
Start:  AIC= 779.07
 Ozone ~ 1

          Df Sum of Sq    RSS   AIC
+ Temp     1     59434  62367   707
+ Wind     1     45694  76108   729
+ Solar.R  1     14780 107022   767
<none>                  121802   779

Step:  AIC= 706.77
 Ozone ~ Temp

          Df Sum of Sq    RSS   AIC
+ Wind     1     11378  50989   686
+ Solar.R  1      2723  59644   704
<none>                   62367   707
- Temp     1     59434 121802   779

Step:  AIC= 686.41
 Ozone ~ Temp + Wind

          Df Sum of Sq   RSS   AIC
+ Solar.R  1      2986 48003   682
<none>                 50989   686
- Wind     1     11378 62367   707
- Temp     1     25119 76108   729

Step:  AIC= 681.71
 Ozone ~ Temp + Wind + Solar.R

          Df Sum of Sq   RSS   AIC
<none>                 48003   682
- Solar.R  1      2986 50989   686
- Wind     1     11642 59644   704
- Temp     1     19050 67053   717

Call:
lm(formula = Ozone ~ Temp + Wind + Solar.R)

Coefficients:
(Intercept)         Temp          Wind       Solar.R
  -64.34208      1.65209      -3.33359       0.05982

>
```

The final model is the one we already found using the previous methods. The stepwise procedure pays off when you have many independent variables, and you want to see how all the possible models compare to each other. We will continue to use the `airqual` dataset in the next section, so do not detach yet.

SELECTING A REGRESSION MODEL USING MACHINE LEARNING

In this section, we will learn to use ML to derive a model using a training dataset and test it using another dataset. Therefore, first split the datasets using package `caTools` as before; use `require` just in case `caTools` is not loaded

```
require(caTools)
split <- sample.split(airqual)
train.aq <- subset(airqual, split == "TRUE")
test.aq <- subset(airqual, split == "FALSE")
```

verify dimensions

```
> dim(train.aq)
[1] 74   6
> dim(test.aq)
[1] 37   6
>
>
```

We will employ the R package `caret`, which stands for Classification and Regression Training. We need to install and load.

```
install.packages("caret")
library(caret)
```

use function `train` of `caret`, specifying `data=train.aq` to select the training set, and `lm` as method

```
lm.train <- train(Ozone ~ Solar.R + Temp + Wind, data=train.aq, method = "lm")
```

query the summary and realize that this is equivalent to what we got by just invoking lm in the previous section; the coefficients are similar but solar radiation seems to be less significant

```
> summary(lm.train)

Call:
lm(formula = .outcome ~ ., data = dat)

Residuals:
    Min      1Q  Median      3Q     Max
-40.325 -14.773  -4.794  12.590  96.118

Coefficients:
            Estimate Std. Error t value Pr(>|t|)
(Intercept) -63.77507   32.60140  -1.956 0.054432 .
Solar.R       0.05351    0.03203   1.671 0.099220 .
Temp          1.65656    0.36122   4.586 1.93e-05 ***
Wind         -3.31177    0.90740  -3.650 0.000502 ***
---
Signif. codes:  0 '***' 0.001 '**' 0.01 '*' 0.05 '.' 0.1 ' ' 1

Residual standard error: 23.28 on 70 degrees of freedom
Multiple R-squared:  0.5584,    Adjusted R-squared:  0.5394
F-statistic:  29.5 on 3 and 70 DF,  p-value: 1.925e-12

>
```

You may wonder why we use this function of `caret` instead of the direct call to `lm`. The main advantage of calling it from `train` of `caret` is that we can readily change the method to use a different model and perform other functions on the model.

Now, we predict from the test dataset, being careful to use `newdata` argument, but employing the coefficients trained by ML

```
y <- predict(lm.train,newdata=test.aq)
```

we build a plot for visualization of the result (Figure 9.6)

```
x <- test.aq$Ozone
plot(x,y,xlab="Measured",ylab="Estimated",
xlim=c(0,150),ylim=c(0,150),    main="Ozone estimated by trained MLR model")
lines(x,x)
```

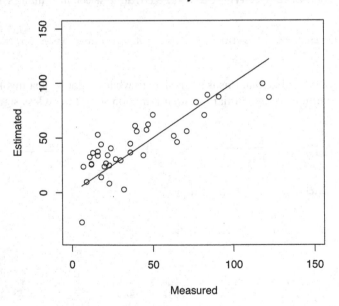

FIGURE 9.6   Ozone estimation using MLR build by ML.

CLASSIFICATION AND REGRESSION TREES

Now, we make use of changing the `method` argument in `train` to classification and regression, invoking function `rpart` which stands for Recursive Partitioning and Regression Trees.

```
tree <- train(Ozone ~ Solar.R + Temp + Wind, data = train.aq, method = "rpart")
```

A convenient tool to visualize the tree is package `rpart.plot`, let us install and load

```
install.packages("rpart.plot")
library(rpart.plot)
```

and use it to make a plot of tree component `finalModel` (Figure 9.7)

```
rpart.plot(tree$finalModel)
```

To interpret the result, we call for a summary

```
>summary(tree)
```

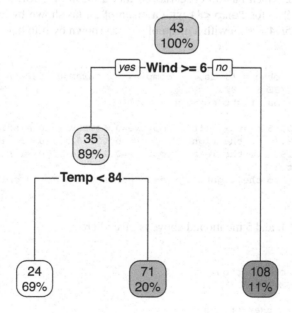

FIGURE 9.7   Decision tree for the air quality dataset.

which is lengthy and thus we only discuss some segments. In the following part, we see the values of cp (complexity parameter) and relative error

```
          CP nsplit rel error
1 0.44970204      0 1.0000000
2 0.29488956      1 0.5502980
3 0.02220341      2 0.2554084


Variable importance
Wind Temp
  50   50
```

In this part, we have information on Node 1 at the top or root of the tree (Figure 9.7). The mean of Ozone is ~43 and will split into 66 observations (66/74=89%) for Wind ≥6 and 8 observations (8/74=11%) for Wind <6 with a mean of 108 (as shown later for info on Node 3)

```
Node number 1: 74 observations,      complexity param=0.449702
  mean=42.63514, MSE=1160.448
  left son=2 (66 obs) right son=3 (8 obs)
  Primary splits:
      Wind    < 6    to the right, improve=0.4497020, (0 missing)
      Temp    < 83.5 to the left,  improve=0.4351213, (0 missing)
      Solar.R < 153  to the left,  improve=0.2190049, (0 missing)
  Surrogate splits:
      Temp < 91.5 to the left,  agree=0.932, adj=0.375, (0 split)
```

Next, we study Node 2 which has 66 observations and a mean of Ozone is ~35, it splits into 51 observations (51/74=69%) for Temp <84 with a mean of 24 (as shown by info below for Node 4) and 15 observations (15/74=20%) with a mean of ~71 (as shown by info below for Node 5)

```
Node number 2: 66 observations,      complexity param=0.2948896
  mean=34.68182, MSE=621.823
  left son=4 (51 obs) right son=5 (15 obs)
  Primary splits:
      Temp    < 83.5 to the left,  improve=0.6170297, (0 missing)
      Wind    < 8.3  to the right, improve=0.2363621, (0 missing)
      Solar.R < 153  to the left,  improve=0.2083347, (0 missing)
  Surrogate splits:
      Wind < 6.6  to the right, agree=0.788, adj=0.067, (0 split)
```

The mean for nodes 3, 4, and 5 mentioned above is given here

```
Node number 3: 8 observations
  mean=108.25, MSE=776.9375

Node number 4: 51 observations
  mean=24.05882, MSE=162.6436

Node number 5: 15 observations
  mean=70.8, MSE=494.8267

>
```

We conclude that 11% of observations with high values of Ozone (mean 108) are classified by higher wind, Wind ≥ 6, while the remainder of observations are separated by cooler temperature, Temp <84, into 69% of the total with low values of Ozone (mean 24), leaving 20% with medium values of Ozone (mean 71).

At this point, we can detach dataset airqual by

```
> detach (airqual)
```

SUPERVISED CLASSIFICATION OF REMOTE SENSING IMAGE

We will use National Land Cover Database (NLCD) layers as a reference dataset to learn supervised classification; this allows you to focus on the classification process and not on how to generate the reference dataset. Normally, a reference dataset is laborious to develop, involves field work, and uses additional information from other sources. Land cover class for 2019 is available from the website maintained by collaborating agencies of the USA federal government (MRLC 2022). NLCD legends relevant to the exercise are in the range of 11–95 and have the simplified description given in Table 9.3. For a more complete description, see MRLC (2023). This exercise is based on the procedures reported by Ghosh and Hijmans (2019).

For the purposes of this exercise, you can download the archive nt-ml.zip from the RTEM repository. This archive contains two files nt-lc-reclass.csv and NLCD-NT-2019.tif that you should copy to your data/nt folder. The first one is a simplification of the classes given in Table 9.3 such that classes 21 and 23 were assigned to 22, 43 and 52 to 41, and 71–81. Open this file with a text editor and you will see it is CSV with three columns, class or legend, label, and description.

```
Class, Label, Description
11,Water, Open water
22,DevLo, Developed Open/Low/Medium Intensity combines 21->22 23->22
24,DevHi, Developed High Intensity
31,Barre, Barren land
41,DeFor, Deciduous/mixed forest and shruhb combines 43->41 52->41
42,EvFor, Evergreen forest
81,Grass, Pasture/Hay combined with grass 71->81
82,Crop, Cultivated crops
90,WdWet, Woody wetlands
95,HbWet, Emergent Herbaceous Wetlands
```

TABLE 9.3
NLCD Legends and Description

Legend	Description
11	Open water
12	Perennial ice/snow
21	Developed, open space – areas with a mixture of some constructed materials, but mostly vegetation in the form of lawn grass
22	Developed, low intensity – areas with a mixture of constructed materials and vegetation, Impervious 20%–49%. Commonly include single-family housing units
23	Developed, medium intensity – areas with a mixture of constructed materials and vegetation. Impervious surfaces account for 50%–79% of the total cover
24	Developed high intensity – highly developed areas where people reside or work in high numbers
31	Barren land (Rock/Sand/Clay) Vegetation <15%
41	Deciduous forest – 75% species shed leaves with season
42	Evergreen forest
43	Mixed Forest – neither deciduous nor evergreen
52	Shrub/Scrub – includes forests in early succession
71	Grassland/Herbaceous – may include grazing
81	Pasture/Hay – grasses and legumes mixtures
82	Cultivated crops – annual and perennial
90	Woody wetlands – forest with soil periodically saturated with water
95	Emergent herbaceous wetlands – herbaceous cover soil periodically saturated with water

The second file `NLCD-NT-2019.tif` was obtained from an NLCD 2019 land cover tiff `NLCD _ 2019 _ Land _ Cover _ L48 _ 20210604` from the USGS, it was reclassified to the legend values just given above, reprojected from Conus Albers to UTM Zone 14, and was clipped to the NT area that we worked with in Lab session 8.

Let us read the first file and build a matrix with columns equal to the `Class`, which we will use to reclassify the image, use `[,-3]` to skip reading the third column

```
# classes defined
class <- read.table("data/nt/nt-lc-reclass.csv", header=T, sep=",")[,-3]
rcl.mat <- cbind(class$Class,class$Class)
```

Now, read the tif image as raster and reclassify to have the data correspond to the classes

```
# read nLCD file
nlcd <- raster('data/nt/NLCD-NT-2019.tif')
nlcd <- reclassify(nlcd,rcl=rcl.mat)
```

And look at `nlcd`

```
> nlcd
class       : RasterLayer
dimensions  : 1167, 1167, 1361889   (nrow, ncol, ncell)
resolution  : 30, 30   (x, y)
extent      : 664995, 700005, 3671805, 3706815   (xmin, xmax, ymin, ymax)
crs         : +proj=utm +zone=14 +datum=WGS84 +units=m +no_defs
source      : memory
names       : layer
values      : 11, 95   (min, max)
>
```

You know how to interpret this information from the work we did in Lab session 8. However, for the work we want to do now, we need to understand a bit more of the raster object `nlcd`, which is structured in `slots`. To find out the slots of `nlcd`, you can use `slotNames`

```
> slotNames(nlcd)
 [1] "file"     "data"     "legend"   "title"    "extent"   "rotated"
 [7] "rotation" "ncols"    "nrows"    "crs"      "history"  "z"
>
```

and these can be addressed with the @ symbol. We are interested in slot "`data`" and thus we can further look at its structure

```
> slotNames(nlcd@data)
 [1] "values"    "offset"     "gain"       "inmemory"   "fromdisk"
 [6] "isfactor"  "attributes" "haveminmax" "min"        "max"
[11] "band"      "unit"       "names"
>
```

We are also interested in slot "`values`" that should contain the classes; we can look at the first 100 and realize it has the NLCD legends simplified to the classes we defined

```
> nlcd@data@values[1:100]
  [1] 81 22 81 81 81 81 81 81 81 81 81 81 81 81 81 81 81 81 81 81 81 81 81 81 81
 [26] 81 81 81 81 81 81 81 81 81 81 81 81 81 81 81 81 81 81 81 81 81 81 81 81 81
 [51] 81 81 81 81 81 81 81 82 82 82 82 82 82 82 82 82 82 81 81 81 82 82 82 82
 [76] 82 82 82 82 82 82 82 82 82 82 82 82 82 82 82 82 82 82 82 82 82 82 82 82 82
>
```

If you drill into the slot attributes, you realize it is empty

```
> nlcd@data@attributes
list()
>
```

To obtain an image visualization that shows the classes, we need to assign levels of a data frame to the values and to do so we use `ratify` which creates the Raster Attribute Table (RAT) as we did in Lab session 8

```
nlcd.rat <- ratify(nlcd)
```

Look at the new raster and you will see attributes which have levels named `ID`

```
> nlcd.rat
class       : RasterLayer
dimensions  : 1167, 1167, 1361889  (nrow, ncol, ncell)
resolution  : 30, 30  (x, y)
extent      : 664995, 700005, 3671805, 3706815  (xmin, xmax, ymin, ymax)
crs         : +proj=utm +zone=14 +datum=WGS84 +units=m +no_defs
source      : memory
names       : layer
values      : 11, 95  (min, max)
attributes  :
         ID
 from:  11
   to :  95

>
```

By the way, you can drill into the slot `attributes` of the slot `data` and realize that now it has information in the form of a list

```
> nlcd.rat@data@attributes
[[1]]
     ID
     10    11
2    22
3    24
4    31
5    41
6    42
7    81
8    82
9    90
10   95

>
```

Now, we are going to attach the `class$Label` to the raster as we did in Lab session 8

```
class.vis <- class
names(class.vis) <- c("ID","Label")
# attach Label to levels
levels(nlcd.rat) <- list(class.vis)
nlcd.rat
```

Recall also that we are using `ID` for the first column because that is the name in the RAT and that to visualize an image based on classes, we can use function `levelplot` of package `rasterVis`. Install this package, if you did not in Lab 8, and load it

```
library(rasterVis)
```

and now use `levelplot` to obtain the image shown in Figure 9.8

```
levelplot(knr.rat, main="Unsupervised Classes")
```

Our goal is to build a CART model and for that purpose we will sample the image to build a training set

```
set.seed(99)
samp.nlcd <- sampleStratified(nlcd, size = 200, na.rm = TRUE, sp = TRUE)
```

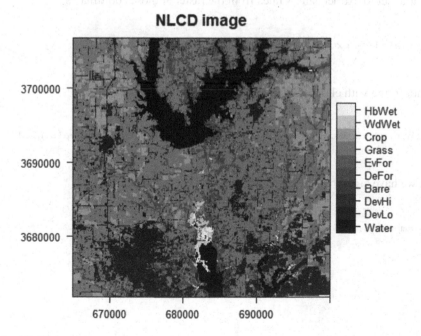

FIGURE 9.8   NLCD image plotted as class.

Verify you have all classes in the sample, note we use @ for slot

```
> unique(samp.nlcd@data[,2])
 [1] 11 22 24 31 41 42 81 82 90 95
>
```

Explore this object so that you can understand its components

```
slotNames(samp.nlcd)
# example
samp.nlcd@coords
```

change name from layer to class to make more readable

```
names(samp.nlcd) <- c("cell", "class")
```

Just before we build the model, let us create and export the point dataset so we could use it later in Chapter 10 when we study geographic information system (GIS) software.

```
ref.train.nlcd <- data.frame(samp.nlcd@coords,samp.nlcd@data$class)
names(ref.train.nlcd) <- c("x","y","val")
write.table(ref.train.nlcd,"data/nt/refnlcd.csv",sep=",",quote=F, row.names=F)
```

Use the spatial set to extract band values from the raster at those coordinates

```
sampvals.nlcd <- extract(nt, samp.nlcd, df = TRUE)[,-1]
```

create a data frame with class and band values

```
sampdata.nlcd <- data.frame(class=samp.nlcd@data$class, sampvals.nlcd)
```

check that we have eight variables, one for class and seven bands

```
> dim(sampdata.nlcd)
[1] 2000    8
>
```

check that the variables have the name for class and the band names

```
> names(sampdata.nlcd)
[1] "class"        "CoastalAerosol" "Blue"        "Green"
[5] "Red"          "NIR"            "SWIR1"       "SWIR2"
>
```

You should already have loaded rpart, but can use require for package rpart just in case it needs to be loaded

```
require(rpart)
```

now we build the CART classifier, noting we use the sampled data

```
cart <- rpart(as.factor(class)~., data=sampdata.nlcd, method = 'class')
```

apply a par function before plotting to ensure it fits in graphics device

```
par(xpd = NA)
```

and plot to obtain Figure 9.9

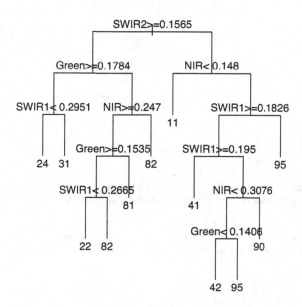

FIGURE 9.9 CART for NT image classification based on NLCD.

```
plot(cart, uniform=TRUE, main="Training CART")
text(cart, cex = 0.8)
```

The nodes are split based on just four bands, Green, NIR (near IR), SWIR1, and SWIR2 (shortwave in two bands), so there are three other bands that do not become a part of the model. The leaves or terminal nodes have the land cover classes according to ID. Since we trained the model, we can predict using the entire dataset

```
pr2019 <- predict(nt, cart, type='class')
```

and use the results to create a data frame with predicted and reference values making sure we remove no data (NA) values

```
prd <- pr2019@data@values
ref <- nlcd@data@values
p <- na.omit(unique(prd))
r <- na.omit(unique(ref))
xy <- data.frame(cbind(prd,ref))
```

armed with the resulting data frame, we generate a confusion matrix and analyze it

```
tb <- table(xy$prd,xy$ref)
err.mat(tb)
```

look at the results

```
> err.mat(tb)
$Po
[1] 0.514

$Pe
[1] 0.204

$kappa
[1] 0.389

$prd.acc
    11    22    24    31    41    42    81    82    90    95
 0.930 0.386 0.442 0.554 0.437 0.504 0.442 0.687 0.227 0.573

$usr.acc
    11    22    24    31    41    42    81    82    90    95
 0.978 0.349 0.224 0.012 0.680 0.052 0.770 0.308 0.178 0.088

$err.co
    11    22    24    31    41    42    81    82    90    95
 0.022 0.651 0.776 0.988 0.320 0.948 0.230 0.692 0.822 0.912
```

```
$err.om
    11     22     24     31     41     42     81     82     90     95
 0.070  0.614  0.558  0.446  0.563  0.496  0.558  0.313  0.773  0.427

>
```

The confusion matrix for this classification results in $Po = 0.514$, $Pe = 0.204$, $\kappa = 0.389$, which does not indicate good accuracy. The errors of commission are all elevated 20%–90% except for water at 2.2%. Similarly, the errors of omission are all elevated 30%–77% except for water at 7%.

It is desirable to mark the terminal nodes with labels instead of ID. To do this, first reassign the levels

```
# change name to keep the original
sampdata <- sampdata.nlcd
# change from NLCD code to label
sampdata$class < factor(sampdata.nlcd$class)
for(i in 1:length(class$Label)){
 levels(sampdata$class)[levels(sampdata$class)==
as.character(class$Class[i])] = class$Label[i]
}
```

And then redo the CART and its plot to obtain Figure 9.10, which shows labels for the leave nodes.

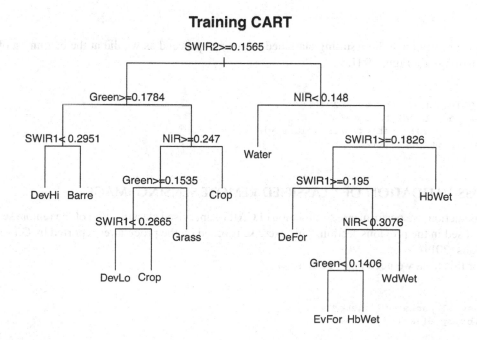

FIGURE 9.10 CART with labels for the classes.

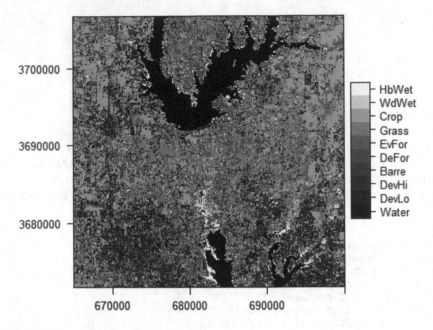

FIGURE 9.11   CART predicted image.

```
# cart with labels
cart.lab <- rpart(as.factor(class)~., data=sampdata, method = 'class')
# do this before plotting to ensure it fits in graphics device
par(xpd = NA)
# plot
plot(cart.lab, uniform=TRUE, main="Training CART")
text(cart.lab, cex = 0.8)
```

We wish to visualize the resulting classified image and proceed as we did at the beginning of this section to obtain Figure 9.11.

```
pr2019.rat <- ratify(pr2019)
# attach Label to levels
levels(pr2019.rat) <- list(class.vis)
levelplot(pr2019.rat)
```

CROSS-VALIDATION OF CLASSIFIED REMOTE SENSING IMAGE

In this section, we perform cross-validation of CART supervised classification of the remote sensing raster used in the previous section. This exercise is based on the procedures reported by Ghosh and Hijmans (2019).

For this task, we need package `dismo`

```
install.packages("dismo")
library(dismo)
```

we select five subsets using `k-fold` and note each subset has 400 observations

```
set.seed(99)
> j <- kfold(sampdata, k = 5, by=sampdata$class)
> table(j)
j
  1   2   3   4   5
400 400 400 400 400
```

We will be calculating several objects for each iteration and therefore we allocate lists, arrays, and matrices to store the results; note the matrices are for ten classes and five iterations

```
x <- list();tab <- list();po <- array(); kappa <- array()
prd.acc <- matrix(ncol=10,nrow=5)
usr.acc <- matrix(ncol=10,nrow=5)
```

and now loop for each `k` to create training and test datasets, run CART on the training set, predict using the test set, create a data frame with predicted and reference, store these in a list, store the confusion matrix analysis in a list, and the metrics in arrays and matrices.

```
for (k in 1:5) {
  train <- sampdata[j!= k, ]
  test <- sampdata[j == k, ]
  cart <- rpart(as.factor(class)~., data=train, method = 'class')
  pclass <- predict(cart, test, type='class')
  #create a data.frame using the reference and prediction
  class.df <- data.frame(test$class, as.character(pclass))
  names(class.df) <- c("Ztest","Zref")
  x[[k]] <- class.df
  tab[[k]] <- err.mat(table(class.df))
  po[k] <- err.mat(table(class.df))$Po
  kappa[k] <- err.mat(table(class.df))$kappa
  prd.acc[k,]<- err.mat(table(class.df))$prd.acc
  usr.acc[k,]<- err.mat(table(class.df))$usr.acc
}
```

We can examine the results; for example, producer accuracy which is a matrix for which rows are the k runs and columns are the classes. We can form a data frame so that we can print the names

```
> prd.acc.df <- data.frame(prd.acc); names(prd.acc.df) <- class$Label
> prd.acc.df
  Water DevLo DevHi Barre DeFor EvFor Grass  Crop WdWet HbWet
1 0.804 0.548 0.458 0.561 0.500 0.630 0.431 0.625 0.765 0.449
2 0.881 0.400 0.576 0.600 0.389 0.400 0.450 0.400 0.619 0.594
3 0.900 0.553 0.625 0.590 0.304 0.471 0.455 0.529 0.410 0.378
4 0.826 0.500 0.543 0.529 0.500 0.568 0.442 0.473 0.459 0.523
5 0.800 0.465 0.605 0.605 0.556 0.435 0.361 0.600 0.684 0.547
>
```

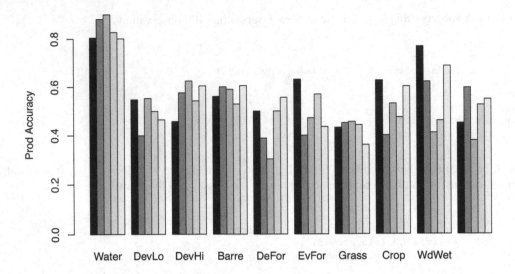

FIGURE 9.12   Producer accuracy by k-fold cross-validation.

Note the best is for water, which is 0.8 or above for all runs. For a graphical display, we can use barplot to obtain Figure 9.12

```
barplot(prd.acc,beside=T,names=class$Label,ylab="Prod Accuracy")
```

We could also calculate the average and standard deviation of the overall accuracy Po

```
> mean(po);sd(po)
[1] 0.5378
[1] 0.02407696
>
```

Another option for analysis is to consolidate all data frames into one using function `do.call` and run confusion analysis on the combined data frame

```
y <- do.call(rbind, x)
y <- data.frame(y)
tb <- table(y)
err.mat(tb)
```

the results of the confusion analysis are

```
> err.mat(tb)
$Po
[1] 0.538

$Pe
[1] 0.1
```

```
$kappa
[1] 0.487

$prd.acc
   11    22    24    31    41    42    81    82    90    95
0.839 0.495 0.566 0.577 0.425 0.492 0.430 0.516 0.541 0.494

$usr.acc
   11    22    24    31    41    42    81    82    90    95
0.940 0.470 0.490 0.525 0.440 0.440 0.475 0.660 0.360 0.580

$err.co
   11    22    24    31    41    42    81    82    90    95
0.060 0.530 0.510 0.475 0.560 0.560 0.525 0.340 0.640 0.420

$err.om
   11    22    24    31    41    42    81    82    90    95
0.161 0.505 0.434 0.423 0.575 0.508 0.570 0.484 0.459 0.506

>
```

Note that the Po of the combined data frame is the same as the average of the values of Po for all runs.

EXERCISES

Exercise 9.1 Applying Bayes' rule.

Apply Bayes' rule to the well water contamination example with $P[A]=0.2$, false positive error of 0.07 and false negative error of 0.03. Calculate the probability of a contaminated well when the test result is negative and of uncontaminated well when the test result is positive.

Exercise 9.2 Ml using naïve Bayes' classifier.

Run the naïve Bayes' classifier using the example file demonstrating how to generate and use training and evaluation datasets.

Exercise 9.3 Binomial test.

Apply the binomial test to the example in the guide with probability of success 0.2. First use 5 successes out of 20 counts, then change to 8 successes out of 20.

Exercise 9.4 Testing Proportions.

Apply the proportion test and Chi-square test to the example in the guide, which assumes we obtain 10, 12, 13 grass pixels (hits) out of 80, 90, 85 examined for each area.

Exercise 9.5 Contingency analysis.

Apply contingency analysis to the examples in the guide. First, use the data in Table 9.1 and then cross-tabulate Ozone and Temp variables of the airquality dataset.

Exercise 9.6 Confusion matrix analysis.

Apply confusion matrix analysis of the example in the guide, that is assume the error matrix corresponds to the data shown in Table 9.2.

Exercise 9.7 Multiple linear regression.

Apply multiple regression analysis to the air quality dataset following the example in the guide. Generate a first model, then practice drop and add operations, and stepwise regression.

Exercise 9.8 Multiple linear regression using ML.

Apply the `train` function of package `caret` to the multiple regression analysis to the air quality dataset following the example in the guide.

Exercise 9.9 Running CART.

Apply CART to the air quality dataset following the example in the guide. Interpret the resulting tree.

Exercise 9.10 Supervisied Calssification of remote sensing image.

Perform supervised classification of the `nt` raster dataset using NLCD as a reference and following the example in the guide. Interpret the resulting tree. Perform cross-validation and discuss the results.

REFERENCES

Acevedo, M.F. 2024. *Real-Time Environmental Monitoring: Sensors and Systems - Textbook, Second Edition*. Boca Raton, FL: CRC Press, Taylor & Francis Group, 392 pp.

Ghosh, A., and R.J. Hijmans. 2019. *Remote Sensing Image Analysis with R*. Berlin/Heidelberg, Germany: Springer, 48 pp.

MRLC. 2022. *NLCD 2019 Land Cover (CONUS)*. Accessed January 2023. https://www.mrlc.gov/data/nlcd-2019-land-cover-conus.

MRLC. 2023. *National Land Cover Database Class Legend and Description*. Accessed January 2023. https://www.mrlc.gov/data/legends/national-land-cover-database-class-legend-and-description.

R Project. 2023. *The Comprehensive R Archive Network*. Accessed January 2023.

10 Databases and Geographic Information Systems

INTRODUCTION

In this session, we will work with databases and database management systems including geographic information systems (GIS). For simplicity, we will learn SQLite connecting to it via R using the package RSQLite. We will work with SQL commands as implemented in SQLite, to create a database, create tables, and insert data. Of particular importance to monitoring, we will see how to create a database from flat files. We also learn the same type of skills using SQLiteStudio, which has a user-friendly GUI. After completion of these exercises, we learn GIS using QGIS using a few examples based on raster and point layers, including how to add raster and vector layers, and conduct operations on these layers, such as nearest neighbor analysis and terrain analysis. We end the lab session with the process of creating a spatial database using SpatiaLite in QGIS.

MATERIALS

READINGS

For theoretical background, you can use Chapter 10 of Acevedo, M.F. 2024. *Real-Time Environmental Monitoring: Sensors and Systems - Textbook, Second Edition* which is a companion to these guides (Acevedo 2024). Other bibliographical references are cited throughout the guide.

SOFTWARE

- R, for data analysis (R Project 2023)
- RSQLite package for R
- SQLiteStudio (SQLiteStudio 2022)
- QGIS (QGIS 2022)

DATA FILES (AVAILABLE FROM THE RTEM GITHUB REPOSITORY)

- Archive windheight.zip for database exercises.
- Archive NT-NLCD-DEM.zip for GIS exercises.

SUPPLEMENTARY SUPPORT MATERIAL

Supplementary support material including additional screenshots, images, and procedures are available from the publisher eResources web page provided for this book.

EXAMPLE: WIND SPEED AND PLANNING A DATABASE

As discussed in Lab session 7, at the UNT Discovery Park campus, we have a weather station consisting of a tower and Campbell Scientific meteorological sensors read by a Campbell Scientific datalogger CR3000 (Campbell Scientific Inc. 2018), and two anemometers, one of them near the top (10 m above the ground) and another a 2 m above the ground. The datalog file (see an example in 7–15 of lab 7 guide) from the CR3000 shows a header containing the variable names.

DOI: 10.1201/9781003184362-10

For wind, the variables corresponding to 10 m are WS _ mph _ Max, WS _ mph _ Min, WS _ mph _ Avg, WS _ mph _ Std, WindDir. The ones corresponding to 2 m are WS _ mph _ 2m _ Max, WS _ mph _ 2m _ Min, WS _ mph _ _ 2m _ Avg, WS _ mph _ 2m _ Std.

We have already learned in Lab 7 to read these files, therefore, to abbreviate the file reading work during this lab session and able to focus on the concept of databases, you can use three files from the RTEM GitHub repository contained in archive windheight.zip. Extract files Wind2m.csv, Wind10m.csv, and TimeStamp.csv to your working directory folder labs/data. Wind speed is given in mph, and wind direction is referenced to $N=0°$ and increasing counterclockwise.

Using a text editor, e.g., Geany or Notepad++, examine the three files TimeStamp.csv, Wind10m.csv, and Wind2m.csv. These files have 27,519 records according to the line counter of Notepad++ or Geany. We will examine the first few records of each. First, TimeStamp.csv consists of RecordID and Timestamp

```
RecordID,Timestamp
1,10/27/2018 10:00
2,10/27/2018 10:10
3,10/27/2018 10:20
4,10/27/2018 10:30
```

then we see that Wind2m.csv has RecordID, followed by maximum, minimum, average, and standard deviation

```
RecordID,WS2mMax,WS2mMin,WS2mAvg,WS2mStd
1,1.909,0,0.112,0.37
2,4.928,0,0.884,1.32
3,6.94,0,3.184,1.346
4,7.443,1.909,4.97,1.19
```

and finally, we see that Wind10m.csv has the same variables as for 2 m but includes direction as well

```
RecordID,WS10mMax,WS10mMin,WS10mAvg,WS10mStd,WDir10m
1,4.625,0,2.77,1.009,233.1
2,5.085,0.767,2.742,1.128,231.5
3,7.716,1.863,4.649,1.118,198
4,8.9,3.332,6.466,1.299,173.3
```

Now, we study how these can be used to create a database. First, they all share the column RecordID; thus, this can be used as PK and FK. A reasonable option is to use it as PK in all files, and as FK (referred to the Timestamp file) to the other two files. A possible design is the Entity Relation Diagram (ERD) based on tables given in Figure 10.1 and the Chen diagram of Figure 10.2. See also Chapter 10 of the textbook.

RSQLITE

SQLite is serverless, meaning that it can store the database in a file directory instead of a server, and therefore it is easy to manipulate data and the database all within your file system (SQLite 2023). RSQLite is an R package that links with SQLite. We will assume that in your R working directory

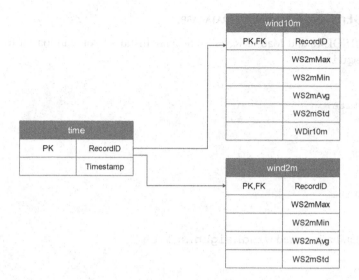

FIGURE 10.1 Simple ERD as tables.

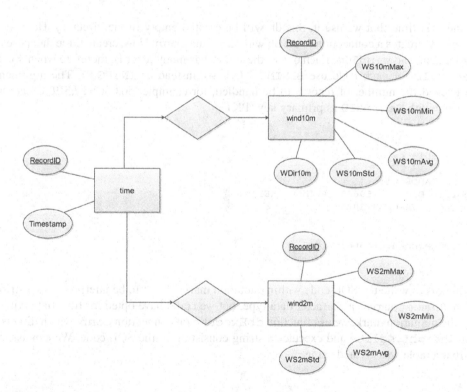

FIGURE 10.2 Simple Chen ERD.

(e.g., `labs`) you have a folder `/data`. Now, we create another folder in labs named `/databases`. In the `labs/data` folder, we store "flat" data files, i.e., files with records in rows such as the datalog files we have encountered thus far, whereas in the `labs/databases` folder we will store the database files, which we will build during the lab session.

INSTALLING RSQLITE AND CREATING A DATABASE

Install package RSQLite and load it. Recall that once installed you can load it for a session using `library` or `require`

```
library(RSQLite)
```

or

```
require(RSQLite)
```

Let us create a database named `windheight.sqlite`

```
# create db
conn <- dbConnect(SQLite(), "databases/windheight.sqlite")
```

Being the first time that we use it, this db will be created empty in the directory. Here, function `dbConnect` creates a connection to the db with the name `conn`. Now, create three database tables: time, wind2m, and wind10m. Each one of these files has many records, therefore when you define the `recordID` for each table use `BIGINT` (size) instead of `INTEGER`. The argument size should exceed the number of records to be handled, for example, 50,000>27,519. Create a table named time with `RecordID` as primary key (PK)

```
createtable <-
"
CREATE TABLE time (
RecordID BIGINT (50000) PRIMARY KEY,
Timestamp VARCHAR(30)
);
"
dbExecute(conn,createtable)
```

The syntax is to write the SQL code within quotation marks " " to be interpreted as a string. We will treat `Timestamp` as a character data type, but we could have opted for time data type. After closing the quotation mark, we use function `dbExecute` on connection `conn`, which directs to the db `windheight.sqlite` and executes a string consisting of the SQL code. We can assure ourselves that a table was created using

```
> dbListTables(conn)
[1] "time"
>
```

The list of the tables indicates that the table was created. Next, create a second table `wind2m`, using the same method, except `RecordID` is a FK in reference to table time as well as a PK. Then, the attributes are declared as real data type,

```
createtable <-
"
CREATE TABLE wind2m (
        RecordID BIGINT (50000) REFERENCES time (RecordID)
                                PRIMARY KEY,
        WS2mMax REAL,
        WS2mMin REAL,
        WS2mAvg REAL,
        WS2mStd REAL
);
"
dbExecute(conn,createtable)
```

Check that we created the table `wind2m` in the database

```
> dbListTables(conn)
[1] "time"   "wind2m"
>
```

Now, create a table for 10 m using the same method

```
createtable <-
"
CREATE TABLE wind10m (
        RecordID BIGINT (100000) REFERENCES time (RecordID)
                                PRIMARY KEY,
        WS10mMax REAL,
        WS10mMin REAL,
        WS10mAvg REAL,
        WS10mStd REAL,
        WDir10m REAL
);
"
dbExecute(conn,createtable)
```

and check that we have the new table

```
> dbListTables(conn)
[1] "time"    "wind10m" "wind2m"
>
```

Now that we have all three tables, we disconnect

```
dbDisconnect(conn)
```

finishing an example of a session creating a database and its tables. Next, we will enter data into the tables in the database.

INPUT DATA TO THE SQLITE DATABASE

In this section, we use the R package RSQLite to input data to the `windheight.sqlite` database and verify that we have indeed entered data using a `dbGetQuery` for each table. In R, a good way to input data from a flat file to a database table is to first create a data frame from the file, and then write the data to the database. Let us see how to do this.

Reconnect to the db

```
conn <- dbConnect(SQLite(), "databases/windheight.sqlite")
```

Create a `data.frame` for timestamp by using `read.table` which we have used often in these lab guides

```
time <- read.table("data/TimeStamp.csv",sep=",",header=TRUE)
```

Once we have the data frame, we write it to the table in the db

```
dbWriteTable(conn, "time", time, overwrite=TRUE)
```

We do the same for the 2 m table

```
wind2m <- read.table("data/Wind2m.csv",sep=",",header=TRUE)
dbWriteTable(conn, "wind2m",wind2m,overwrite=TRUE)
```

and for the 10 m table

```
wind10m <- read.table("data/Wind10m.csv",sep=",",header=TRUE)
dbWriteTable(conn, "wind10m",wind10m,overwrite=TRUE)
```

Check your data input with `dbGetQuery` for each. Shown here are just short segments of the output, and as you can see the output of a `dbGetQuery` is a data frame. First, work on table time,

```
x <- dbGetQuery(conn, "SELECT * FROM time")
x
....
27504     27504     3/28/2019 6:50
27505     27505     3/28/2019 6:55
27506     27506     3/28/2019 7:00
27507     27507     3/28/2019 7:05
27508     27508     3/28/2019 7:10
```

Then, `wind2m`

```
x <- dbGetQuery(conn, "SELECT * FROM wind2m")
x
19983   19983   8.620   4.593   6.706   1.099
19984   19984   9.460   2.748   6.292   1.368
19985   19985   8.110   4.089   6.063   0.989
19986   19986   7.947   3.251   5.274   1.065
19987   19987   8.950   3.419   5.766   1.333
19988   19988   8.280   3.083   5.727   1.259
```

and finally wind 10 m

```
x <- dbGetQuery(conn, "SELECT * FROM wind10m")
x
16658   16658   12.950   6.423   10.140   1.608   3.287
16659   16659   12.470   6.160    8.810   1.752  14.720
16660   16660   13.130   5.984    9.140   1.920 349.100
16661   16661   13.440   7.256   10.460   1.457 348.200
16662   16662   14.310   7.256   11.290   1.655   4.222
```

This concludes our example of DDL (data definition language). Now that we have data in the db tables, we will show in the next section how to perform queries which is an example of DML (data manipulation language).

QUERIES

Use RSQLite's function `dbGetQuery` to perform the following queries and store the results as data frames.

Query 1: Find all records of average wind speed at 10 m that exceed 10 mph

```
w10avg.gt.10 <- dbGetQuery(conn, "SELECT WS10mAvg FROM wind10m where WS10mAvg >10")
```

Query 2: Find all records of average wind speed at 10 m that are <5 mph

```
w10avg.lt.5 <- dbGetQuery(conn, "SELECT WS10mAvg FROM wind10m where WS10mAvg <5")
```

Query 3: Find all records of timestamp and average wind speed at 2 m that exceed 10 mph. For more than one table, you can use a JOIN.

```
query <-"
SELECT Timestamp,
       WS2mAvg
  FROM time
       INNER JOIN
       wind2m
       ON time.RECORDID=wind2m.RECORDID
 WHERE WS2mAvg > 10;
"
t.w2avg.gt.10 <- dbGetQuery(conn,query)
```

Query 4: Find all records of timestamp, average wind speed at 2 m and at 10 m such that wind speed at 10 m exceed 10 mph, and wind direction is southerly between SW and SE. Use two JOIN, one for each table pair.

```
query <- "
SELECT Timestamp,
       WS10mAvg,
       WS2mAvg, WDir10m
  FROM time
       INNER JOIN
       wind10m
       ON time.RECORDID=wind10m.RECORDID
       INNER JOIN
       wind2m
       ON time.RECORDID=wind2m.RECORDID
 WHERE WS10mAvg > 10 and Wdir10m between 135 and 225;
"
t.w2avg.w10.avg.t.10.s <- dbGetQuery(conn,query)
```

And disconnect

```
dbDisconnect(conn)
```

USING QUERY RESULTS FOR ANALYSIS

Now, we use R to plot wind speed at 10 m vs wind speed at 2 m, when wind speed at 10 m exceeds 10 mph, and wind is southerly between SW and SE (Figure 10.3).

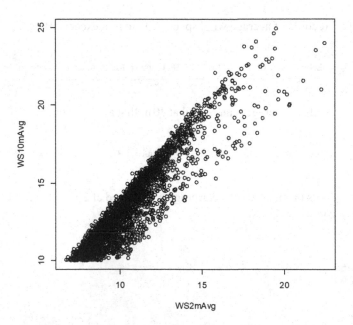

FIGURE 10.3   Wind speed at 10 m vs. wind speed at 2 m when wind speed at 10 m exceeds 10 mph, and wind is southerly between SW and SE.

```
X <- t.w2avg.w10.avg.t.10.s
names(X); attach(X)
plot(WS2mAvg,WS10mAvg)
```

Now, we can detach the dataset

```
detach(X)
```

SIMPLE SQLITE DATABASE CREATED FROM R BY INSERT

First, connect to a database called `example.sqlite`. Being the first time that we use it, this db will be created empty in the directory.

```
# connect to a db, it will create it if does not exist
conn <- dbConnect(SQLite(), "databases/example.sqlite")
```

Recall that function `dbConnect` creates a connection to the db. Now, we will create tables. First, create a table named `location`

```
# create tables
dbExecute(conn,"CREATE TABLE location (
  StationID INTEGER PRIMARY KEY,
  StationName CHAR(45),
  StationLocation CHAR(45));")
```

As you can see, function `dbExecute` acts on connection conn, which directs to the db example. sqlite, and executes a string consisting of SQL code. This code creates a table named location with three columns. One of these, `StationID`, is the primary key.

We can assure ourselves that a table was created using

```
> dbListTables(conn)
[1] "location"
>
```

Next, create a second table, using the same method

```
dbExecute(conn,"CREATE TABLE equipment (
  StationID INTEGER REFERENCES location(StationID)
                    PRIMARY KEY,
  StationEquipment CHAR(45),
  Manufacturer CHAR(45));")
```

Here, `StationID` is a primary key, but it is also a foreign key, which comes from the table location. Check that the table was created

```
> dbListTables(conn)
[1] "equipment" "location"
>
```

Now that we have two tables, we insert some values into each table. Again, we use function dbEx-ecute and the SQL code uses INSERT

```
dbExecute(conn,"INSERT INTO location VALUES (1,'DP','Discovery Park');")
dbExecute(conn,"INSERT INTO location VALUES (2,'GBC','Greenbelt');")
dbExecute(conn,"INSERT INTO location VALUES (3,'EESAT','Main Campus');")
dbExecute(conn,"INSERT INTO equipment VALUES (1,'CR Datalogger','Campbell');")
dbExecute(conn,"INSERT INTO equipment VALUES (2,'SBC-WSN','LowPowerLabs');")
dbExecute(conn,"INSERT INTO equipment VALUES (3,'UVDatalogger','SolarLight');")
```

At this point, we have a db with data, and three records in each table. This concludes our example of DDL, and now that we have data in the db tables, we will show how to perform queries, which is an example of DML.

Let us disconnect from the database, so that it is closed it for now.

```
dbDisconnect(conn)
```

Check your directory labs/databases and you will see that example.sqlite has some content since it is more than 0 bytes.

We will now query the database, using SQL code embedded in function dbGetQuery. First, reconnect to the database so that you can access the tables

```
conn <- dbConnect(SQLite(), "databases/example.sqlite")
```

For example, we can get the entire content of location

```
x <- dbGetQuery(conn,"select * from location;")
```

inspect x

```
> x
  StationID StationName StationLocation
1         1          DP   Discovery Park
2         2         GBC        Greenbelt
3         3       EESAT      Main Campus
>
```

We have obtained a data.frame in R. You already know how to handle a data.frame from previous lab sessions of this manual. For example, the second column is station name

```
> x[,2]
[1] "DP"      "GBC"      "EESAT"
>
```

Let us try a predicate

```
x <- dbGetQuery(conn,"select * from location where StationName like 'D%';")
```

we get

```
  StationID StationName StationLocation
1         1          DP  Discovery Park
>
```

We conclude this exercise. Remember to disconnect

```
dbDisconnect(conn)
```

SQLITESTUDIO

SQLiteStudio is a GUI that facilitates using SQLite. It can be downloaded from (SQLiteStudio 2022) and a tutorial is available online (GC Digital Fellows 2022). Download SQLiteStudio, install it, and open a new session.

ADDING A DATABASE TO SQLITESTUDIO

Start with Database menu item at the left and use Add database. At the dialog box, browse to your labs/databases folder to find your database windheight.sqlite and proceed to add it. We will review the structure of each table to verify that data types and keys have the configuration we specified in the previous exercises. Use the navigation tree on the left-hand side. Start with time, under the Structure tab, check the attributes and data type. You can make changes if needed; for instance, if RecordID shows as INTEGER, we click on it and edit in the dialog window to BIGINT using size 50,000. So, now we have BIGINT (50000) in the Structure tab. Press the green check checkmark to commit these changes. The result is exemplified in Figure 10.4
Repeat the process for the other two tables and recall pressing commit for any changes made. Now, verify that you have data in each table using the Data tab as exemplified in Figure 10.5 for wind at 10 m.

QUERIES USING SQLITESTUDIO

Now, use SQLiteStudio to perform the same four queries we did using RSQLite and export the result of each query to csv, HMTL, and XML files storing them in your /labs/data folder. Go to Tools and Open SQL editor.
Query 1: Find all records of average wind speed at 10 m that exceed 10 mph. Type the query in the editor box and press the blue arrowhead to run (Figure 10.6).
Now, use the export icon (Figure 10.7) and you will see several format options, CSV, HTML, XML

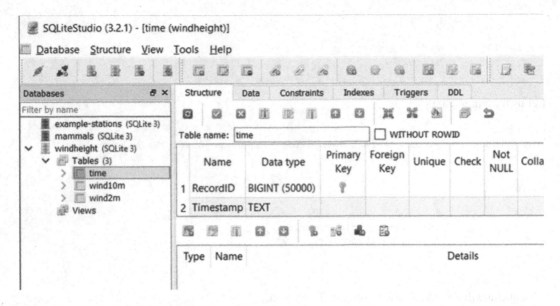

FIGURE 10.4   Checking attributes and data type using SQLiteStudio.

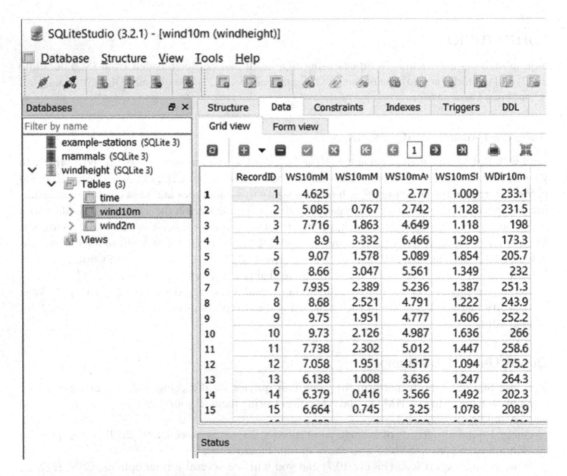

FIGURE 10.5   Example of verifying data contents using the Data tab.

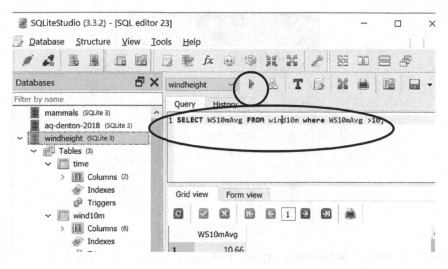

FIGURE 10.6 Example of a query using the Query tab in SQLiteStudio.

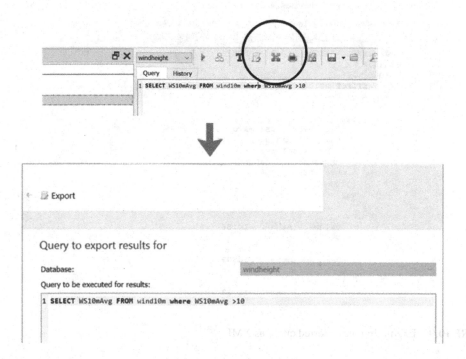

FIGURE 10.7 Exporting the result of a query.

We can export to CSV and analyze using R as we have done in previous sections. Interestingly, we can export as HTML and open it in a browser (Figure 10.8). This would be helpful when communicating your results to others by displaying it on a website.

Also, of interest is to export the query result as XML since it would be helpful in building an XML DB; we can examine the XML file using a text editor (Figure 10.9). Note the tag structure where the `<column>` tags are the attributes or columns of the table and the `<row>` tags are the instances of the table.

You can follow the same procedure for Queries 2, 3, and 4.

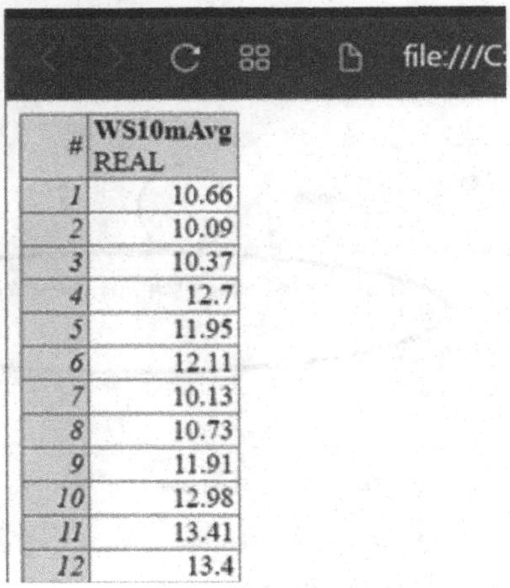

#	WS10mAvg REAL
1	10.66
2	10.09
3	10.37
4	12.7
5	11.95
6	12.11
7	10.13
8	10.73
9	11.91
10	12.98
11	13.41
12	13.4

FIGURE 10.8   Displaying the exported query as HTML.

```
C:\acevedo\courses\rtem\labs-dev\data\query1.xml - Notepad++                    —    □    ×
ile  Edit  Search  View  Encoding  Language  Settings  Tools  Macro  Run  Plugins  Window  ?
  example dtd      example css      example-stations.xml      exampleSQLse R      query1.xml
  1   <?xml version="1.0" encoding="UTF-8"?>
  2   <results>
  3       <query>
  4           SELECT WS10mAvg FROM wind10m where WS10mAvg &gt;10
  5       </query>
  6       <columns>
  7           <column>
  8               <displayName>WS10mAvg</displayName>
  9               <name>>WS10mAvg</name>>
 10               <table>wind10m</table>
 11               <database></database>
 12               <type>REAL</type>
 13           </column>
 14       </columns>
 15       <rows>
 16           <row>
 17               <value column="0">10.66</value>
 18           </row>
 19           <row>
 20               <value column="0">10.09</value>
 21           </row>
 22           <row>
 23               <value column="0">10.37</value>
 24           </row>
```

FIGURE 10.9   Examining the exported query as XML.

CREATING A SIMPLE SQLITE DATABASE FROM SQLITESTUDIO

In the previous sections, we learned how to add and manipulate an existing DB to SQLiteStudio. In this section, we go through the process of creating a database using SQLiteStudio. Start with the `Database` menu item at the left and use `Add database`. At the dialog box, fill names for the file and for the list, e.g., example-stations (Figure 10.10). Checkmark "Databases" in the View menu to be able to see the database created; use the folder icon next to the file name box and navigate to the folder where you have your databases.

Connect to the database using the icon in the upper right corner. Then, create tables as illustrated in the following four figures. In Figure 10.11, we create a column for the new table.

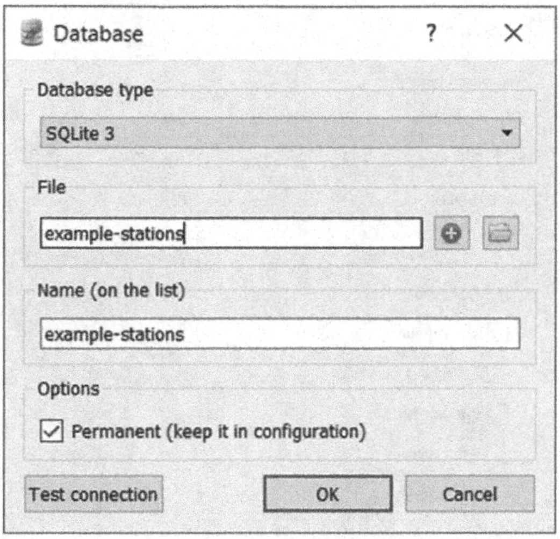

FIGURE 10.10 Creating a new DB.

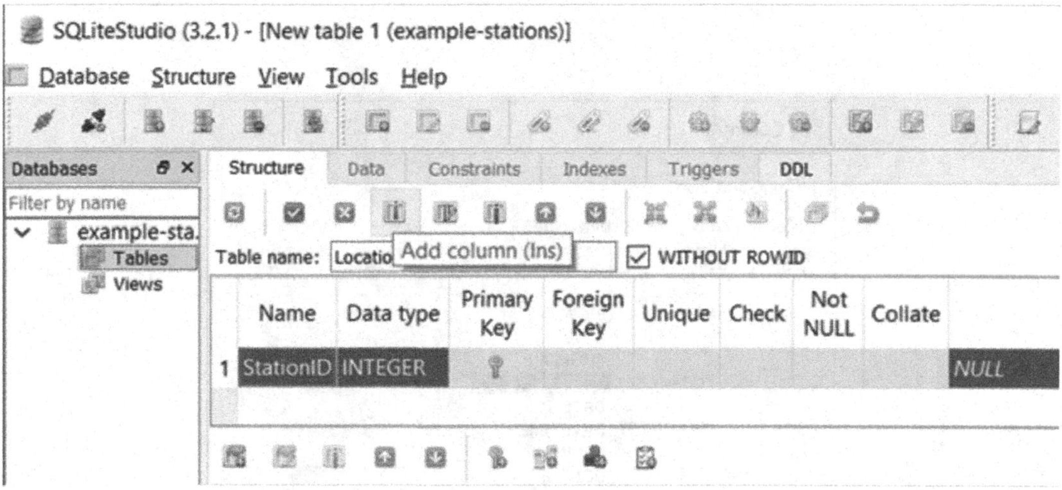

FIGURE 10.11 Creating columns in a new table.

In Figure 10.12, we assign the column as PK. Then, in Figure 10.13, we are creating another column and assigning data type character. Figure 10.14 shows you can start to see the structure being developed in the `Databases Panel` on the left and the columns in the `Structure` tab.

Press the check mark to commit edits. You are shown the corresponding SQL code executed. You can also use the DDL tab to inspect the executed code (Figure 10.15).

Use the same process to create table `Equipment`. Create column `Station ID` that will be a primary key as well as a foreign key (Figure 10.16).

Press the `Configure` button to input settings of the foreign key to get a dialog window in which you can fill the table and column fields (Figure 10.17).

And now in the `Structure` tab you would see the attributes, their data type, and which ones are keys (Figure 10.18). Once we complete these steps, we are ready to add data by going to the data tab and use the plus sign icon.

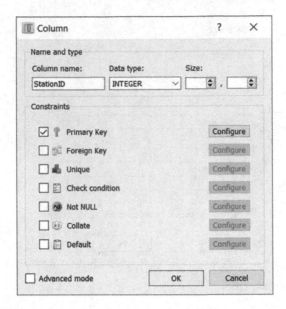

FIGURE 10.12 Creating a column as a primary key.

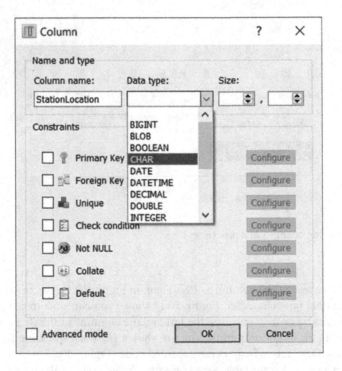

FIGURE 10.13 Creating another column data type string of characters.

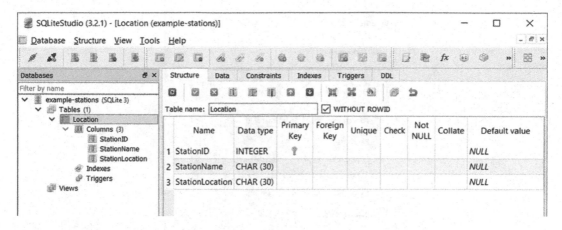

FIGURE 10.14 New table with three columns or attributes.

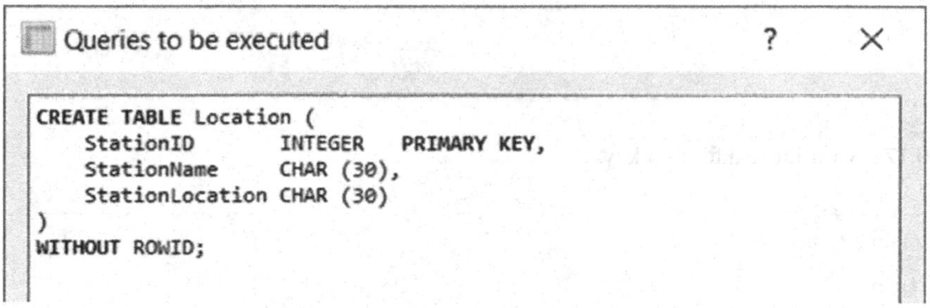

```
Queries to be executed                              ?    ×

CREATE TABLE Location (
    StationID       INTEGER    PRIMARY KEY,
    StationName     CHAR (30),
    StationLocation CHAR (30)
)
WITHOUT ROWID;
```

FIGURE 10.15 SQL code that executed to create this table and its columns.

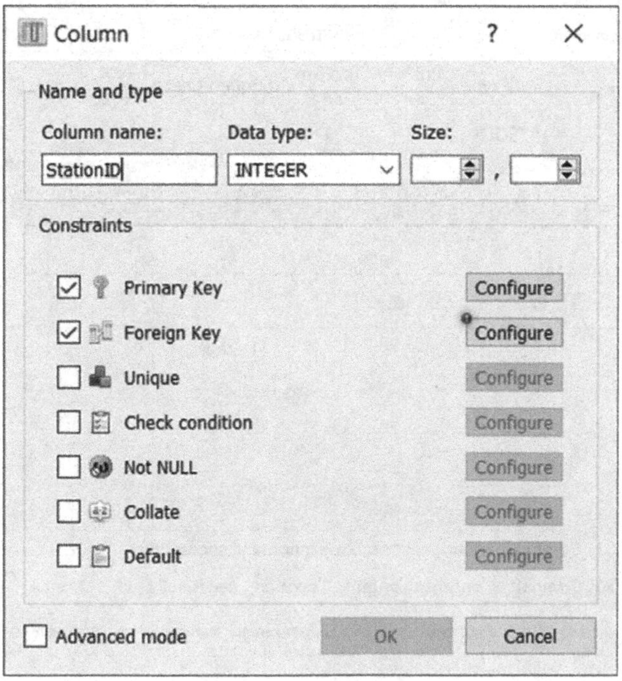

Column   ?   ×

Name and type

Column name: Data type: Size:

StationID   INTEGER   ,

Constraints

☑ 🔑 Primary Key   Configure
☑ 🗝 Foreign Key   Configure
☐ 🔒 Unique   Configure
☐ ☑ Check condition   Configure
☐ 🚫 Not NULL   Configure
☐ 🌐 Collate   Configure
☐ 📋 Default   Configure

☐ Advanced mode   OK   Cancel

FIGURE 10.16 Station ID will be the primary and foreign key.

FIGURE 10.17 Configure a foreign key.

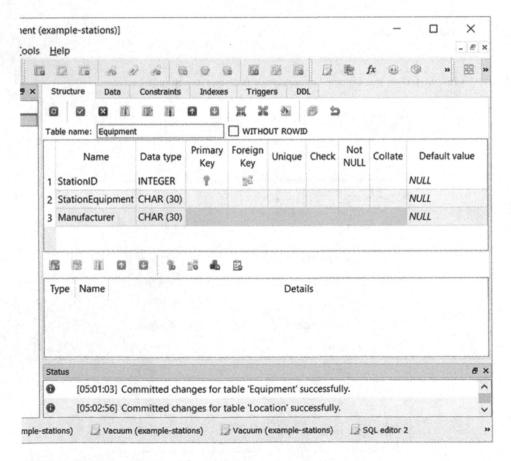

FIGURE 10.18 Table including keys.

Figure 10.19 shows this step for table `Location` and Figure 10.20 shows it for table `Equipment`. Once we have added data, we can perform a query, go to `Tools` menu, and open the SQL editor to type the query (Figure 10.21)

For example,

```
select * from location;
```

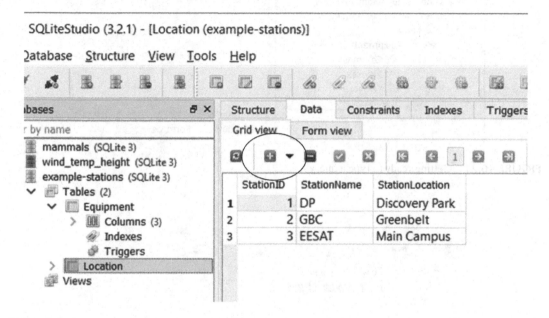

FIGURE 10.19 Add data to Location table. Note the plus sign+icon.

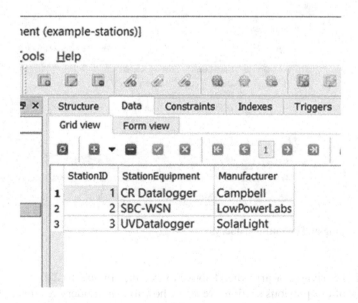

FIGURE 10.20 Add data to equipment.

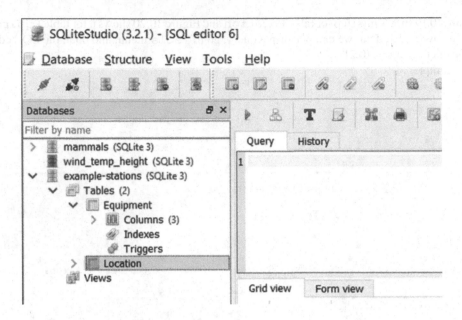

FIGURE 10.21   Getting ready to type a query.

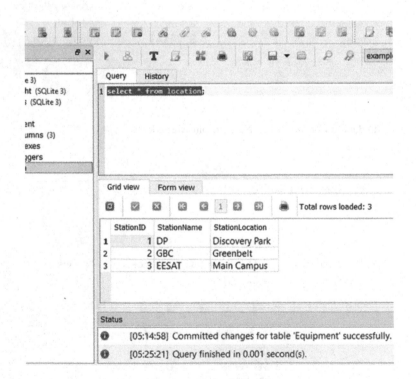

FIGURE 10.22   Typing and executing a query.

Type it and press the triangle or arrowhead above to execute (Figure 10.22)

From the work of our previous section, we know how to export query results and interpret them. Thus, now we take time to learn how to export a DB as a whole or in parts.

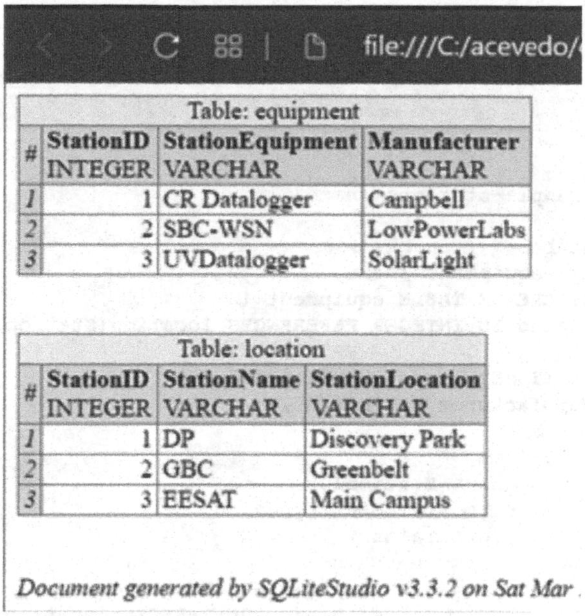

Table: equipment

#	StationID INTEGER	StationEquipment VARCHAR	Manufacturer VARCHAR
1	1	CR Datalogger	Campbell
2	2	SBC-WSN	LowPowerLabs
3	3	UVDatalogger	SolarLight

Table: location

#	StationID INTEGER	StationName VARCHAR	StationLocation VARCHAR
1	1	DP	Discovery Park
2	2	GBC	Greenbelt
3	3	EESAT	Main Campus

Document generated by SQLiteStudio v3.3.2 on Sat Mar

FIGURE 10.23 Exported database as HTML.

Exporting a Database from SQLiteStudio

Once we create a database in SQLiteStudio, we can export it, or only certain objects of the database. Once exported with the desired format, we can use it for a variety of purposes. For example, opening the HTML file in a browser (Figure 10.23) or examine XML code with an editor (Figure 10.24). Note tags for DDL, constraint, type, and definition.

Disconnect from the database when done using icon in upper left-hand side corner just below the Database menu item (Figure 10.25).

GEOGRAPHIC INFORMATION SYSTEMS: USING QGIS

Download QGIS (QGIS 2022) and install. From the RTEM repository download data archive `NT-NLCD-DEM.zip` which has several files, one of these `NLCD-NT-2029.tif` corresponds to National Land Cover Database (NLCD) land cover classes for the North Texas area which we used in Lab 9 as a reference for supervised classification. Extract the `tif` file to your `labs/data/nt/` if you have not yet. This `tif` file was originally downloaded from the USGS NLCD 2019 Products (ver. 2.0, June 2021) (Dewitz and U.S. Geological Survey 2021).

Adding GIS Raster Layers from Raster tif Files

Start QGIS and go to `Layer | Add layer | Add Raster layer` and browse to your `labs/data/nt` and find the `NLCD-NT-2019.tif` file. Click add to close the dialog window. Now you will see the image on the QGIS image area or canvas (Figure 10.26). At the bottom banner, you will see coordinates, scale, etc. and as you move around the image, the coordinates would update.

Now repeat the process to add the `nt20210618 _ ndvi.tif` file that was created in Lab 8 and which will cover the NLCD image as shown in Figure 10.27. Navigate to the list of layers on the bottom left-hand side, you can toggle the layers on and off, using the checkbox on the layers to add or remove from the canvas display (Figure 10.28). In this figure, both NLCD-NT-2019 and

FIGURE 10.24 Exported XML document for DB.

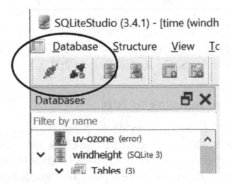

FIGURE 10.25 Connect and disconnect icons.

FIGURE 10.26 Raster `NLCD-NT-2019.tif` added as a raster layer.

FIGURE 10.27 NDVI image added.

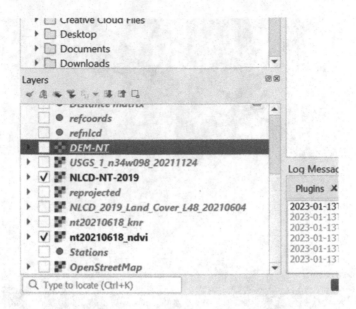

FIGURE 10.28   List of layers showing checkboxes to toggle them on and off.

nt20210618_ndvi are checked. As an example, uncheck `nt20210618 _ ndvi` and you will see the NLCD layer and vice versa.

We can produce a histogram by going to the left-hand side list of layers, right click on the `ndvi` layer, go to properties, then histogram, and then compute histogram (Figure 10.29).

We can see a sharp peak around −0.02 that corresponds to the open water pixels and a broader peak around 0.3 that corresponds to vegetation. A useful tool is an *identifier* (shown as a circle with an *i* and a cursor) (Figure 10.30), which can be used to query information about a pixel directly on the layer. For example, Figure 10.31 displays the value and coordinates of a pixel that was clicked on the `ndvi` layer image.

A convenient feature of QGIS is the ability to roll the mouse button to zoom in and out. For example, zoom in to the Greenbelt Corridor area to the north of Lake Lewisville headwaters to obtain Figure 10.32. You can see more details, such as the Clear Creek joining the Elm Fork of The Trinity River, and the area with higher normalized difference vegetation index (NDVI) values

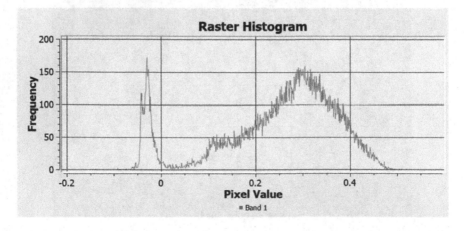

FIGURE 10.29   Histogram of the NDVI raster layer.

FIGURE 10.30   Identify features tool.

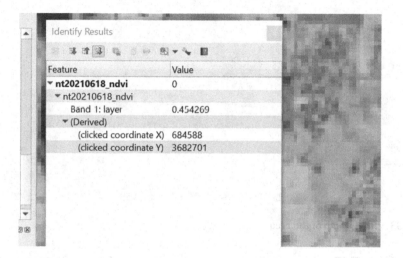

FIGURE 10.31   Example of identifier result.

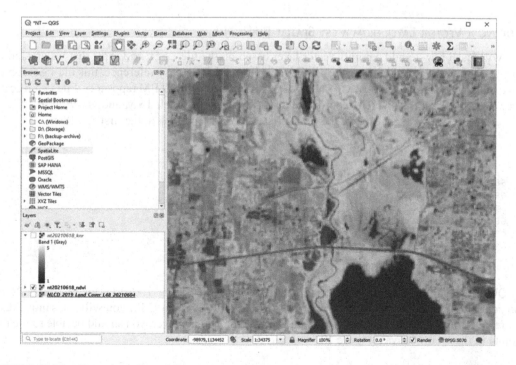

FIGURE 10.32   Zoom in to Greenbelt area in the NDVI layer.

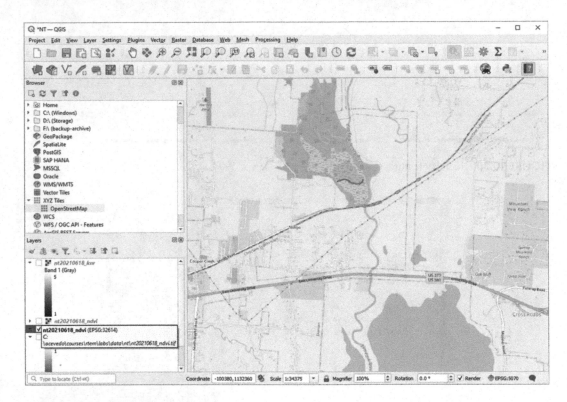

FIGURE 10.33 Street map of the area.

(see Lab 8). Go to XYZ tiles in the left navigation tree and Open Street Map to obtain Figure 10.33. Now, we have a view of the streets and roads of the area in map form.

ADDING A VECTOR LAYER FROM A CSV SPATIAL POINT FILE

We will learn how to add a points vector layer based on the file refnlcd.csv that we created in Lab 9 when we sampled land cover to build a CART model; for easy reference, this file refnlcd. csv is also in the archive NT-NLCD-DEM.zip available from the RTEM repository. File refnlcd. csv has 2000 records of UTM coordinates and NLCD class, labeled x,y, and val. Add this file to labs/data/nt. We can explore the file with a text editor and look at the first few lines.

```
x,y,val
683970,3678180,11
684570,3673320,11
678240,3697440,11
684600,3675600,11
675120,3703980,11
679290,3694590,11
```

In QGIS, we can add this layer by using Layer | Add Layer | Add Delimited Text Layer. In the dialog window, browse to this file, select file format CSV, you will see sample data corresponding to the first few lines of the file. Once you add the layer, you should be able to see the points in the QGIS canvas (Figure 10.34).

FIGURE 10.34   Points vector layer added from a CSV file.

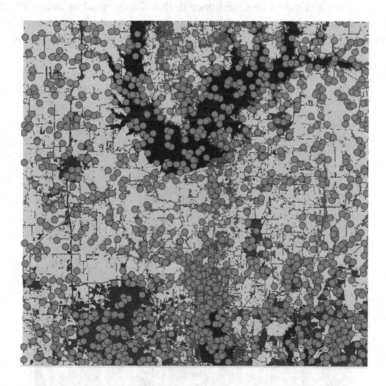

FIGURE 10.35   Sampled points on the NLCD raster layer.

Navigate around the points and use the identifier tool to query several points and obtain the class that corresponds to the point. Navigate to the list of layers on the bottom left-hand side, and check both this layer and the NLCD layer, to visualize both together (Figure 10.35). Note that the sampling points are denser toward the south in the urban areas and Lake Lewisville headwaters.

Raster Analysis: Terrain Calculations

In this section, we will look at an example of raster layer calculations using QGIS, based on a digital elevation model (DEM). The example will be slope aspect calculated from a DEM. For the example, we will use file `DEM-NT.tif` contained in the archive `NT-NLCD-DEM.zip` available from the RTEM repository and which you downloaded for the previous exercise. File `DEM-NT.tif` was obtained by clipping an original image `USGS _ 1 _ n34w098 _ 20211124.tif` downloaded from USGS (2022). Extract the `DEM-NT.tif` file to your `labs/data/nt/`. Then using QGIS, go to `Layer | Add layer | Add Raster layer` and browse to your `labs/data/nt` and find the `DEM-NT.tif` file and press Add, and Close the dialog window. You would have the image on the QGIS canvas (Figure 10.36).

Go to the list of layers on the bottom left-hand side (Figure 10.37), find the DEM-NT and you will see its check marked and has values between 156.4 and 271.6 m of elevation and note that the darker pixels correspond to the lowest elevation, and by looking at the image you would realize these are the lakes. Right click on the DEM-NT layer and select `Properties` (Figure 10.38).

Using the left navigation tool of this dialog window, select `Histogram`, and compute `Histogram` to obtain Figure 10.39. Other than the spikes likely correspond to the lakes, we conclude that elevation is quite uniformly distributed around the image up to about 240 m.

Now we compute an aspect layer by going to `Raster|Analysis|Aspect` and press Run to obtain the aspect image shown in Figure 10.40. For interpretation, go to the Layer navigation, right click on `Aspect`, select `Properties`, and once in that dialog window, select `Histogram`, and compute `Histogram` to obtain Figure 10.41. Note that many pixels are in the 100°–200° range, which would be mostly SE, while NW (~270°–360°) facing slopes are less frequent.

FIGURE 10.36   DEM-NT layer added.

FIGURE 10.37   Layers navigation panel.

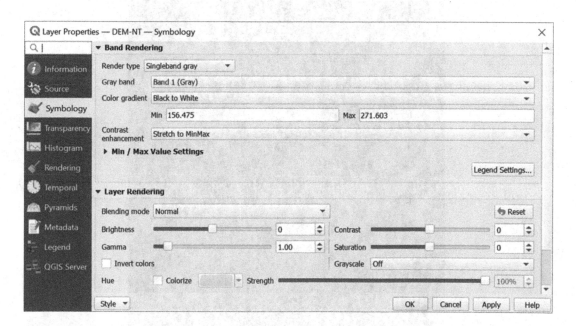

FIGURE 10.38   DEM-NT layer properties.

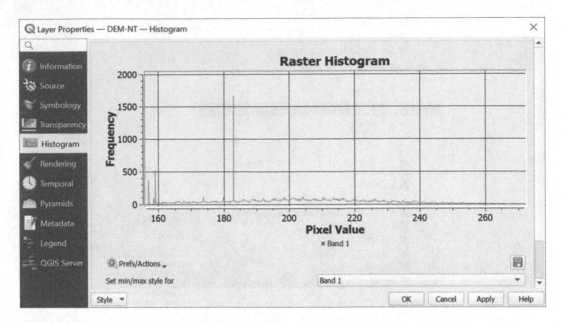

FIGURE 10.39 Histogram of the DEM-NT layer.

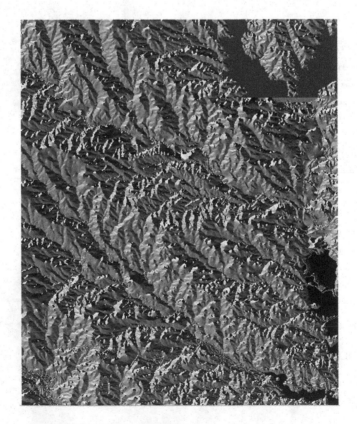

FIGURE 10.40 Aspect slope for the DEM-NT raster.

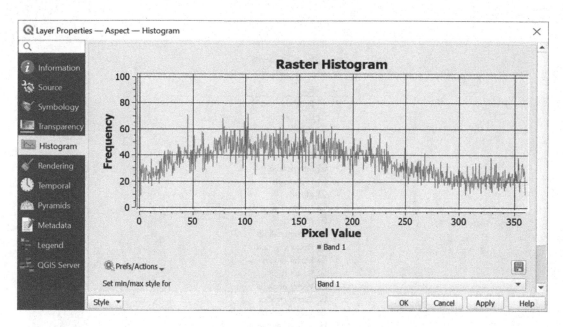

FIGURE 10.41   Aspect layer histogram.

Vector Analysis

As an example of a vector point operation, we generate a point vector layer of 100 points at random within the extent of the NLCD points vector layer. We can do this by going to Vector | Research Tools| Random points in Extent but for the sake of learning how to access a variety of processing tools go to Processing| Toolbox and the menu of options that appear at the right-hand side (Figure 10.42). There are many options that you can browse from this tree. However, when you know what you want to look for, the easiest is to type a word or two in the search box. For instance, if you type "random points" you will see an option for Random points in Extent.

Selecting that option leads to a dialog window where you can select the extent from "calculate from layer", use refnlcd and enter 100 points. Figure 10.43 shows the new layer; using an identifier tool, a click on each point will display the id of the point. You can toggle on both this new layer and refnlcd to realize that these points are much sparse and not so clustered in the southern section of the layer.

We will calculate distance and perform nearest neighbor analysis on this new points layer. Go to the Processing Toolbox, search for distance, and select Distance matrix. In the dialog window, select the input and target point layer as the random point layer you generated. Once completed, in the resulting layer when you query a point you will see the distances to the other points as the attribute of the point. Go back to the Processing Toolbox and select Nearest Neighbour Analysis. Use the distance matrix layer as the input layer, and the results will be displayed in the Log tab of the dialog window.

```
{'EXPECTED_MD': 1723.5889058058535,
 'NN_INDEX': 1.0509986205108715,
 'OBSERVED_MD': 1811.4895623297946,
 'POINT_COUNT': 100,
 'Z_SCORE': 0.9756393577990425}
```

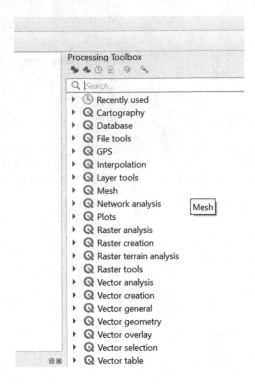

FIGURE 10.42   Processing toolbox.

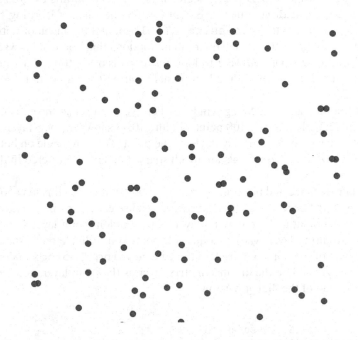

FIGURE 10.43   Example of points layer with point coordinates generated at random within the extent of the refnlcd points layer.

VECTOR AND RASTER ANALYSIS

By checking both the point layer of the last section and the `ndvi` layer, you can visualize both together (Figure 10.44).

Point coordinates can be used to extract the values of a raster layer into a new points layer such that each point has the value of the corresponding pixel of the raster layer. Go to `Processing Toolbox`, search for sample raster, and select `Sample raster values`, for input use the 100 random points layer and for raster use the NDVI raster. The resulting vector layer now has in addition to the coordinates the NDVI values for the points.

CREATING A SPATIAL DATABASE USING QGIS

In the previous sections, we have seen examples of raster and vector operations using QGIS. In this section, we learn to create a spatial database using QGIS.

Go to `SpatiaLite` in the left navigation tree. Right click and select `Create Database` and save in `labs/data/nt` as `nt.sqlite` as illustrated in Figure 10.45.

Now, we will add a layer to this database. You can use the feather icon or go to `Layer | Add Layer | Add SpatiaLite Layer`. A dialog window should pop up (Figure 10.46). At the dialog window (Figure 10.47), we will input `Stations` in the layer name box, `Point` in the geometry type box, keep the EPSG:4326 - WGS 84 reference system, in the `New Field` box, enter `Station _ name`, use `Text data` for Type, and click add to Fields list. In advanced options, select `autoincrement primary key`.

FIGURE 10.44 Points layer visualized together with NDVI layer.

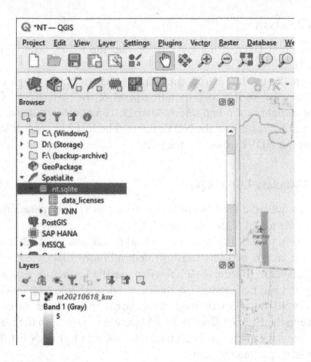

FIGURE 10.45   Creating SpatiaLite database.

FIGURE 10.46   Adding a layer to the SpatiaLite database.

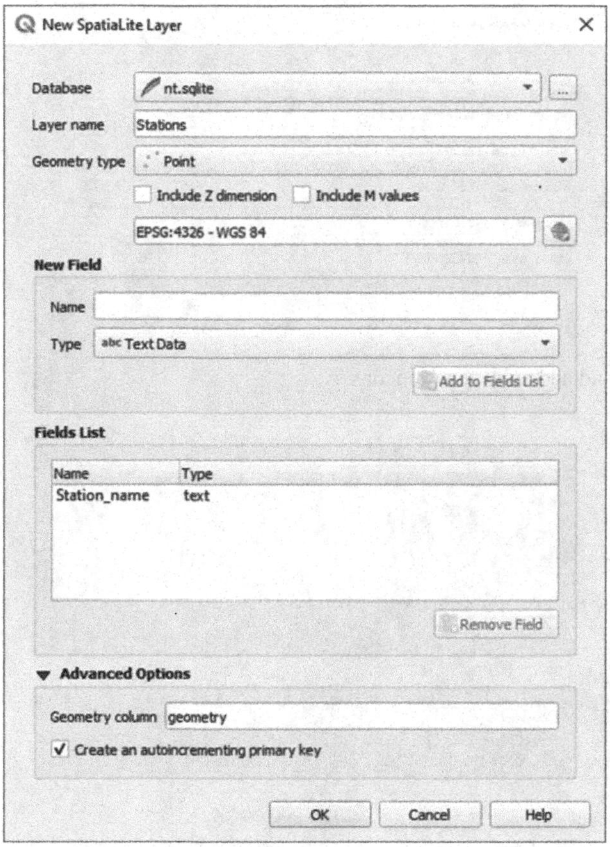

FIGURE 10.47   Adding a table to the layer: start with names.

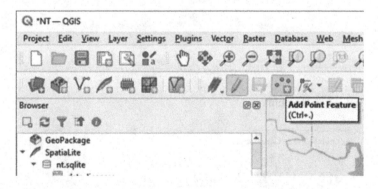

FIGURE 10.48   Fields added.

Add successively Station type, Instruments, Elevation, and Install date as shown in Figure 10.48. Press ok. Now use the pencil icon to start editing and the icon for Add point feature (Figure 10.49). Add a point near to the Clear Creek and Elm Fork confluence. Press ok and the point will be added to the map (Figure 10.50). Add another for a Wireless Sensor Network (WSN) station (Lab 6) just east of the previous one (Figure 10.51) resulting in Figure 10.52. Right click on Stations and save the layer. In this section, we learned how to create a spatial DB and some skills adding vector features such as point data.

Stations - Feature Attributes   [x]

pkuid	Autogenerate	✕	✓
Station_name	Weather Station	✕	
Station_type	*NULL*		
Instruments	CSI 1000	✕	
Elevation	3	✕	
Install_date	7/28/2004	✕	▾

OK   Cancel

FIGURE 10.49   Edit and add point feature icons.

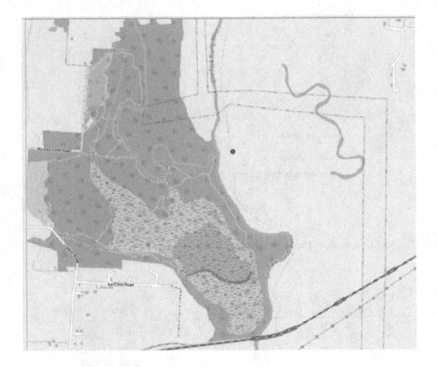

FIGURE 10.50   Dialog to add a point.

Stations - Feature Attributes   [x]

pkuid	Autogenerate	✕	✓
Station_name	Gateway	✕	
Station_type	WSN	✕	
Instruments	Moteino USB and Raspberry Pi	✕	
Elevation	4	✕	
Install_date	3/8/2009	✕	▾

OK   Cancel

FIGURE 10.51   Point added to map.

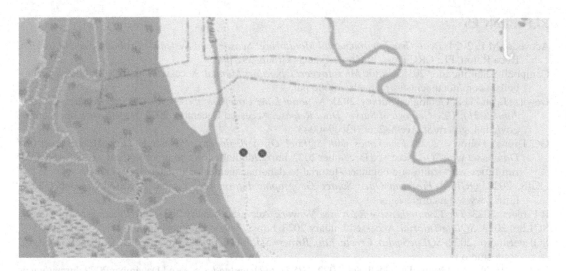

FIGURE 10.52 Add another point for a WSN station.

EXERCISES

Exercise 10.1 Database design.
Use the wind speed dataset to plan a database. Develop ER diagrams and tables.

Exercise 10.2 Creating and using databases using RSQLite.
Install RSQLite and create a database. Input data to the sqlite database. Perform queries and use the query results for analysis. Practice how to create a simple sqlite database created from R by insert.

Exercise 10.3 Using SQLiteStudio.
Install SQLiteStudio and add your database. Perform queries using SQLiteStudio. Create a simple sqlite database created from SQLiteStudio. Export the database example-stations from SQLiteStudio as XML.

Exercise 10.4 Raster and vector layers using GIS.
Use QGIS and follow the steps in the guide to add raster layers from the NLCD and NDVI tif files, visualize, and add a street view map. Use QGIS and follow steps in the guide to add a point vector layer from the refnlcd csv file and visualize it.

Exercise 10.5 Raster analysis in GIS.
Use QGIS and follow steps in the guide to calculate the slope aspect for the DEM-NT layer. Interpret the aspect layer.

Exercise 10.6 Point vector layer analysis in GIS.
Use QGIS and follow the steps in the guide to generate a random set of 100 points, calculate distance, perform nearest neighbor analysis, and extract values from the NDVI raster layer.

Exercise 10.7 Spatial databases in QGIS.
Use QGIS and follow the steps in the guide to add a SpatiaLite database and add points layer.

REFERENCES

Acevedo, M.F. 2024. *Real-Time Environmental Monitoring: Sensors and Systems - Textbook, Second Edition.* Boca Raton, FL: CRC Press, Taylor & Francis Group, 392 pp.

Campbell Scientific Inc. 2018. *CR300 Micrologger: Operators Manual.* Accessed March 2021. https://s.campbellsci.com/documents/us/manuals/cr3000.pdf.

Dewitz, J., and U.S. Geological Survey. 2021. *National Land Cover Database (NLCD) 2019 Products (ver. 2.0, June 2021): U.S. Geological Survey Data Release.* Accessed December 2022. https://www.sciencebase.gov/catalog/item/5f21cef582cef313ed940043.

GC Digital Fellows. 2022. *Fun Times with SQLite! Or, a Beginner's Tutorial to Data Management and Databases with SQL.* Accessed December 2022. https://digitalfellows.commons.gc.cuny.edu/2016/04/08/fun-times-with-sqlite-or-a-beginners-tutorial-to-data-management-and-databases-with-sql/.

QGIS. 2022. *QGIS A Free and Open Source Geographic Information System.* Accessed December 2022. https://www.qgis.org/en/site/.

R Project. 2023. *The Comprehensive R Archive Network.* Accessed January 2023. http://cran.us.r-project.org/.

SQLite. 2023. *SQLite Tutorial.* Accessed January 2023. https://www.sqlitetutorial.net/.

SQLiteStudio. 2022. *SQLiteStudio. Create, Edit, Browse SQLite Databases.* Accessed December 2022. https://sqlitestudio.pl/.

USGS, The National Map - Data Delivery. 2022. *GIS Data Download.* Accessed December 2022. https://www.usgs.gov/the-national-map-data-delivery/gis-data-download.

11 Atmospheric Monitoring

INTRODUCTION

We will practice analysis of atmospheric monitoring data emphasizing (1) analysis and modeling of the dynamics of CO_2 concentration and Earth's temperature, (2) ultraviolet (UV) radiation monitoring and analysis, (3) monitoring of atmospheric gases, focusing on total column ozone as an example, which relates to the UV radiation received at the Earth surface, (4) visualizing vertical profiles of atmospheric gases focusing on ozone, and (5) air quality, using ground level or tropospheric ozone as an example.

MATERIALS

READINGS

For theoretical background, you can use Chapter 11 of Acevedo, M.F. 2024. *Real-Time Environmental Monitoring: Sensors and Systems - Textbook, Second Edition* which is a companion to these guides (Acevedo 2024). Other bibliographical references are cited throughout the guide.

SOFTWARE (LINKS PROVIDED IN THE REFERENCES)

- R system (R Project 2023)
- RSQLite package for R
- SQLiteStudio (SQLiteStudio 2022)
- R package `renpow`
- R package `tibble`

DATA FILES (AVAILABLE FROM THE RTEM GITHUB REPOSITORY)

- Archive `atmospheric.zip`

SUPPLEMENTARY SUPPORT MATERIAL

Supplementary support material including additional screenshots, images, and procedures are available from the publisher's eResources web page provided for this book.

INCREASING ATMOSPHERIC CO_2 CONCENTRATION

An important piece of our knowledge of planetary carbon dynamics comes from the measurement of atmospheric CO_2 concentrations recorded at Mauna Loa, Hawaii (Vaughan et al. 2001; Lovett et al. 2007). A visit to the website of NOAA's Global Monitoring Division (NOAA 2020) will inform us of recent values of monthly average of CO_2 concentration in parts per million (ppm). For example, in July 2020 was 414.38, which is ~3 ppm up from 411.74 ppm for the same month the previous year, July 2019.

Concentration in ppm express dry air mole fraction defined as the number of molecules of CO_2 divided by the number of all molecules in air, including CO_2 itself, after water vapor has been removed (NOAA 2020). The July 2020 value of 414.38 ppm represents a mole fraction of 0.000414. On the website, we can see a graph of CO_2 in ppm as monthly average and its trend (seasonal

DOI: 10.1201/9781003184362-11

correction) for the last five years of record. The trend is calculated by a moving average of seven adjacent seasonal cycles centered on the month to be corrected (NOAA 2019). The last five years trend (left-hand side plot) changes from 402.5 to about 411.3 ppm in five years, which is an average increase of approximately 9/5 ~1.8 ppm/year. Besides the graphs, the website offers the data for download by using the Data tab.

To facilitate the work in this guide, we can use the co2 _ mm _ mmlo.csv file in the atmo-spheric.zip archive of the RTEM repository which has a file with data to 2020. Download this file to folder data in your labs working folder. If you want, you can download more recent data by downloading files from the Mauna Loa CO_2 monthly mean data and store in your labs/data folder as co2 _ mm _ mlo.csv instead of the file in the RTEM repository. First, look at the file using Notepad++ or Geany.

```
year,month,decimal date,average,interpolated,trend,ndays
1958,3,1958.2027,315.70,314.43,-1,-9.99,-0.99
1958,4,1958.2877,317.45,315.16,-1,-9.99,-0.99
1958,5,1958.3699,317.51,314.71,-1,-9.99,-0.99
1958,6,1958.4548,317.24,315.14,-1,-9.99,-0.99
1958,7,1958.5370,315.86,315.18,-1,-9.99,-0.99
1958,8,1958.6219,314.93,316.18,-1,-9.99,-0.99
1958,9,1958.7068,313.20,316.08,-1,-9.99,-0.99
1958,10,1958.7890,312.43,315.41,-1,-9.99,-0.99
```

The file has 52 lines of useful information before the first record (March 1958) in line 52. We will use columns 4 and 5 that have the values of average and interpolation. Using function read.table, we read the downloaded datafile into a dataset or data frame, which is organized by columns and can be manipulated by R functions.

```
CO2.mo <- read.table("data/co2_mm_mlo.csv",skip=52,header=F,sep=',')[,4:5]
names(CO2.mo) <- c("Monthly Avg","Trend")
```

Notice that we skip the first 52 lines of the file and only use columns 4 and 5. We named the two variables accordingly instead of using the header in line 52 for simplicity. The first ten records show us that we have correctly imported the data into a dataset

```
> CO2.mo[1:10,]
   Monthly Avg   Trend
1       315.71  314.62
2       317.45  315.29
3       317.50  314.71
4       317.10  314.85
5       315.86  314.98
6       314.93  315.94
7       313.20  315.91
8       312.66  315.61
9       313.33  315.31
10      314.67  315.61
>
```

We now convert to a time series starting month 3 of 1958 with monthly frequency

```
CO2.mo.ts <- ts(CO2.mo, start=c(1958,3),frequency=12)
```

The first few lines show that we now have the data stamped by a month and year.

```
> CO2.mo.ts
         Monthly Avg    Trend
Mar 1958      315.71   314.62
Apr 1958      317.45   315.29
May 1958      317.50   314.71
Jun 1958      317.10   314.85
Jul 1958      315.86   314.98
Aug 1958      314.93   315.94
Sep 1958      313.20   315.91
Oct 1958      312.66   315.61
Nov 1958      313.33   315.31
```

Now, we plot and obtain Figure 11.1.

```
ts.plot(CO2.mo.ts,type="l",lty=2:1,lwd=1:2,xlab="Year",ylab="CO2 (ppm)")
legend('topleft',leg=names(CO2.mo),lty=2:1,lwd=c(1,2))
```

Figure 11.1 illustrates the CO_2 trajectory for the entire record of measurement (since March 1958) using the data downloaded from this website in August 2020. From the dataset, we plot similar graphs as shown on the website. The dashed line represents the monthly average values (centered on the middle of each month), which fluctuate up and down during the year according to the seasons.

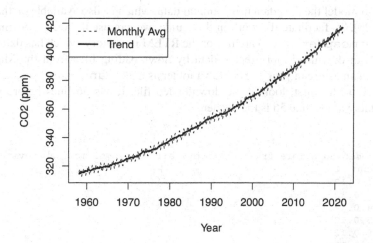

FIGURE 11.1   Monthly mean CO_2 at Mauna Loa. Entire record from 1958 to 2020 plotted using R function `ts.plot`. (Data downloaded from NOAA (2020).)

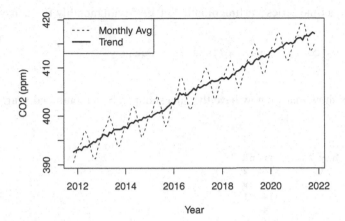

FIGURE 11.2 Monthly mean CO_2 at Mauna Loa, zooming in for period 2010–2020, plotted using R function `ts.plot`. (Data downloaded from NOAA (2020).)

Removing the average of this seasonal cycle yields a solid line that shows a clear accelerating increase during the entire record.

Now, we will plot the same variables using a time window of the most recent ten years (2010–2020) so that we can visualize the graph with more details. We do this by using a time window of the time series from 2010 month 7, `c(2010,7)` to 2020 month 7, `c(2020,7)`, which produces Figure 11.2. You would modify the start and end according to the date that you download the data if you opt to obtain data from the website.

```
recent10yrs <- window(CO2.mo.ts,start=c(2010,7), end=c(2020,7))
ts.plot(recent10yrs,type="l",lty=2:1,lwd=1:2,xlab="Year",ylab="CO2 (ppm)")
legend('topleft',leg=names(CO2.mo),lty=2:1,lwd=c(1,2))
```

NONLINEAR FIT BY TRANSFORMATION

Let us see how to model the CO_2 data using annual data, which is also available on the NOAA website (NOAA 2020). To facilitate the work in this guide, we can use the `co2 _ annmean _ mlo. csv` file in the `atmospheric.zip` archive of the RTEM repository, which has data up to 2020. If you want, you can download more recent data by downloading files from the Mauna Loa CO_2 annual mean data and store in your `labs/data` folder as `co2 _ annmean _ mlo.csv` instead of the RTEM GitHub file. First, look at the downloaded file. It has 56 lines before you get to the columns with data values; line 56 is the header

```
# CO2 expressed as a mole fraction in dry air, micromol/mol, abbreviated as ppm
year,mean,unc
1959,315.98,0.12
1960,316.91,0.12
1961,317.64,0.12
1962,318.45,0.12
1963,318.99,0.12
1964,319.62,0.12
1965,320.04,0.12
1966,321.37,0.12
```

We read the data file into a data frame, calculate time as a difference in years, and calculate the natural logarithm of the ratio of CO_2 concentration at year t to CO_2 concentration at $t = 0$ (initial value at 1959),

```
CO2.yr <-
read.table("data/co2_annmean_mlo.csv",skip=56,sep=',',header=F)[,1:2]
names(CO2.mo) <- c("Year","Annual Avg")
t <- CO2.yr[,1]-CO2.yr[1,1]
y <- log(CO2.yr[,2]/CO2.yr[1,2])
```

and run a linear regression with zero intercept to obtain the slope

```
lm(y ~ 0+ t)
```

The results indicate that the slope is 0.00399 per year and that this is a good estimate (adjusted R^2 is 0.9922, which is close to 1, and the p-value is negligible, 2.2×10^{-16}).

```
> summary(lm(y ~ 0+ t))

Call:
lm(formula = y ~ 0 + t)

Residuals:
      Min        1Q      Median        3Q         Max
-0.0182460 -0.0132109 -0.0089623  0.0003345  0.0268480

Coefficients:
   Estimate Std. Error t value Pr(>|t|)
t 3.990e-03  4.479e-05   89.07   <2e-16 ***
---
Signif. codes:  0 '***' 0.001 '**' 0.01 '*' 0.05 '.' 0.1 ' ' 1

Residual standard error: 0.01247 on 61 degrees of freedom
Multiple R-squared:  0.9924,  Adjusted R-squared:  0.9922
F-statistic:  7934 on 1 and 61 DF,  p-value: < 2.2e-16

>
```

This means a 0.39% per year rate coefficient, which translates to a doubling time of $0.693 / 0.0039 \approx 177$ years. Doubling CO_2 concentration with respect to a reference year is often used as a scenario for climate change assessment and modeling.

However, plotting this log of ratios,

```
plot(t,y, type="l",ylab="ln(ratio)",xlab="Time (Years)",lwd=c(2,1))
abline(lm(y ~ 0+ t),lty=2)
legend('topleft',leg=c("Data", "Model"),lty=1:2,lwd=c(2,1))
```

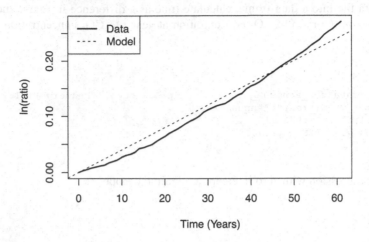

FIGURE 11.3 Determining possible exponential increasing trend of CO_2.

we realize that it cannot be considered a straight line (Figure 11.3). Therefore, our first guess of a linear rate of change is not a good enough approximation, and the rate may be a nonlinear function of the concentration. This implies that the prediction of doubling time would be off. The trajectory before $t \sim 46$ years (~2005) indicates under prediction and thereafter increases faster.

NONLINEAR REGRESSION BY OPTIMIZATION

We just saw that simple exponential growth does not account for data on atmospheric CO_2 increase. Now, we introduce nonlinear regression, which consists of a numerical procedure to minimize the error in the fit of a function to the data. A convenient way of doing this is to use function `nls` of R. We need to start the optimization algorithm from an initial guess for the coefficients. Generally, this requires some additional knowledge of the system. Alternatively, we use the values obtained by transformation as first guess for the algorithm. The initial guess is declared in a `start= list(...)` argument to `nls`.

Use the CO_2 annual data and apply the doubly exponential model. Update the estimate of doubling time. First, use quick trial and error to see that k_1 must be around 0.1 and k_2 around 0.01 so that we have initial estimates of the rate coefficients.

```
t <- CO2.yr[,1]- CO2.yr[1,1]
y <- CO2.yr[,2]; y0 <- y[1]
dexp <- nls(y~ y0*(exp(k1*exp(k2*t))-exp(k1)+1),start=list(k1=0.1,k2=0.01))
summary(dexp)
```

we can look at the nonlinear regression summary

```
> summary(dexp)

Formula: y ~ y0 * (exp(k1 * exp(k2 * t)) - exp(k1) + 1)
```

```
Parameters:
   Estimate Std. Error t value Pr(>|t|)
k1 0.171791   0.004596   37.38   <2e-16 ***
k2 0.014036   0.000313   44.84   <2e-16 ***
---
Signif. codes:  0 '***' 0.001 '**' 0.01 '*' 0.05 '.' 0.1 ' ' 1

Residual standard error: 0.8753 on 60 degrees of freedom

Number of iterations to convergence: 6
Achieved convergence tolerance: 3.892e-06

>
```

Note that the coefficients have values $k_1 = 0.171$ and $k_2 = 0.014$. Now, we predict the values of CO_2 using these coefficients and plot for comparison

```
yest <- predict(dexp)
plot(t,y, type="l",ylab="CO2 (ppm)",xlab="Time (Years)",lwd=1,ylim=c(300,400))
lines(t,yest, lty=2,lwd=1)
legend('topleft',leg=c("Data", "Model"),lty=1:2,lwd=c(1,1))
```

The result shown in Figure 11.4 illustrates a good fit to the data. Then, calculate doubling time using

```
> # doubling time
> k1=0.17; k2=0.014
> log(log(2+exp(k1)-1)/k1)/k2
[1] 108.9817
>
```

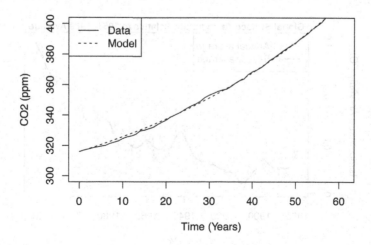

FIGURE 11.4 Modeling CO_2 increase using the doubly exponential.

which means our estimate is now ~109 years instead of 177 years as obtained by the simple exponential. Looking back in time, we may guess that 109 years ago (first decade of the last century)
the concentration would have been half of current values or about 200 ppm. Looking ahead in
time, double CO_2 concentration with respect to 1959 ($\sim 2 \times 315 = 730$ ppm) would be obtained in
year 2068. These are predictions based only on the trend and do not consider modifications due to
changing emissions.

GLOBAL TEMPERATURE: INCREASING TREND

NASA's Global Climate Change website (NASA 2023) has data on global surface temperature since
1880. Global surface temperature refers to the average over land and ocean. The record is expressed
as an anomaly or difference relative to the 1951–1980 average temperature. The dataset up to year
2019 is in file global-temp-anomaly.txt contained in archive atmospheric.zip, which
you can download from the GitHub RTEM repository. The first few records are

```
Land-Ocean Temperature Index (C)
---------------------------------

Year No_Smoothing  Lowess(5)
-------------------------------
1880       -0.16        -0.08
1881       -0.07        -0.12
1882       -0.10        -0.16
1883       -0.16        -0.19
1884       -0.28        -0.23
1885       -0.32        -0.25
```

The first column is the year; the second column is the annual average, and the third column is the
five-year average. If you want, you can download a more recent file from the site (NASA 2020)
instead of the one available from the RTEM GitHub repository.

You can use the following lines of R code to read the data file global-temp-anomaly.txt,
create a dataset, convert to time series, and plot as shown in Figure 11.5.

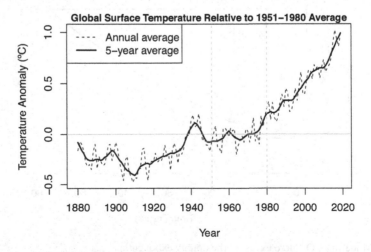

FIGURE 11.5 Global average temperature of Earth during 1880–2020 as anomaly with respect to the 1951–
1980 average. Plotted using R function ts.plot. (Data downloaded from NASA (2023).)

```
gt <- read.table("data/global-temp-anomaly.txt",skip=5,col.
names=c("Year","Avg","Avg5yr"))
gt.ts <- ts(gt[,2:3],start=1880, deltat=1)
ts.plot(gt.ts,type="l",lty=2:1,lwd=1:2,xlab="Year",ylab="Temperature Anomaly (°C)")
abline(h=0, col='gray')
abline(v=1951,lty=2,col='gray'); abline(v=1980,lty=2,col='gray')
legend("topleft",lty=2:1,,c("Annual average","5-year average"), lwd=c(1,2))
title("Global Surface Temperature Relative to 1951-1980 Average",cex.main=0.9)
```

This graph includes the annual average and the five-year average and shows a clear increasing trend and positive anomaly after 1980, that is in the last 40 years, raising to 0.99°C above the 1951–1980 average. The ten warmest years in the examined record have occurred in the period 2000–2020 (except 1998).

Does the global temperature data show an exponential increase? We can proceed as we did for CO_2. We add an arbitrary positive value (+1.0) to the anomaly to make all anomaly values positive

```
t <- gt[,1]-gt[1,1]
gtp <- gt[,3]+1
y <- log(gtp/gtp[1])
summary(lm(y ~ 0+ t))
```

The results indicate an estimate of $k = 0.0025$ per year, with poor R^2 (0.43) but significant (negligible p-value).

```
> summary(lm(y ~ 0+ t))

Call:
lm(formula = y ~ 0 + t)

Residuals:
     Min       1Q   Median       3Q      Max
-0.52122 -0.23618 -0.15218  0.02203  0.41489

Coefficients:
   Estimate Std. Error t value Pr(>|t|)
t 0.0025657  0.0002463   10.42   <2e-16 ***
---
Signif. codes:  0 '***' 0.001 '**' 0.01 '*' 0.05 '.' 0.1 ' ' 1

Residual standard error: 0.2343 on 139 degrees of freedom
Multiple R-squared:  0.4385,  Adjusted R-squared:  0.4344
F-statistic: 108.5 on 1 and 139 DF,  p-value: < 2.2e-16

>
```

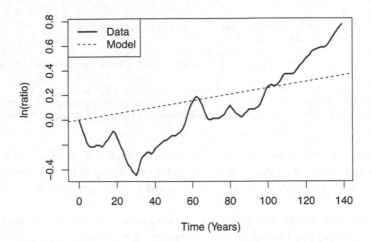

FIGURE 11.6 Determining possible exponential increase of global temperature.

This means a 0.25% per year rate coefficient, which translates to a doubling time of $0.693 / 0.0025 \approx 277$ years. However, we know the R^2 is not very good. Moreover, by plotting this log of ratios,

```
plot(t,y, type="l",ylab="ln(ratio)",xlab="Time (Years)",lwd=c(2,1))
abline(lm(y ~ 0+ t),lty=2)
legend('topleft',leg=c("Data", "Model"),lty=1:2,lwd=c(2,1))
```

we realize that a straight line is not at all a good estimate (Figure 11.6). Therefore, the rate may be a nonlinear function of temperature. This implies that the prediction of doubling time would be off, and the doubling time could be shorter. The trajectory after $t \sim 100$ years (~1980) indicates a faster increase.

Apply the doubly exponential to the global temperature data.

```
# global temp data
gt <- read.table("data/global-temp-anomaly.txt")
names(gt) <- c("Year", "Annual Avg", "5-year Avg")
# time in years
t <- gt[,1]-gt[1,1]
# shift up to make positive
gtp <- gt[,3]+1
y <- gtp; y0 <- y[1]
# non linear doubly exp
dexp <- nls(y~ y0*(exp(k1*exp(k2*t))-exp(k1)+1),start=list(k1=0.1,k2=0.01))
```

In this case, the coefficient values are $k_1 = 0.0068$, $k_2 = 0.034$ as we can see from the summary

```
> summary(dexp)

Formula: y ~ y0 * (exp(k1 * exp(k2 * t))) - exp(k1) + 1)
```

```
Parameters:
    Estimate Std. Error t value Pr(>|t|)
k1 0.006800   0.002298   2.958  0.00364 **
k2 0.034515   0.002581  13.372  < 2e-16 ***
---
Signif. codes:  0 '***' 0.001 '**' 0.01 '*' 0.05 '.' 0.1 ' ' 1

Residual standard error: 0.1363 on 138 degrees of freedom

Number of iterations to convergence: 19
Achieved convergence tolerance: 3.621e-06

>
```

Estimate and plot

```
yest <- predict(dexp)
# shift back down by 1
ya <- y-1; yaest <- yest-1
plot(t,ya, type="l",ylab="Temperature Anomaly (°C)",xlab="Time (Years)",lwd=1,
ylim=c(-1,1))
lines(t,yaest, lty=2,lwd=1)
legend('topleft',leg=c("Data", "Model"),lty=1:2,lwd=c(1,1))
abline(h=0, col='gray')
```

to get Figure 11.7. Then calculate doubling time

```
# doubling time
k1=0.0068; k2=0.0345
log(log(2+exp(k1)-1)/k1)/k2
> log(log(2+exp(k1)-1)/k1)/k2
[1] 134.1803
>
```

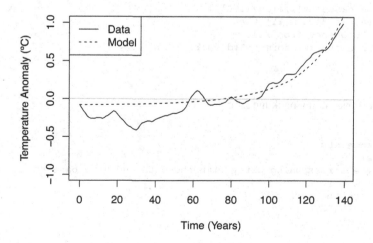

FIGURE 11.7   Modeling temperature increase using the doubly exponential.

which means ~134 years instead of ~277 years we obtained when using the simple exponential. To put this in perspective, we would double the current anomaly (~+0.99°C) of 2020 entering the next century.

UV RADIATION MONITORING

Besides measurements of total irradiance as learned in Lab session 7, it is of interest to measure other aspects related for example, to the spectral distribution of solar radiation. We installed a UV-B detector (UV biometer) on the roof of EESAT building, UNT main campus, which operated during 1999–2017. For convenience, in 2013 the original datalogger was interfaced to a Raspberry Pi using serial communications. The data are given in Minimal Erythema Dose per hour (MED/ hours). MED is "…the amount of UV radiation that will produce minimal erythema (sunburn or redness caused by engorgement of capillaries) of an individual's skin within a few hours following exposure" (Heckman et al. 2013). Sample files are included in the `atmospheric.zip` archive posted in the RTEM GitHub repository.

In this archive, a file named `uv-data-30min2006-2007.csv` contains UV data collected from the UV biometer. Store this file under `labs/data`. Examine the file using Notepad++ or Geany. It is formatted for two detectors, but this file has data for only one detector. The first fee records show

```
Date,Time,SUV_Det1,UVA_Det2,Temp1,Temp2
10.07.2006,10:30,    0.857,    0.000,    25,    0
10.07.2006,11:00,    1.121,    0.000,    25,    0
10.07.2006,11:30,    1.368,    0.000,    25,    0
10.07.2006,12:00,    1.608,    0.000,    25,    0
10.07.2006,12:30,    1.800,    0.000,    25,    0
10.07.2006,13:00,    1.584,    0.000,    25,    0
10.07.2006,13:30,    1.481,    0.000,    25,    0
```

The timestamp is given in two columns, one for date and the other for time. We will merge and apply `strptime` to form a POSIX time stamp that will be combined with the third column as a time series

```
# read uv data
uv <- read.table("data/uv-data-30min2006-2007.csv", sep=",", header=TRUE)
# merge first 2 cols into 1
tstamp <- array()
for(i in 1:dim(uv)[1])
tstamp[i] <- paste(uv[i,1],uv[i,2])
# note format is "%d.%m.%Y" then
# read time sequence from file
tt <- strptime(tstamp, format="%d.%m.%Y %H:%M",tz="")
xx <- uv[,3]
```

We make a time series using package `xts`

```
# load package xts
library(xts)
# create time series using array with the data and time base tt
x.t <- xts(xx,tt)
# give name to the series
names(x.t) <- c("UV")
```

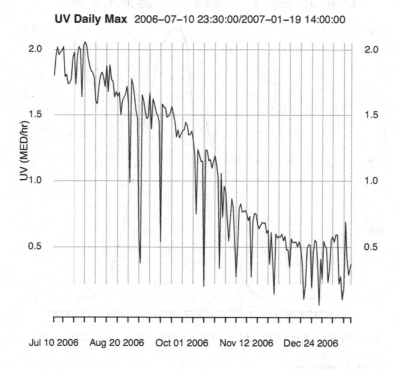

FIGURE 11.8 Daily maximum of UV measured at roof of EESAT building UNT campus.

Having the time series, we can calculate the maximum for each day and plot it for the period 2006–2007 vs. time (Figure 11.8)

```
# daily max
x.d.max <- apply.daily(x.t,max)
# plot the entire series of the max daily
plot(x.d.max,ylab="UV (MED/hr)", main="UV Daily Max",lwd=1)
mtext("UV Roof EESAT UNT",3,-1)
```

We can focus on the dynamics within a day by splitting the time series in days

```
x.t.days <- split(x.t, f="days")
```

and look at a summer day, say July 12, or day 3 in the record, and plot to obtain Figure 11.9

```
# plot day i
# july 12 2006
i=3
plot(x.t.days[[i]], ylab="UV (MED/hr)", main=paste("UV Hourly ", "day=",i))
mtext("UV Roof EESAT UNT",3,-1)
```

Similarly, for a day in the winter, say Jan 18 or day 193 to obtain Figure 11.10

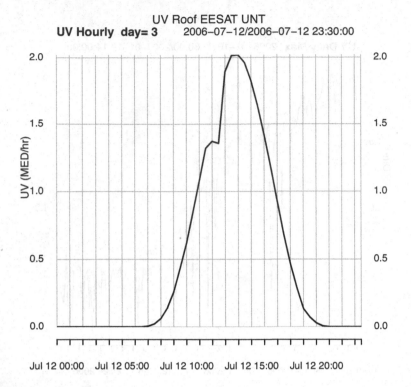

FIGURE 11.9   UV in a summer day.

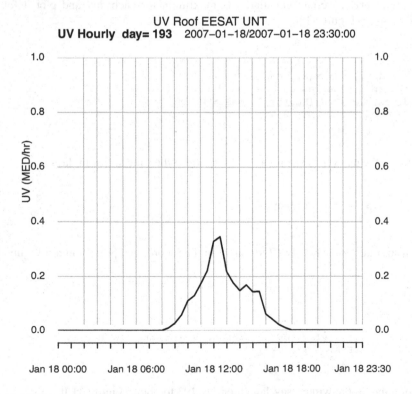

FIGURE 11.10   UV in a winter day.

```
# jan 18 2007
i=193
plot(x.t.days[[i]],ylim=c(0,1),ylab="UV (MED/hr)", main=paste("UV Hourly ",
"day=",i))
mtext("UV Roof EESAT UNT",3,-1)
```

ATMOSPHERIC GASES: TOTAL COLUMN OZONE UNT MONITOR

At the roof of EESAT building, UNT main campus, we also operated a total column ozone monitor that we developed during 1999–2014 based on fiber optics and using optical absorption spectroscopy (OAS) and differential OAS (DOAS) methods (Nebgen 2006; Jerez 2011). This instrument was then adapted to be deployed outdoors, and using an Arduino and a Raspberry Pi (Faschingbauer et al. 2014). Sample files are included in the `atmospheric.zip` archive posted in the RTEM repository. In this archive, the file named `TCO-2011-04-05-2012-11-10.csv` has daily data from 2011 and 2012 collected from this instrument and compared to NASA satellite-borne TOMS (Total Ozone Mapping Spectrometer).

Store the file under `labs/data` and examine it with Notepad++ or Geany. The first few records are

```
TIMESTAMP,UNT-EESAT,NASA-TOMS
"2011-04-05 00:00:00",283.0,292.0
"2011-04-06 00:00:00",302.0,290.0
"2011-04-07 00:00:00",293.0,295.0
"2011-04-08 00:00:00",280.0,277.0
"2011-04-09 00:00:00",259.0,NA
"2011-04-11 00:00:00",306.0,NA
"2011-04-12 00:00:00",277.0,280.0
"2011-04-13 00:00:00",287.0,NA
"2011-04-14 00:00:00",339.0,315.0
```

We can read the data file,

```
# read Total Column Ozone  data
tco <- read.table("data/TCO-2011-04-05-2012-11-10.csv", sep=",", header=TRUE)
tt <- strptime(tco[,1], format="%Y-%m-%d %H:%M:%S",tz="")
xx <- tco[,2:3]
```

create a time series

```
# load package xts
library(xts)
# create time series using array with the data and time base tt
x.t <- xts(xx,tt)
```

give name to the variables in the series

```
names(x.t)
```

select a time window

```
# select March 2011 to Dec 2011
w.t <- x.t["2011-3/2011-12"]
```

and plot to obtain Figure 11.11.

```
plot(w.t,ylab="Ozone (ppm)", multi.panel=T, ylim=c(200,400), main="Total
Column", col=1, lwd=1)
mtext("Roof EESAT UNT",3,-3)
```

You can observe that there is good correspondence between the two sets of values. The time series obtained from the ground instrument fills gaps in the TOMS time series.

TOTAL COLUMN OZONE: USING DATA AVAILABLE ON THE WEB

In this section, we use products of Environment and Climate Change Canada, Toronto (n.d.). World Meteorological Organization-Global Atmosphere Watch Program (WMO-GAW)/World Ozone and Ultraviolet Radiation Data Centre (WOUDC), retrieved from the WOUDC (2023) website. Select Total Ozone plots, station Punta Arenas, Chile, and "All" for instruments and year. Select for

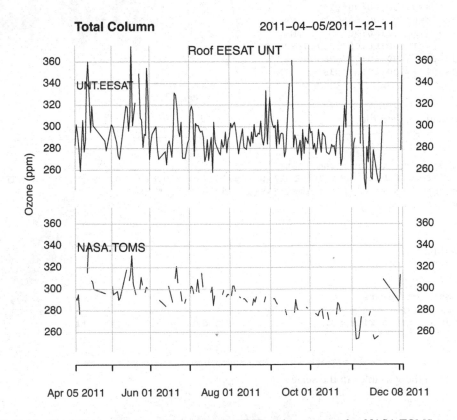

FIGURE 11.11   Total Column Ozone measured by the UNT station compared to NASA TOMS.

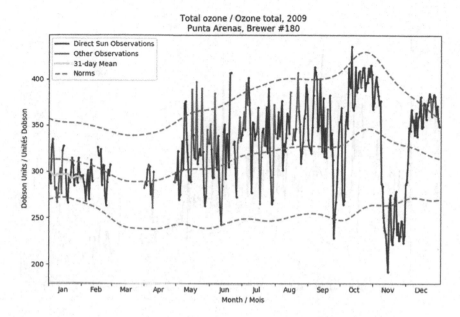

FIGURE 11.12 Total ozone. Punta Arenas, Chile, 2009. (Credit: Environment and Climate Change Canada, Toronto (n.d.). WMO-GAW/WOUDC. Retrieved January 2023 from WOUDC (2023).)

example Year 2009, which is obtained with a Brewer instrument. See Figure 11.12. You can observe increased ozone in October, a spring month in the southern hemisphere (opposite to the spring in the northern hemisphere).

Here, Ozone is given in Dobson units (DUs) defined in terms of what the thickness of a layer of pure ozone would be at the ground if the total column of ozone was reduced to this layer at standard conditions of temperature and pressure; 100 DU correspond to an ozone layer of 1 mm at 1 atm and 0°C (Graedel and Crutzen 1993).

Now, select a mid-latitude site in the northen hemisphere site. Say, Fort Collins, Colorado, USA, available for 1965 using a Dobson instrument. See Figure 11.13. See how the highest values still occur in the spring (March–April for the northen hemisphere).

VERTICAL PROFILES

Profiles of atmospheric concentration of various gases are measured by sondes launched from the ground and that also measure meteorological variables at increasing altitude (km) or decreasing air pressure (hPa) intervals. In this section, we also use products of Environment and Climate Change Canada, Toronto (n.d.). WMO-GAW/WOUDC, retrieved from the WOUDC (2023) website.

Select Ozonesonde plots to obtain data of ozone as a function of altitude. Choose South Pole and one sounding for 1987 in the summer. The partial pressure of ozone (leftmost line) is maximum at about 20 km altitude and has a value of about 15 mPa. The tropopause is at about 8 km (Figure 11.14).

Choose a mid-latitude northern hemisphere, say Palestine, Texas, USA, and one sounding for 1975 in the winter. The partial pressure of ozone (leftmost line) is maximum at about 20 km altitude and has a value of about 14 mPa. The tropopause is at about 17 km (Figure 11.15).

In addition to retrieving products, the WOUDC (2023) offers data download that we can use to take a closer look at the data or do our plots and analysis. As an example, we will retrieve a vertical profile dataset and process it using *R*. In this case, from the home page you can go to Data Search / Download, select Ozonesonde, country USA, station Hilo (HI) (109), instrument

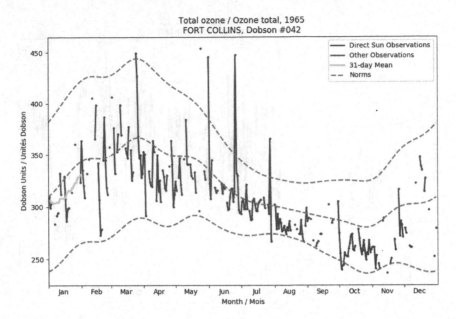

FIGURE 11.13   Total ozone. Fort Collins, Colorado, USA, 1965. Dobson instrument. (Credit: Environment and Climate Change Canada, Toronto (n.d.). WMO-GAW/WOUDC. Retrieved January 2023 from WOUDC (2023).)

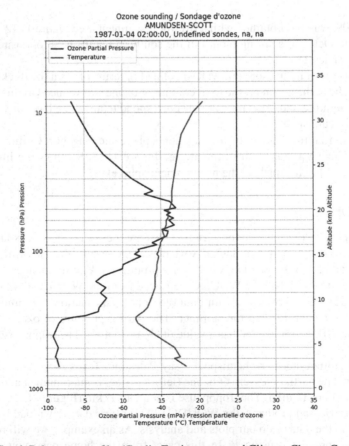

FIGURE 11.14   South Pole ozone profile. (Credit: Environment and Climate Change Canada, Toronto (n.d.). WMO-GAW/WOUDC. Retrieved January 2023 from WOUDC (2023).)

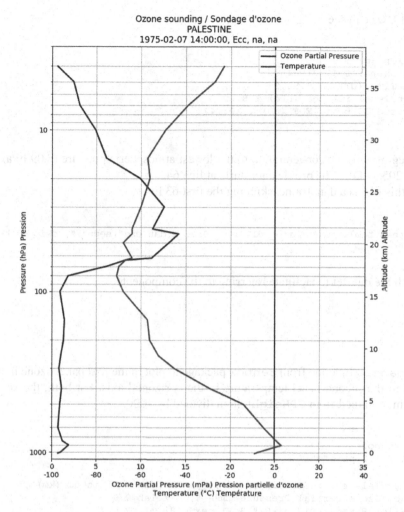

FIGURE 11.15 Ozone profile obtained at Palestine, Texas, USA. (Credit: Environment and Climate Change Canada, Toronto (n.d.). WMO-GAW/WOUDC. Retrieved January 2023 from WOUDC (2023).)

Ecc, and go to Search, and pick 2017-01-04 as an example. You would download a file `hilo _ 20170104 _ V05.1 _ R.dat.csv` and store in `labs/data`. This file contains data on ozone and temperature profiles along with meteorological conditions obtained by a sonde launched at Hilo, Hawaii, on 2017-01-04. Examining the contents of this file, we see from lines 1 to 62 important information about the dataset, including the origin and investigators, for example, lines 33–37

```
#DATA_GENERATION
Date,Agency,Version,ScientificAuthority
2017-08-10,NOAA-CMDL,05.1,Bryan Johnson (NOAA/ESRL Global Monitoring
Division. USA)
* SHADOZ Principal Investigator    : Anne M. Thompson (NASA/GSFC, USA)
* Station Principal Investigator(s): Bryan Johnson (NOAA/ESRL Global
Monitoring Division, USA)
```

In lines 15–19, we can see

```
* Elevation (m) : 11.0
* Launch Date : 2017-01-04
* Launch Time (UT) : 18:58:59
* Highest level reached (hPa) : 8.09
* Integrated O3 until EOF (DU) : 205.47
```

Note the integrated ozone concentration to the lowest atmospheric pressure (8.09 hPa) reached by the sonde is 205.47 DU. The profile data starts at line 64.

We read this file as a data frame, skipping the first 63 lines

```
prof <- read.table("data/hilo_20170104_V05.1_R.dat.csv",sep=",", skip=63,header=T)
```

and we attach the dataset to facilitate referring to the components.

```
attach(prof)
```

and use the `panels` function from `renpow` package to plot in the first panel ozone in the horizontal axis and in the second panel temperature in the horizontal axis. For both, the vertical axis is altitude in km, obtained from `GPHeight` in m divided by 1000.

```
require(renpow)
panels(6,6,1,2,pty="m")

plot(O3PartialPressure,GPHeight/1000,type="l",ylab="Altitude (km)",
    xlab="Ozone Partial Pressure (mPa)", cex.lab=0.8)
mtext("Hilo, Hawaii,2017-01-04",3,-1,cex=0.8)

plot(Temperature,GPHeight/1000,type="l",ylab="Altitude (km)",
    xlab="Temperature (°C)", cex.lab=0.8)
mtext("Hilo, Hawaii,2017-01-04",3,-1,cex=0.8)
```

The result is shown in Figure 11.16. You will note how temperature starts increasing and ozone increasing at about 17 km. As always remember to detach when you are done.

```
detach(prof)
```

AIR QUALITY: USING DATA AVAILABLE ON THE WEB

In this part of the lab session, you will download air quality data from the web, use R to create databases, use SQLite to perform queries, export, and use R to plot and perform statistics. For this purpose, we will use the EPA website Interactive Map of Air Quality Monitors (U.S. EPA 2023). Select click to launch the `AirData Map App` and inspect the dialog window with a "To use this map" to list of operations you can do.

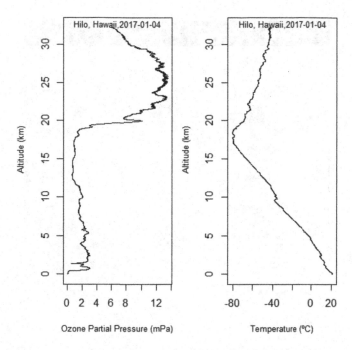

FIGURE 11.16 Ozone and temperature profiles for Hilo, Hawaii, USA, 2017-01-0. (Data downloaded from Environment and Climate Change Canada, Toronto (n.d.). WMO-GAW/WOUDC. Retrieved January 2023 from WOUDC (2023).)

FIGURE 11.17 Select a data layer.

Select `layer` (Figure 11.17) and choose `Ozone-Active` (Figure 11.18). Find Denton, Texas, and select station at "`Denton Airport South`". Inspect the popup window and select daily data for 2018 (Figure 11.19).

Download file `daily_48_121_0034_2018.csv` to your `labs/data` folder. Inspect the file using Notepad++ or Geany. The first line is a header

```
"State Code","County Code","Site Number","Parameter
Code","POC","Latitude","Longitude","Datum","Parameter Name","Duration
Description","Pollutant Standard","Date (Local)","Year","Day In Year
(Local)","Units of Measure","Exceptional Data Type","Nonreg Observation
Count","Observation Count","Observation Percent","Nonreg Arithmetic
Mean","Arithmetic Mean","Nonreg First Maximum Value","First Maximum
Value","First Maximum Hour","AQI","Daily Criteria Indicator","Tribe
Name","State Name","County Name","City Name","Local Site Name","Address","MSA
or CBSA Name","Data Source"
```

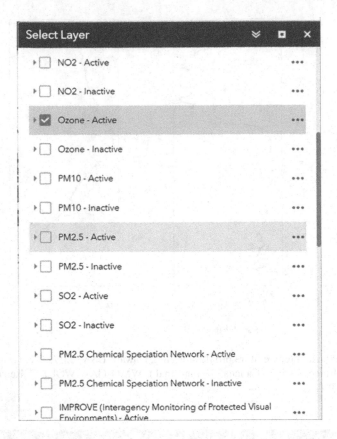

FIGURE 11.18 Finding monitors and variables.

FIGURE 11.19 Finding Denton Airport station and available datasets.

Note that there are spaces and parentheses in the labels. Starting in the second line each record is like this

```
"48","121","0034","44201","1","33.219069","-97.196284","WGS84","Ozone","1
HOUR","Ozone 1-hour 1979","2018-01-01","2018","001","Parts per million","None
","","24","100.0000","","0.027083","","0.032","15",".","Y","","Texas","Denton
","Denton","Denton Airport South","Denton Airport South","Dallas-Fort Worth-
Arlington, TX","AQS Data Mart"
```

Note that you have a mix of string or character variables together with numeric. Our goal is to import the dataset into RSQLite. Thus, we use R to read the file, preserving spaces in names (check.names=FALSE), then convert spaces to underscore, and getting rid of parenthesis. These extra steps are needed so that we do not have spaces, dots, or parenthesis in column names for SQLite.

```
aqsfile <- "data/daily_48_121_0034_2018.csv"
aqsdf <- read.table(aqsfile, header=TRUE, sep=",",check.names=FALSE)
names(aqsdf) <- gsub(" ", "_", names(aqsdf))
names(aqsdf) <- gsub("\\(", "", names(aqsdf))
names(aqsdf) <- gsub("\\)", "", names(aqsdf))
```

Next, we subset the data frame

```
set1 <- c(1:3,6:7,27:34)
set2 <- c(4,9:26)
aqsdf[,set2]
names(aqsdf[,set2])
```

verify the converted names

```
> names(aqsdf[,set2])
 [1] "Parameter_Code"              "Parameter_Name"
 [3] "Duration_Description"        "Pollutant_Standard"
 [5] "Date_Local"                  "Year"
 [7] "Day_In_Year_Local"          "Units_of_Measure"
 [9] "Exceptional_Data_Type"       "Nonreg_Observation_Count"
[11] "Observation_Count"           "Observation_Percent"
[13] "Nonreg_Arithmetic_Mean"      "Arithmetic_Mean"
[15] "Nonreg_First_Maximum_Value"  "First_Maximum_Value"
[17] "First_Maximum_Hour"          "AQI"
[19] "Daily_Criteria_Indicator"
>
>
```

Convert to sqlite db with name db/aq-denton-2018.sqlite in your working directory

```
# using SQlite
library(RSQLite)
db <- dbConnect(SQLite(), "db/aq-denton-2018.sqlite")
dbWriteTable(db, "aqdenton2018", aqsdf[,set2], overwrite=TRUE)
dbDisconnect(db)
```

Now use SQLiteStudio, add a database and read from file db/aq-denton-2018.sqlite. Inspect the structure and the data (Figure 11.20). Verify that all is correct.

Select tools, open SQL editor, and perform a query (see Figure 11.21)

```
SELECT Day_In_Year_Local,Arithmetic_Mean, First_Maximum_Value,First_Maximum_Hour
FROM aqdenton2018
WHERE Parameter_Name = "Ozone" AND
Duration_Description = "1 HOUR";
```

Now, we get ready to export the results of the query (Figure 11.22) Check the format and options (Figure 11.23)

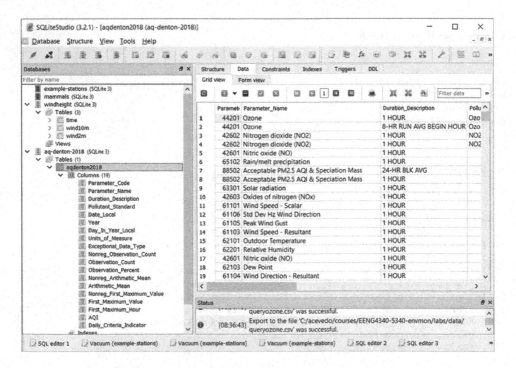

FIGURE 11.20 Using SQLStudio to inspect the database.

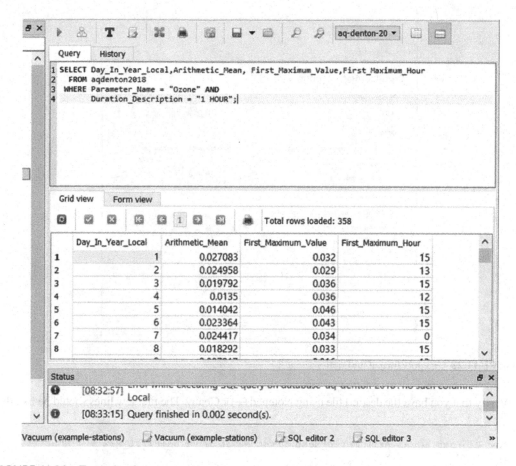

FIGURE 11.21 Example of query.

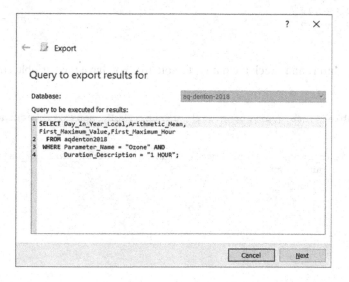

FIGURE 11.22 Exporting the query.

FIGURE 11.23 Export format and options.

Verify that you have the desired file using Notepad++ or Geany. The first few lines should be like this

```
Day_In_Year_Local,Arithmetic_Mean,First_Maximum_Value,First_Maximum_Hour
1,0.027083,0.032,15
2,0.024958,0.029,13
3,0.019792,0.036,15
4,0.0135,0.036,12
5,0.014042,0.046,15
6,0.023364,0.043,15
```

Now, read the file into R and check the names to note that you have the variables in positions 1 to 4,

```
x <- read.table("data/queryozone.csv", header=TRUE, sep=",",check.names=FALSE)
names(x)
> names (x)
[1] "Day_In_Year_Local"   "Arithmetic_Mean"       "First_Maximum_Value"
[4] "First_Maximum_Hour"
>
```

We can also use `tibble` to check the result. Use `require(tibble)` if needed

```
> as_tibble(x)
# A tibble: 358 × 4
   Day_In_Year_Local Arithmetic_Mean First_Maximum_Value First_Maximum_Hour
           <int>           <dbl>              <dbl>                 <int>
 1               1         0.0271               0.032                    15
 2               2         0.0250               0.029                    13
 3               3         0.0198               0.036                    15
 4               4         0.0135               0.036                    12
 5               5         0.0140               0.046                    15
 6               6         0.0234               0.043                    15
 7               7         0.0244               0.034                     0
 8               8         0.0183               0.033                    15
 9               9         0.00792              0.016                    12
10              10         0.0168               0.026                    15
# … with 348 more rows
# i Use 'print(n = ...)' to see more rows
>
```

Use `require(renpow)` if needed to make sure is loaded and use its function `panels` to make two horizontal panels

```
panels(6,6,2,1)
```

and use `matplot` for a graph of avg and first max and plot() for the Hour first max (see Figure 11.24)

```
matplot(x[,1],x[,2:3],type="l",xlab="Day",ylab="Ozone(ppm)")
legend("topright",lty=1:2,col=1:2,leg=c("Avg","1st Max"))
plot(x[,1],x[,4],type="p",xlab="Day",ylab="Hour 1st Max")
```

This section should have helped you realize that there are many layers of air quality data available for analysis, become acquainted how to retrieve air quality data, how to build your own databases using the data, how to make queries, and process the results.

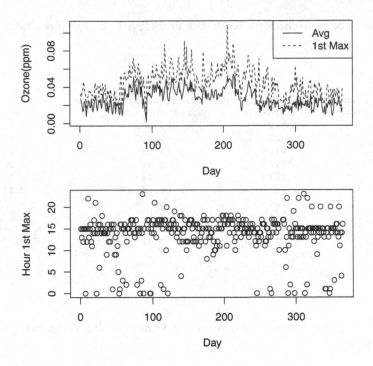

FIGURE 11.24   Ozone plots using R.

EXERCISES

Exercise 11.1 Nonlinear regression using R.

   Model annual CO_2 and global temperature data using nonlinear regression. Show your code and results.

Exercise 11.2 UV monitoring data.

   Analyze UV vs. time data acquired at the roof of EESAT building UNT campus as described in the lab guide.

Exercise 11.3 Total column ozone monitoring.

   Analyze total column ozone data acquired by the UNT monitor as described in the lab guide.

Exercise 11.4 Atmospheric ozone vertical profiles.

   Create graphs of total ozone dynamics and ozone profiles using data from WOUDC as described in the guide.

Exercise 11.5 Air quality databases.

   Create and query an air quality database using RSQLite and SQLiteStudio as described in the lab guide.

REFERENCES

Acevedo, M.F. 2024. *Real-Time Environmental Monitoring: Sensors and Systems - Textbook, Second Edition.* Boca Raton, FL: CRC Press, Taylor & Francis Group, 392 pp.

Faschingbauer, A., J. Stumberg, and T. Eminger. 2014. Instrument Panel Design – Automated Ozone Monitor. Senior Design, BSEE, Electrical Engineering, University of North Texas.

Graedel, T.E., and P.J. Crutzen. 1993. *Atmopsheric Change: An Earth System Perspective.* New York: W.H. Freeman, 446 pp.

Heckman, C.J., R. Chandler, J.D. Kloss, A. Benson, D. Rooney, T. Munshi, S.D. Darlow, C. Perlis, S.L. Manne, and D.W. Oslin. 2013. Minimal Erythema Dose (MED) testing. *Journal of Visualized Experiments: JoVE* 75:e50175. doi: 10.3791/50175.

Jerez, C. 2011. Measuring atmospheric ozone and nitrogen dioxide concentration by differential optical absorption spectroscopy. PhD dissertation (Environmental Sciences), University of North Texas.

Lovett, G.M., D.A. Burns, C.T. Driscoll, J.C. Jenkins, M.J. Mitchells, L. Rustad, J.B. Shanley, G.E. Likens, and R. Haeuber. 2007. Who needs environmental monitoring? *Frontiers in Ecology and the Environment* 5(5):253–260.

NASA. 2023. *Vital Signs Ocean Warming.* Accessed January 2023. https://climate.nasa.gov/vital-signs/ocean-warming/.

NASA. 2020. *Global Climate Change. Vital Signs of the Planet.* Accessed August 2020. http://climate.nasa.gov/vital-signs/global-temperature/.

Nebgen, G. 2006. Automated low cost instrument for measuring total column ozone. Ph.D. Dissertation, Environmental Science, University of North Texas.

NOAA. 2019. *Trends in Atmospheric Carbon Dioxide.* NOAA, Earth System Research Laboratory, Global Monitoring Division. Accessed September 2019. http://www.esrl.noaa.gov/gmd/ccgg/trends/.

NOAA. 2020. *Trends in Atmospheric Carbon Dioxide.* NOAA, Earth System Research Laboratory, Global Monitoring Division. Accessed August 2020. http://www.esrl.noaa.gov/gmd/ccgg/trends/.

R Project. 2023. *The Comprehensive R Archive Network.* Accessed January 2023. http://cran.us.r-project.org/.

SQLiteStudio. 2022. *SQLiteStudio. Create, Edit, Browse SQLite Databases.* Accessed December 2022. https://sqlitestudio.pl/.

U.S. EPA. 2023. *Interactive Map of Air Quality Monitors.* Accessed January 2023. https://www.epa.gov/outdoor-air-quality-data/interactive-map-air-quality-monitors.

Vaughan, H., T. Brydges, A. Fenech, and A. Lumb. 2001. Monitoring long-term ecological changes through the Ecological Monitoring and Assessment Network: Science-based and policy relevant. *Environmental Monitoring and Assessment* 67:3–28.

WOUDC. 2023. *World Ozone and Ultraviolet Radiation Data Centre, Data Products.* Accessed January 2023. https://woudc.org/data/products/.

12 Water Monitoring

INTRODUCTION

This lab session analyzes datasets on water quantity and quality monitoring, including surface water and groundwater. For surface water quantity, the datasets relate to water level and flow monitoring, starting with current conditions available on the web for stream stage, stream discharge, and hydrographs, as well as rating curves. Then, we learn historical data retrieval using an R package such that retrieved data go immediately into an R data frame. For groundwater, the datasets are also collected from the web and include well water level below the surface, and these are used to practice time series analysis and prediction using autoregressive (AR) models. For water quality, the guide emphasizes datasets collected by datasondes immersed in streams and lakes.

MATERIALS

READINGS

For theoretical background, you can use Chapter 12 of Acevedo, M.F. 2024. *Real-Time Environmental Monitoring: Sensors and Systems - Textbook, Second Edition* which is a companion to these guides (Acevedo 2024). Other bibliographical references are cited throughout the guide.

SOFTWARE

- R system (R Project 2023)
- RSQLite package for R
- SQLiteStudio (SQLiteStudio 2022)
- `dataRetrieval` package for R (USGS 2023a; De Cicco et al. 2022)
- R package tibble

DATA FILES (AVAILABLE FROM THE RTEM GITHUB REPOSITORY)

- Archive `2018 _ Upper _ Pecan _ Creek.zip`
- Archive `MBOWN-238-1985-2023.zip`

SUPPLEMENTARY SUPPORT MATERIAL

Supplementary support material including additional screenshots, images, and procedures are available from the publisher eResources web page provided for this book.

SURFACE WATER QUANTITY MONITORING: CURRENT CONDITIONS AND STATISTICS

In this section, we use streamflow data from the U.S. Geological Survey (USGS) available from USGS (2023d) (Figure 12.1). This example was developed for data corresponding to the week of April 13–18, 2019. Whenever selecting *current* conditions on the website, keep in mind that the data will vary according to the date that you are conducting this exercise.

Select `Current Streamflow` from the left navigation list or from the maps. This would take you to a window titled "`Map of real-time streamflow compared to historical`

DOI: 10.1201/9781003184362-12

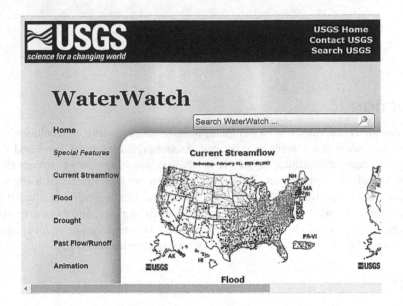

FIGURE 12.1 Current Streamflow on USGS Water Watch web page (USGS 2023d).

streamflow for the day of the year (United States)", where you can search for the stream gage or station of interest (Figure 12.2) by state, region, clicking on the map, or by USGS stream gage number.

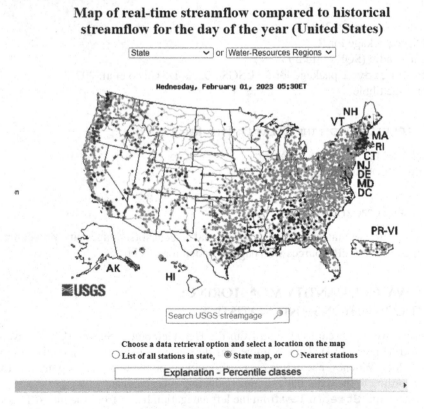

FIGURE 12.2 Web page to select stream gage (USGS 2023d).

WaterWatch Streamflow Map
Choose a region and then click "GO" to view a regional map
(Warning: It may take several minutes to process)

Map of real-time streamflow compared to historical streamflow for the day of the year

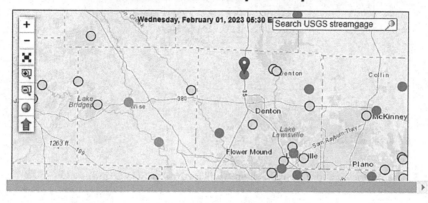

FIGURE 12.3   Location Clear Creek (USGS 2023d).

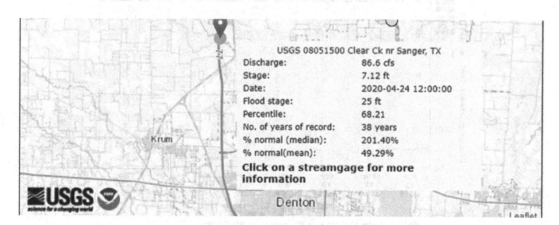

FIGURE 12.4   Details for the station (USGS 2023d).

As an example, we will work with a stream gage near Sanger, Texas, USA, that monitors Clear Creek. Its number is 08051500; you can enter this number in the `Search USGS Streamgage` box to find it. See the result in Figure 12.3. Display the information of the gage showing a time stamp. Note that the number of years of records at the time was 38 (Figure 12.4).

Now select the hydrograph tab (Figure 12.5). Click on the graph to obtain an image of the hydrograph. This example corresponds to April 12–18, 2019 (Figure 12.6). Data will vary according to the date you conduct this exercise. The flow is given in cubic feet per second, abbreviated as cfs; 1 cfs is 0.0283 m³/s. The flow in April 11 and 12 is about 50 cfs (~1.4 m³/s), which is close to the median daily for the 38 Years of record, shown by triangle marks. The flow rose sharply to a peak on April 13 to about 1000 cfs (~28 m³/s) then receded after a small peak to 100 cfs (~2.8 m³/s) and

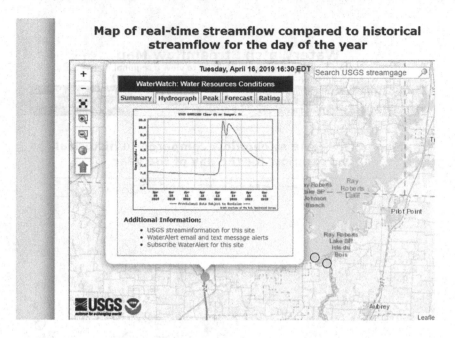

FIGURE 12.5   Real-time data for Clear Creek on April 16, 2019 (USGS 2023d).

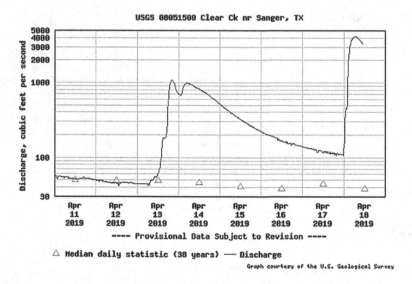

FIGURE 12.6   Hydrograph (water flow or discharge vs. time) for Clear Creek April 12–18, 2019 (USGS 2023d).

then in April 18, 2019, rose to about 4000 cfs (~113 m³/s). These peaks correspond to rainstorms over the watershed.

Next, let us produce a figure of the water level or gage height from April 12–18, 2019 (Figure 12.7). The peak heights on April 13 and 18, 2019 were about 10 ft (~3.1 m) and 17 ft (~5.2 m), respectively, which corresponds to the peak flows discussed above.

For comparison, we produce a hydrograph for a week with no precipitation. For example, April 17–24, 2020 (Figure 12.8) shows a steady decline of flow from about 120 to 90 cfs (3.4 to 2.5 m³/s), which is still much higher than the median daily.

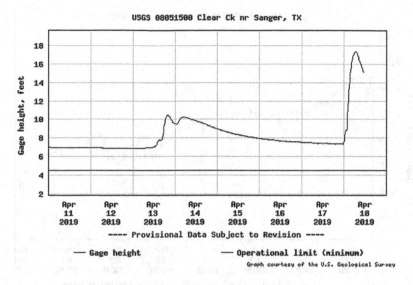

FIGURE 12.7   Gage height (water level) vs. time for Clear Creek April 12–18, 2019 (USGS 2023d).

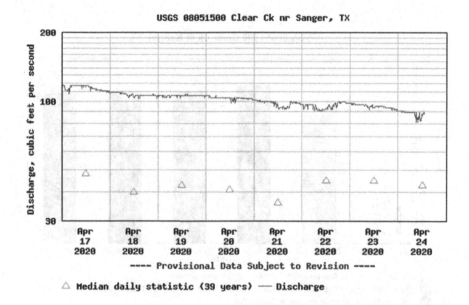

FIGURE 12.8   Hydrograph for a week with no precipitation, April 17–24, 2020 (USGS 2023d).

Likewise, the water level or gage height in this period (Figure 12.9) of no rain shows a decline from ~7.3 to ~7.1 ft, or ~2.21to 2.15 m, which was 0.1 m in five days.

Now, we use the peak tab to produce flooding charts for stage (gage height) and discharge (flow). Compare the stage and flow on the day you obtained results to the stage and flow in April 1990 that reached 29.9 ft (~9 m) (Figure 12.10) and 100,000 cfs (~2830 m³/s) (Figure 12.11).

Recall that a rating curve is the result of curve fit of streamflow from gage height (water level). Once calibrated, the flow can be estimated from measured height, because level or height is easier to monitor real-time. Let us find the rating curve for the Clear Creek stream gage. From the left navigation menu of the website select Toolkit, and then choose Rating Curve, at the dialog

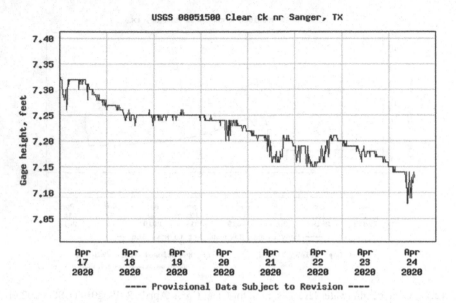

FIGURE 12.9   Water level or gage height (USGS 2023d).

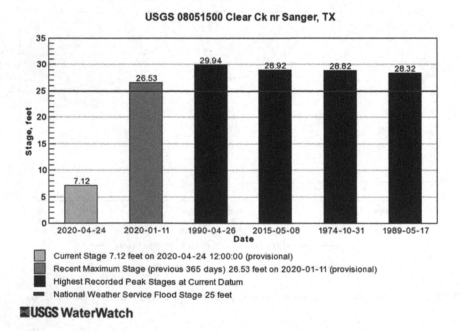

FIGURE 12.10   Peak stage statistics (USGS 2023d).

box enter the site number (from the previous work note that Clear Creek is 08051500). Use the default for the remainder entries and press Go. The curve is displayed as a graph. See the example in Figure 12.12. The rating curve provided is labeled provisional because the flow vs. height relation changes over time, as the river channel characteristics that control this relation vary.

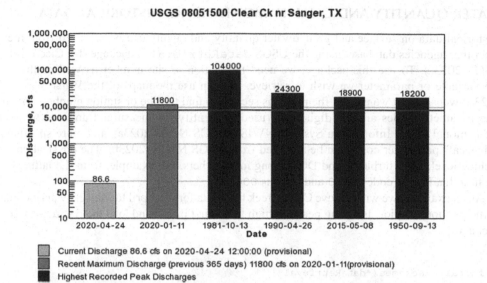

FIGURE 12.11 Peak flow statistics (USGS 2023d).

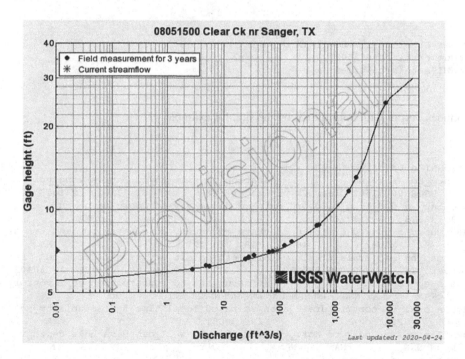

FIGURE 12.12 Rating curve (USGS 2023d).

WATER QUANTITY AND QUALITY MONITORING: HISTORICAL DATA

Historical data on surface and groundwater quantity and quality can be retrieved from the USGS and other agencies databases using the USGS `dataRetrieval` R package (De Cicco et al. 2022; USGS 2023a). To use this package, we need the station or site number as well as the code for the variable or parameter we wish to retrieve. We can use the maps of the Water Watch (USGS 2023d) website that we used in the previous section to find the site or station numbers for streams. The parameter codes are five-digit codes used to identify the measurand and can be found via the National Water Information System (NWIS) (USGS NWIS 2023a), and more specifically the "Physical" parameter codes can be accessed from USGS NWIS (2023b). The physical parameters include level, flow, turbidity, and DO, among many others. For example, from the latter web page we find that flow is code `00060` and stage is `00065`.

As an example, we will retrieve Clear Creek flow data for the April to May 2019 period identified in the previous section as having peaks of high flow. First install and load the `dataRetrieval` R package

```
install.packages("dataRetrieval")
library(dataRetrieval)
```

We already know from the previous section that the site number for Clear Creek USGS stream gage is `08051500` and from the previous paragraph we know that flow is parameter code `00060`. Use the site number and run `readNWISite` on the sie number to obtain info about the site

```
# Clear Creek near Sanger, TX
siteNumber <- "08051500"
readNWISsite(siteNumber)
```

that produces output with information on the site; for example

```
> readNWISsite(siteNumber)
  agency_cd  site_no                station_nm site_tp_cd lat_va long_va
1     USGS 08051500 Clear Ck nr Sanger, TX            ST 332010  971045
  dec_lat_va dec_long_va coord_meth_cd coord_acy_cd coord_datum_cd
1   33.33623   -97.17946             M            F          NAD27
  dec_coord_datum_cd district_cd state_cd county_cd country_cd land_net_ds
1              NAD83          48       48       121         US          NA
     map_nm map_scale_fc alt_va alt_meth_cd alt_acy_va alt_datum_cd   huc_cd
1 Sanger, TX      24000 565.52           X       0.24       NAVD88 12030103
  basin_cd topo_cd              instruments_cd construction_dt inventory_dt
1     <NA>    <NA> NNNNYNNNNNNNNNNNNNNNNNNNNNNNNNN              NA           NA
  drain_area_va contrib_drain_area_va tz_cd local_time_fg reliability_cd
1           295                   295   CST             Y           <NA>
  gw_file_cd nat_aqfr_cd aqfr_cd aqfr_type_cd well_depth_va hole_depth_va
1   NYNNNNNN        <NA>    <NA>         <NA>            NA            NA
  depth_src_cd project_no
1         <NA> 8653-00140
>
```

Recall that NA stands for no data. Let us query more info on parameter codes 00060, 00061, and 00065

```
> readNWISpCode(c("00060","00061","00065"))
   parameter_cd parameter_group_nm
43       00060              Physical
44       00061              Physical
48       00065              Physical
                                            parameter_nm casrn
43                    Discharge, cubic feet per second
44 Discharge, instantaneous, cubic feet per second
48                                    Gage height, feet
                          srsname parameter_units
43   Stream flow, mean. daily             ft3/s
44 Stream flow, instantaneous             ft3/s
48               Height, gage                ft
>
```

We want to retrieve instantaneous flow, which is service "iv", for a time window encompassing the months of April and May of 2019 and using Central Time (America/Chicago) corresponding to the location of the site.

```
instFlow <- readNWISdata(
   sites = siteNumber, service = "iv", parameterCd = "00060",
   startDate = "2019-04-01T00:00Z", endDate = "2019-06-01T12:00Z",
   tz = "America/Chicago"
)
```

Look at a few records to confirm the retrieval

```
> instFlow[1:5,]
   agency_cd  site_no        dateTime X_00060_00000 X_00060_00000_cd
1      USGS 08051500 2019-03-31 18:00:00          64.5                A
2      USGS 08051500 2019-03-31 18:15:00          64.5                A
3      USGS 08051500 2019-03-31 18:30:00          64.5                A
4      USGS 08051500 2019-03-31 18:45:00          64.5                A
5      USGS 08051500 2019-03-31 19:00:00          65.7                A
             tz_cd
1 America/Chicago
2 America/Chicago
3 America/Chicago
4 America/Chicago
5 America/Chicago
>
```

Note the data are collected every 15 minutes. Out of all columns we will extract the time stamp and the flow that is columns 3 and 4

```
qCC <- instFlow[,3:4]
names(qCC) <- c("TimeStamp","Flow")
```

verify that we correctly extracted the timestamp and flow columns

```
> qCC[1:5,]
             TimeStamp Flow
1 2019-03-31 18:00:00 64.5
2 2019-03-31 18:15:00 64.5
3 2019-03-31 18:30:00 64.5
4 2019-03-31 18:45:00 64.5
5 2019-03-31 19:00:00 65.7
>
```

We will convert the flow in cfs to m^3/s

```
# convert to m3/s
qq <- (qCC[,2])*0.0283
```

Our goal is to obtain a time series, thus encoding the timestamp

```
# read time sequence from file
tt <- strptime(qCC[,1], format="%Y-%m-%d %H:%M:%S", tz="")
```

and use package xts, executing require ensuring it is loaded and converting to time series

```
require(xts)
q.t <- xts(qq,tt)
names(q.t) <- c("Flow(m3/s)")
```

verify a few records

```
> q.t[1:5,]
                      Flow(m3/s)
2019-03-31 18:00:00      1.82535
2019-03-31 18:15:00      1.82535
2019-03-31 18:30:00      1.82535
2019-03-31 18:45:00      1.82535
2019-03-31 19:00:00      1.85931
>
```

And plot to produce Figure 12.13 which shows a sequence of peak flows during this period

```
plot(q.t,main="Clear Creek",ylab="Flow(m3/s)",lwd=1)
```

Some of the early peaks shown in this figure correspond to the peaks we investigated in the previous section. Using a similar procedure, we can retrieve other parameters for this gage such as stream stage.

As you can appreciate from this section, the R package dataRetrieval offers the possibility of programming and automating retrieving data from a variety of water bodies and for a multitude of parameters.

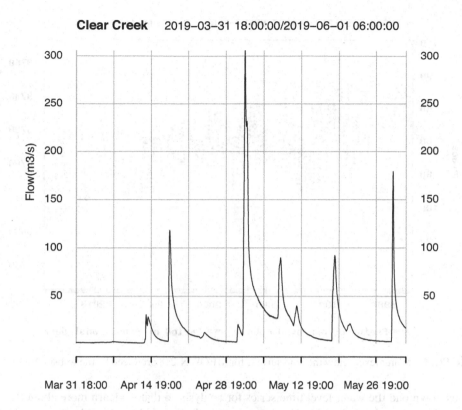

Clear Creek 2019–03–31 18:00:00/2019–06–01 06:00:00

Mar 31 18:00   Apr 14 19:00   Apr 28 19:00   May 12 19:00   May 26 19:00

FIGURE 12.13 Flow time series for Clear Creek showing a succession of peaks during April–May 2019. Data retrieved using `dataRetrieval` R package (USGS 2023a).

GROUNDWATER QUANTITY MONITORING

In the US, the USGS keeps groundwater monitoring data available online (USGS 2023c) with current conditions at selected sites based on automated recording equipment at a fixed interval, transmitted to the USGS every hour, and reported daily. Also available are historical data, we could, of course, use the `dataRetrieval` package as outlined in the previous section, but for the sake of exemplifying another approach, we will look on the website for well USGS 315712106361803 MBOWN-238 near Santa Teresa, New Mexico, and El Paso, Texas, USA. Search for this well on the web site; if you have difficulties, reference (USGS 2023b) provides a direct link. Once there you will see general information such as

```
USGS 315712106361803 MBOWN-238 - JL-49-04-476 (CWF-2C)
El Paso County, Texas
Hydrologic Unit Code 13030102
Latitude  31°57'13", Longitude 106°36'19" NAD27
Land-surface elevation 3,773.50 feet above NGVD29
The depth of the well is 300.0 feet below land surface.
The depth of the hole is 300.0 feet below land surface.
This well is completed in the Rio Grande aquifer system (S100RIOGRD) national
aquifer.
```

You will see data available including water level data since 1985-06-10 and water quality data including inorganics, nutrients, organics, pesticides, and radio isotopes, at selected years. On that page, you can see a time series graph of water level below the land surface (Figure 12.14) showing a declining trend since 1985.

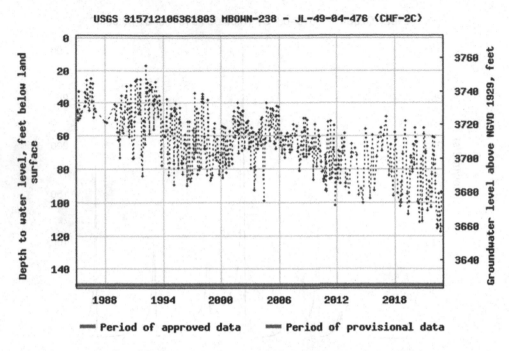

FIGURE 12.14 Water level below the land surface for MBOWN-238 produced by the website USGS (2023b).

Let us download the water level time series for analysis so that we learn more about these data. Click on tab separated data and download as a text file that would have records separated by tabs. It is more convenient to have a csv file. Using a text editor, we can convert tabs to commas using a global replacement. However, you can find these data already as csv in the RTEM GitHub repository in the archive MBOWN-238-1985-2023.zip. You can copy to your labs/data folder. Examine the file MBOWN-238-1985-2023.csv with Geany or the Notepad++

```
# Data for the following 1 site(s) are contained in this file
#    USGS 315712106361803 MBOWN-238 - JL-49-04-476 (CWF-2C)
#-------------------------------------------------------------------------------
#------------
# Data provided for site 315712106361803
#        TS    parameter    statistic    Description
#      100213        72019      00002    Depth to water level, feet below
land surface (Minimum)
#      238867        72019      00001    Depth to water level, feet below
land surface (Maximum)
#      238868        72019      00003    Depth to water level, feet below
land surface (Mean)
#
# Data-value qualification codes included in this output:
#
#    A  Approved for publication -- Processing and review completed.
#    P  Provisional data subject to revision.
#
agency_cd,site_no,datetime,100213_72019_00002,100213_72019_00002_cd,238867_72
019_00001,238867_72019_00001_cd,238868_72019_00003,238868_72019_00003_cd
5s,15s,20d,14n,10s,14n,10s,14n,10s
USGS,315712106361803,1985-06-10,57.64,A,58.57,A,58.06,A
USGS,315712106361803,1985-06-11,54.62,A,57.64,A,57.02,A
```

The records start in line 33 with

```
USGS,315712106361803,1985-06-10,57.64,A,58.57,A,58.06,A
```

And you can infer that the columns are for the agency and site identifier, timestamp by day, and three values separated by a character such as A for approved or P for provisional. The three water level values are minimum, maximum, and mean.

Use R to read the file skipping 32 lines into a dataset gwep.raw

```
gwep.raw <- read.table("data/MBOWN-238-1985-2023.txt",sep=",",skip=32)
```

and check the first few records

```
> gwep.raw[1:5,]
    V1           V2           V3     V4 V5     V6 V7     V8 V9
1 USGS 3.157121e+14 1985-06-10 57.64  A 58.57  A 58.06  A
2 USGS 3.157121e+14 1985-06-11 54.62  A 57.64  A 57.02  A
3 USGS 3.157121e+14 1985-06-12 49.29  A 54.30  A 51.58  A
4 USGS 3.157121e+14 1985-06-13 46.64  A 49.14  A 47.85  A
5 USGS 3.157121e+14 1985-06-14 45.64  A 46.57  A 45.97  A
>
```

We will work with the timestamp in column 3 and the levels in column 4, 6, and 8, store this in a new dataset named gwep

```
gwep <- gwep.raw[,c(3,4,6,8)]
names(gwep) <- c("Date","Level max","Lavel min", "Level avg")
```

and verify that you indeed have the desired results. Use the average in column 4 and convert feet to meters

```
> gwep[1:5,]
        Date Level max Level min Level avg
1 1985-06-10     57.64     58.57     58.06
2 1985-06-11     54.62     57.64     57.02
3 1985-06-12     49.29     54.30     51.58
4 1985-06-13     46.64     49.14     47.85
5 1985-06-14     45.64     46.57     45.97
>
```

look at the summary of this level

```
> summary(avg.m)
   Min. 1st Qu.  Median    Mean 3rd Qu.    Max.    NA's
-37.865 -24.069 -19.291 -19.799 -15.420  -5.075    855
>
```

The deepest level is 37.8 m below the surface and the closest it was to the surface was 5.08 m.

Read a time sequence from the timestamp in column 1

```
tt <- strptime(gwep[,1], format="%Y-%m-%d")
```

we do not specify the time zone for simplicity; the time zone would be the one set in your system. We will make the average level and timestamp into a time series using xts. As before you can ensure it is loaded by

```
require(xts)
```

Once xts is loaded, make the time series

```
x.t <- xts(avg.m,tt)
```

Our next step is to plot the time series. Assuming the surface is at 0 m and the levels below the surface are negative, we plot using xlim in negative values starting at −40, which is close to the deepest value ~−38 m that we found previously

```
plot(x.t,ylim=c(-40,0),main="MBOWN-238",ylab="Avg level below surface (m)",lwd=1)
```

producing Figure 12.15, which confirms the decreasing water level trend.

The plan now is to use the capabilities of xts to get the monthly average of the average level and plot.

```
x.mo <- apply.monthly(x.t, mean)
```

if we check x.mo, we realize that many months have NA values. This is because the mean function returns NA when the data include NAs. To avoid this issue, let us build a function for the mean that returns a value instead of NA, and return NA if all values are NA.

```
mean.na <- function(x){
 y <- mean(x,na.rm=TRUE)
 if(is.nan(y)) y <- NA
 return(y)
}
```

And apply the monthly using this function and plot

```
x.mo <- apply.monthly(x.t, mean.na)
plot(x.mo,ylim=c(-40,0),main="MBOWN-238 - Monthly",ylab="Avg Level below
surface (m)",lwd=1)
```

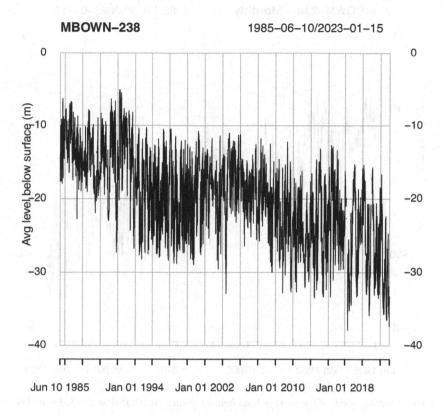

FIGURE 12.15 Average level time series. (Data downloaded from USGS (2023b).)

producing Figure 12.16 is a monthly average of the average water level showing a decreasing trend with a period of higher recharge in the early 2000s.

TIME SERIES PREDICTION

In this section, we will practice AR time series prediction using the dataset of the previous section, in particular, the monthly average. For AR models using the Yule Walker equations, a convenient function is ar.yw that takes as argument a time series object obtained by function ts. Therefore, start by building this object. One way is to use the coredata of the xts object that we built in the previous section

```
xt <- ts(coredata(x.mo)[,1],start=1,delta=1)
```

Now fit an AR model

```
> ar.xt <- ar.yw(xt,na.action=na.pass)
```

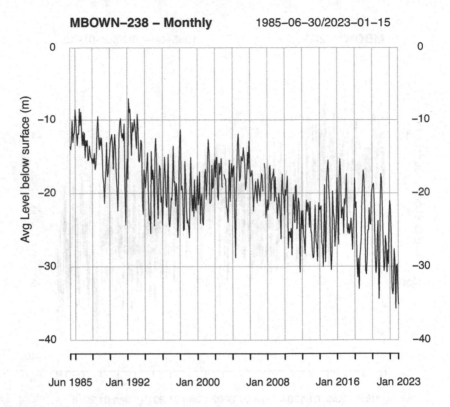

FIGURE 12.16 Monthly series of the average level below the surface. (Data downloaded from USGS (2023b).)

And checking the result we see that it selected an AR(11) model.

```
> ar.xt

Call:
ar.yw.default(x = xt, na.action = na.pass)

Coefficients:
      1        2        3        4        5        6        7        8
 0.7575  -0.2340   0.1495  -0.0701   0.1635  -0.1219   0.1709  -0.1215
      9       10       11
 0.0676  -0.0258   0.1957

Order selected 11  sigma^2 estimated as  9.484
>
```

We will now select a set from the last months in the series, say the last 30 months, to make a 10-month prediction, based on 20 months and test with the remaining 10 months of data. Denote by `tf` the last month in the series, `t2` the month to start testing, and `t1` the previous 20 months before the test.

```
dt.prd <- 20
dt.test <- 10
tf <- length(xt)
t2 <- length(xt)- dt.test +1
t1 <- t2 - dt.prd
```

and use these indices to form the test and prediction series

```
test <- ts(xt[t1:tf])
prd <- ts(xt[t1:t2])
```

We are ready to predict using the AR model, based on the prd series, and looking ahead for the duration of the test series

```
xt.p <- predict(ar.xt,prd,n.ahead=dt.test)
```

The object xt.p contains the predicted values $pred and the standard error $se

```
> xt.p
$pred
Time Series:
Start = 22
End = 31
Frequency = 1
 [1] -30.28983 -27.62436 -26.04351 -26.43995 -26.97662 -26.85114 -26.35069
 [8] -24.57077 -23.92725 -24.25322

$se
Time Series:
Start = 22
End = 31
Frequency = 1
 [1] 3.079642 3.863512 4.002779 4.064804 4.086852 4.138547 4.161155 4.204537
 [9] 4.227032 4.237293

>
```

Thus, we can build series with predicted values and low and high intervals around them

```
mid <- xt.p$pred
up <- xt.p$pred + 2*xt.p$se
low <- xt.p$pred - 2*xt.p$se
```

and plot all series together to get results as in Figure 12.17.

```
ts.plot(ts(test),mid,ylim=c(-40,0),lty=1:2,
        xlab=paste("Months Starting", index(x.mo[t1])),
        ylab="Depth below surface (m)")
lines(up, lty=3)
lines(low,lty=3)
legend("topright",leg=c("Data","Pred Mid","Pred 2xSE"),
lty=c(1,2,3),cex=0.8)
mtext("MBOWN-238",3,-1)
```

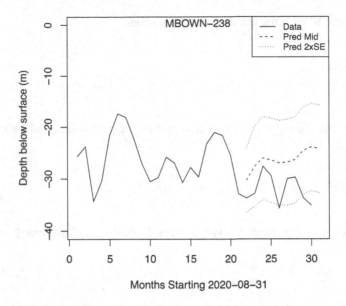

FIGURE 12.17   Prediction of monthly values using an AR(11) model.

Note that the actual values were below the predicted but contained in the interval given by ±2*se*, and show a similar pattern of increasing to a peak, falling to a valley, and increasing to a next peak.

WATER MONITORING: MULTIPROBES

Municipalities in the US monitor water quality for the protection of water resources and report to State and Federal agencies. Water protection includes effluents from wastewater treatment plants and stormwater runoff to streams or lakes. As an example, the City of Denton, Texas, water protection program, operates and maintains water quality instruments such as multiprobe datasondes and automated samplers.

A multi-parameter probe (datasonde) can measure, for example, dissolved oxygen (DO), temperature, total dissolved solids (TDS), electrical conductivity (EC), turbidity, salinity, depth, oxidation reduction potential (ORP), and pH. A multi-parameter probe can be deployed from a buoy or from a floating platform. An example is shown in Figure 12.18.

FIGURE 12.18   Floating platform in Lewisville Lake, deployed by the Ecoplex project, at the water intake of the City of Denton, Texas.

As an example, we will consider data from Lewisville Lake at the intake for the City of Denton water supply, we already used data of a datasonde deployed in Lewisville Lake in Lab 1. In this chapter, we use our time series skills to plot the data. A data file is available from the RTEM repository archive 2018 _ Lake _ Lewisville _ Fall.zip. Extract the content to labs/data. The 2018 _ Lake _ Lewisville _ Fall.csv file in the archive corresponds to the autumn of 2018. Examine the file with Notepad++ or Geany. The first few records are

```
Date (M/D/Y),Time (hh:mm:ss),Temp (degC),SpCond (uS/cm),TDS (mg/L),
Salinity (ppt),DO%,DO (mg/L),Depth (m),pH,Turbidity+ (NTU),Battery (V),,
9/1/2018,0:00:00,30.81,358,232,0.17,128.2,9.55,0.709,8.72,12.3,11.9,,
9/1/2018,0:30:00,30.74,356,231,0.17,126.5,9.43,0.709,8.7,12.6,11.9,,
9/1/2018,1:00:00,30.62,355,230,0.17,124.8,9.33,0.711,8.67,12.3,11.8,,
9/1/2018,1:30:00,30.49,355,231,0.17,123.6,9.25,0.71,8.66,12.1,11.8,,
9/1/2018,2:00:00,30.36,355,230,0.17,122.2,9.17,0.707,8.66,12.2,11.8,,
9/1/2018,2:30:00,30.24,347,226,0.16,121.9,9.17,0.708,8.65,11.5,11.7,,
9/1/2018,3:00:00,30.15,348,226,0.16,119.7,9.01,0.708,8.63,11.4,11.7,,
9/1/2018,3:30:00,30.06,348,226,0.16,117,8.83,0.708,8.61,11.3,11.7,,
```

Here, the first line is shown wrapped so that you can read the entire header. Measurements are on a 30-minute interval. We have eight water quality variables: temperature (°C), EC µS/cm, TDS (mg/L), salinity (ppt), DO%, DO (mg/L), pH, and turbidity (in nephelometric units, NTU). Other variables are date, time, depth, and battery voltage. Note that there are two missing data items at the end of each line.

Use R to read the file, preserve spaces in names (check.names=FALSE), remove the last two columns because they are empty, and remove records with NA by using complete.cases.

```
X <-
read.table("data/2018_Lake_Lewisville_Fall.csv",
header=TRUE,sep=",",check.names=FALSE)
X <- X[,-c(13:14)]
X <- X[complete.cases(X),]
```

The resulting dimension of X after eliminating the NAs can be found by

```
> dim(X)
[1] 3832    12
>
```

or tibble

```
> as_tibble(X)
# A tibble: 3,832 × 12
   Date (M…¹ Time …² Temp …³ SpCon…⁴ TDS (…⁵ Salin…⁶ 'DO%' DO (m…⁷ Depth…⁸    pH
   <chr>    <chr>     <dbl>   <int>   <int>   <dbl> <dbl>   <dbl>   <dbl> <dbl>
 1 9/1/2018 0:00:00   30.8     358     232    0.17  128.    9.55   0.709  8.72
 2 9/1/2018 0:30:00   30.7     356     231    0.17  126.    9.43   0.709  8.7
 3 9/1/2018 1:00:00   30.6     355     230    0.17  125.    9.33   0.711  8.67
 4 9/1/2018 1:30:00   30.5     355     231    0.17  124.    9.25   0.71   8.66
 5 9/1/2018 2:00:00   30.4     355     230    0.17  122.    9.17   0.707  8.66
 6 9/1/2018 2:30:00   30.2     347     226    0.16  122.    9.17   0.708  8.65
 7 9/1/2018 3:00:00   30.2     348     226    0.16  120.    9.01   0.708  8.63
 8 9/1/2018 3:30:00   30.1     348     226    0.16  117     8.83   0.708  8.61
 9 9/1/2018 4:00:00   30.0     349     227    0.16  114.    8.6    0.711  8.58
10 9/1/2018 4:30:00   30.0     349     227    0.16  111.    8.41   0.711  8.56
# … with 3,822 more rows, 2 more variables: 'Turbidity+ (NTU)' <dbl>,
#    'Battery (V)' <dbl>, and abbreviated variable names ¹'Date (M/D/Y)',
#    ²'Time (hh:mm:ss)', ³'Temp (degC)', ⁴'SpCond (uS/cm)', ⁵'TDS (mg/L)',
#    ⁶'Salinity (ppt)', ⁷'DO (mg/L)', ⁸'Depth (m)'
# i Use 'print(n = ...)' to see more rows, and 'colnames()' to see all
variable names
```

This example shows a case when date and time are given in separate columns. Thus, to obtain a full timestamp, paste together date and time strings and convert character string date and time to POSIXlt using strptime.

```
# Date (M/D/Y),Time (hh:mm:ss)
datetime<- paste(X[,1],X[,2])
tt <- strptime(datetime, format="%m/%d/%Y %H:%M:%S",tz="")
# select the water variables
xx <- X[,3:12]
```

We can use tt[1:5] to show a screen capture segment of five lines of the result. Note that the time zone (CST in this case) was attached to the timestamp since the system running this script was set for Central Time.

Load package xts and create a time series.

```
# use package xts
library(xts)
# create time series
x.t <- xts(xx,tt)
names(x.t)
```

Use `x.t[1:5,]` to show a screen capture segment of ten lines of your result

```
> x.t[1:5,]
            Temp (degC) SpCond (uS/cm) TDS (mg/L) Salinity (ppt)   DO%
2018-09-01 00:00:00        30.81            358        232        0.17 128.2
2018-09-01 00:30:00        30.74            356        231        0.17 126.5
2018-09-01 01:00:00        30.62            355        230        0.17 124.8
2018-09-01 01:30:00        30.49            355        231        0.17 123.6
2018-09-01 02:00:00        30.36            355        230        0.17 122.2
                    DO (mg/L) Depth (m)   pH Turbidity+ (NTU) Battery (V)
2018-09-01 00:00:00        9.55      0.709 8.72           12.3       11.9
2018-09-01 00:30:00        9.43      0.709 8.70           12.6       11.9
2018-09-01 01:00:00        9.33      0.711 8.67           12.3       11.8
2018-09-01 01:30:00        9.25      0.710 8.66           12.1       11.8
2018-09-01 02:00:00        9.17      0.707 8.66           12.2       11.8
>
```

Let us work with six variables 1, 2, 6, 7, 8, 9 corresponding to temperature (°C), specific conductance or EC in µS/cm, DO (mg/L), depth (m), pH, and turbidity (NTU).

We can select these variables and plot in two pages using `multi.panel=TRUE` and `yaxis.same=FALSE`. For this purpose, we open a `LewisvilleLake2018.pdf` file and print the two pages to this file in folder output or the name of the folder where you store your graphics results

```
# select Temp, EC, DO,Depth, pH, Turbidity
sel <- c(1,2,6,7,8,9)
# send plots to a pdf
pdf("output/LewisvilleLake2018.pdf")
plot(x.t[,sel],multi.panel=3,yaxis.same=FALSE,main="LewisvilleLake2018",lwd=1,
col=1)
dev.off()
```

The first page of the pdf has the plot for temperature, EC, and DO (Figure 12.19) and the second page of the pdf has graphs for depth, pH, and turbidity (Figure 12.20).

We notice that the low EC and DO values correspond to warmer water (Figure 12.19). Criteria often used for water quality includes having DO above 5 mg/L. We can see that some values are lower, particularly in the summer months. The second page (Figure 12.20) shows spikes in turbidity. These most likely correspond to runoff events due to rainfall events.

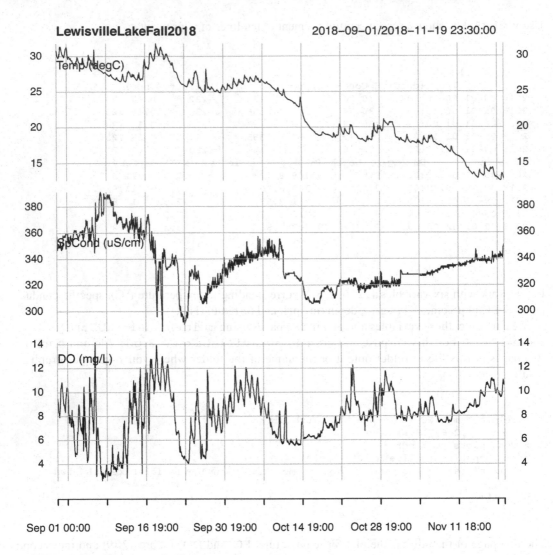

FIGURE 12.19   First page of pdf.

EXERCISES

Exercise 12.1 Surface water current conditions from Water Watch.

Use real-time current conditions flow data from USGS Water Watch for Clear Creek as described in the guide. Produce a figure of the water level and flow for one week. Describe patterns observed. Follow the lab guide to find the rating curve for Clear Creek.

Exercise 12.2 Hydrographs from historical data using R.

Use package `dataRetrieval` to obtain flow data for Clear Creek for April to May 2019 as described in the lab guide. Produce a time series graph using package `xts`.

Exercise 12.3 Groundwater quantity analysis.

Follow the lab guide to produce a time series graph of monthly water level below the surface for well USGS 315712106361803 MBOWN-238.

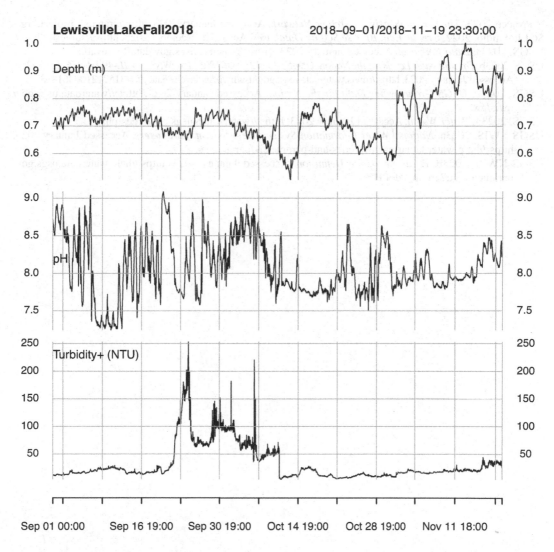

FIGURE 12.20   Second page of pdf.

Exercise 12.4 AR modeling and prediction.

Follow the lab guide to fit an AR time series model to the monthly water level data of the previous exercise (well USGS 315712106361803 MBOWN-238). Perform prediction of water level looking ahead as suggested in the guide. Plot and discuss.

Exercise 12.5 Surface water monitoring using multiprobes.

Using data from the Lewisville lake datasonde produce plots of time series during autumn 2018 for the following: temperature, EC, DO, depth, pH, and turbidity.

REFERENCES

Acevedo, M.F. 2024. *Real-Time Environmental Monitoring: Sensors and Systems - Textbook, Second Edition.* Boca Raton, FL: CRC Press, Taylor & Francis Group, 392 pp.

De Cicco, L.A., R.M. Hirsch, D. Lorenz, W.D. Watkins, and M. Johnson, 2022. Data Retrieval: R packages for discovering and retrieving water data available from Federal hydrologic web services.

R Project. 2023. *The Comprehensive R Archive Network*. Accessed January 2023. http://cran.us.r-project.org/.

SQLiteStudio. 2022. *Create, Edit, Browse SQLite Databases*. Accessed December 2022. https://sqlitestudio.pl/.

USGS. 2023a. *Data Retrieval*. Accessed January 2023. https://rconnect.usgs.gov/dataRetrieval/.

USGS. 2023b. *Groundwater Levels for the Nation, USGS 315712106361803 MBOWN-238- JL-49-04-476 (CWF-2C)*. Accessed January 2023. https://nwis.waterdata.usgs.gov/usa/nwis/gwlevels/?site_no=315712106361803.

USGS. 2023c. *USGS Groundwater Data for the Nation*. Accessed January 2023. https://waterdata.usgs.gov/nwis/gw.

USGS. 2023d. *Water Watch*. Accessed January 2023. https://waterwatch.usgs.gov/index.php.

USGS NWIS. 2023a. *National Water Information System: Help System, Parameters*. Accessed January 2023. https://help.waterdata.usgs.gov/codes-and-parameters/parameters.

USGS NWIS. 2023b. *Parameter Code Definition*. Accessed January 2023. https://help.waterdata.usgs.gov/parameter_cd?group_cd=PHY.

13 Terrestrial Ecosystems Monitoring

INTRODUCTION

This session is devoted to terrestrial ecosystems monitoring, including topics in three major categories: the vadose zone, atmospheric and solar radiation interaction, and vegetation canopy. With respect to the vadose zone, the session covers soil moisture, temperature, and electrical conductivity, emphasizing measurement at various depths in the soil column. To illustrate interactions with the atmosphere and solar radiation, two examples are presented, evapotranspiration and photosynthetically active radiation (PAR). Vegetation canopy analysis is exemplified by analysis of structure via hemispherical photos from the ground, and mapping canopy height by means of laser imaging, detection, and ranging (LiDAR).

MATERIALS

READINGS

For theoretical background, you can use Chapter 13 of Acevedo, M.F. 2024. *Real-Time Environmental Monitoring: Sensors and Systems - Textbook, Second Edition* which is a companion to these guides (Acevedo 2024). Other bibliographical references are cited throughout the guide.

SOFTWARE (LINKS PROVIDED IN THE REFERENCES)

- R system (R Project 2023)
- R Package neonUtilities
- R Package raster
- R package xts

DATA FILES (AVAILABLE FROM THE RTEM GITHUB REPOSITORY)

- Archive soilmoisture.zip
- Archive teros12-datalog.zip
- Archive DP Weather data CR3000 _ 13Feb2019 _ 13Jan2021.zip
- Archive CR3000Tower.zip
- Archive eto.zip
- Archive canopy.zip
- Archive NEON _ par.zip
- Archive NEON _ struct-ecosystem.zip

SUPPLEMENTARY SUPPORT MATERIAL

Supplementary support material including additional screenshots, images, and procedures are available from the publisher eResources web page provided for this book.

DOI: 10.1201/9781003184362-13

SOIL MOISTURE SENSOR CALIBRATION

In Chapter 13 of the companion textbook, we discussed the Decagon Devices EC5 soil moisture probe (Meter 2022a), which is the same probe we presented in relation to the Arduino and WSN nodes in Lab session 5.

In this lab session, we study the calibration of a soil moisture probe using the gravimetric method, which consists of determining soil water content from the difference in weight between a soil sample when wet and the same sample after is dried. For this purpose, start with the wet soil sample and record the output of the sensor; a sample of fixed volume is weighted wet, dried for three days in an oven at 103°C, and then weighted again as dry soil. The difference in weight between wet and dry soil divided by the dry weight is the gravimetric water content (GWC) in g/g. The bulk density (BD) is obtained by dividing the dry weight over the volume. Then, the volumetric water content (VWC) is obtained by multiplying the GWC times the BD.

We will use values derived from Chen (2008) and available as a `calibration-EC5.csv` file in the archive `soilmoisture.zip` from the RTEM repository. The volume is in ml and the weights are in g.

```
Out(mV),Vol(ml),Wet(g),Dry(g)
220.1,35,39.0,39.0
438.0,35,56.4,47.8
471.0,35,51.5,43.5
503.5,35,46.0,39.0
518.6,35,57.0,46.0
541.6,35,58.0,46.0
550.3,35,61.0,48.0
566.1,35,62.0,49.0
580.0,35,60.0,46.0
610.8,35,66.0,49.0
616.0,35,71.8,53.4
625.0,35,75.25,56.9
```

We write R code to read the file,

```
ec5.cal <- read.table("data/calibration-EC5.csv", sep=",", header=TRUE)
names(ec5.cal) <- c("mV","Vol","Wet","Dry")
```

calculate GWC by subtracting dry weight from wet weight and dividing by the dry weight, BD by dividing dry weight by volume, and VWC by multiplying GWC and BD

```
GWC <- (ec5.cal$Wet-ec5.cal$Dry)/ec5.cal$Dry
BD <- ec5.cal$Dry/ec5.cal$Vol
VWC <- GWC*BD
```

Conduct a linear regression of VWC on mV,

```
> ec5.lm <- lm(VWC ~ ec5.cal$mV)
> summary(ec5.lm)
```

```
Call:
lm(formula = VWC ~ ec5.cal$mV)

Residuals:
     Min        1Q     Median        3Q       Max
-0.113335 -0.020031 -0.006222  0.039032  0.068062

Coefficients:
              Estimate Std. Error t value Pr(>|t|)
(Intercept) -0.3360935  0.0762781  -4.406  0.00132 **
ec5.cal$mV   0.0012885  0.0001437   8.968 4.27e-06 ***
---
Signif. Codes:  0 '***' 0.001 '**' 0.01 '*' 0.05 '.' 0.1 ' ' 1

Residual standard error: 0.05279 on 10 degrees of freedom
Multiple R-squared:  0.8894,    Adjusted R-squared:  0.8783
F-statistic: 80.42 on 1 and 10 DF,  p-value: 4.273e-06


>
>
```

The resulting regression has $R^2 = 0.878$ with low p-values for both coefficients. The calibration equation would be $VWC = -0.336 + 0.00129 \times mV$. We can visualize a scatterplot together with the regression line (Figure 13.1).

```
plot(ec5.cal$mV, VWC ,ylab="VWC", xlab="mV")
lines(ec5.cal$mV,ec5.lm$fitted.values)
```

As we have learned before, it is often possible to obtain a better calibration by nonlinear regression; let us try polynomial regression of order 2

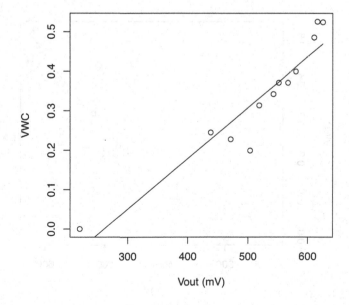

FIGURE 13.1   Calibration of EC-5 output by linear regression.

```
> ec5.poly <- lm(VWC ~ poly(ec5.cal$mV,2))
> summary(ec5.poly)

Call:
lm(formula = VWC ~ poly(ec5.cal$mV, 2))

Residuals:
     Min        1Q    Median        3Q       Max
-0.078558 -0.010612  0.001425  0.010450  0.066011

Coefficients:
                      Estimate Std. Error t value Pr(>|t|)
(Intercept)            0.33417    0.01077  31.016 1.85e-10 ***
poly(ec5.cal$mV, 2)1   0.47345    0.03732  12.686 4.79e-07 ***
poly(ec5.cal$mV, 2)2   0.12384    0.03732   3.318  0.00897 **
---
Signif. Codes:  0 '***' 0.001 '**' 0.01 '*' 0.05 '.' 0.1 ' ' 1

Residual standard error: 0.03732 on 9 degrees of freedom
Multiple R-squared:  0.9503,    Adjusted R-squared:  0.9392
F-statistic: 85.97 on 2 and 9 DF,  p-value: 1.365e-06
>
```

We now have $R^2 = 0.939$ with low p-values for all coefficients. The calibration equation would be $VWC = 0.334 + 0.473 \times mV + 0.124 \times mV^2$. We can visualize a scatterplot together with the regression line (Figure 13.2). Clearly, the dataset would benefit for datapoints in the 300-400 mV range to obtain a better calibration for VWC in the 0.1–0.2 range.

```
VWC.est<- ec5.poly$fitted.values
plot(ec5.cal$mV, VWC ,ylab="VWC", xlab="mV")
lines(ec5.cal$mV, VWC.est, lty=2)
```

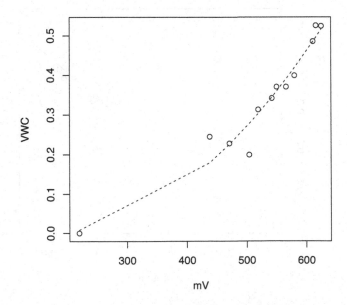

FIGURE 13.2 Polynomial regression to calibrate soil moisture sensor.

SOIL MOISTURE ANALYSIS

We will use a file weather-soil-moisture-GBC-2007-April11-May10.csv file in the archive soilmoisture.zip from the RTEM GitHub repository. This data file has measurements of soil moisture using EC-5 sensors at four different depths 5, 15, 25, and 36 in. (12.4, 38.10, 63.5, 91.44 cm).

```
x <- read.table("data/weather-soil-moisture-GBC-2007-April11-May10.csv",
sep=",", header=TRUE)
sm <- c(8,10,12,14,16)
smleg <- c("Rain","SM1","SM2","SM3","SM4")
```

To create a time series, first generate a POSIX timestamp, excluding timestamps with no data NA,

```
tt.raw <- strptime(x[,1], format="%m/%d/%Y %H:%M",tz="")
tt <- tt.raw[-which(is.na(tt.raw))]
xx <- x[-which(is.na(tt.raw)),sm]
```

and then create the time series using xts. Recall to load package xts or use require(xts)

```
require(xts)
xx.tt<- xts(xx,tt)
```

Plot using xts to obtain Figure 13.3.

```
col.sm <- c('darkgray',1:4)
plot(xx.tt,type="l",lty=1:5,col=col.sm, ylab="Rain(in) and VWC",grid.
col='lightgrey')
addLegend("top",smleg,lty=1:5,col=col.sm,cex=0.8)
```

As one can appreciate from this graph, soil moisture at the top layer responds to rain and then allows some water percolation to deeper layers, depending on moisture content. The rain event of April 23 causes a significant increase in soil moisture at all layers. Now, we will focus on this event by first selecting a timeframe from 12:00 to 23:00 hours on April 23.

```
x.t <- xx.tt["2007-04-23 12:00/2007-04-23 23:00"]
```

and then plot to obtain Figure 13.4.

```
plot(x.t,type="l",lty=1:5,col=col.sm, ylab="Rain(in) and VWC",grid.
col='lightgrey')
addLegend("top",smleg,lty=1:5,col=col.sm,cex=0.8)
```

By inspecting this graph, we conclude that soil moisture responds at the top layer at the beginning of the event when all layers are relatively dry; then as the two deeper layers (SM3 and SM4) saturate around 16:00 hours, only the top two layers show a response to the rain around 18:00 hours.

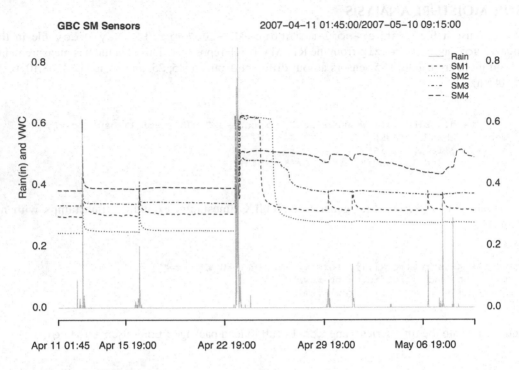

FIGURE 13.3   Rain and four soil moisture sensor response.

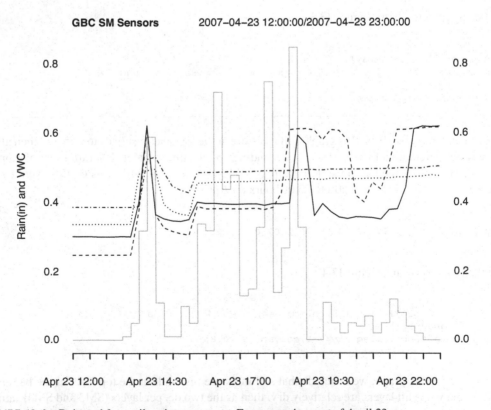

FIGURE 13.4   Rain and four soil moisture sensors. Focus on rain event of April 23.

SOIL CONDUCTIVITY SENSOR NETWORK AND DATA ANALYSIS

In this section, we work on the analysis of sensor network data for VWC, electrical conductivity (EC), and temperature employing as an example a Teros-12 (METER 2022b) sensor network, using the SDI-12 interface. This network is introduced in Chapter 13 of the companion textbook. The sensor network is deployed at the Brackish Groundwater National Desalination Research Facility (Alamogordo, NM) and is part of a project aiming to improve crop yield and soil salinity funded by the INFEWS (Innovations at the Nexus of Food, Energy and Water Systems) program of the National Science Foundation and US Department of Agriculture. The network consists of 16 sensors, one in each one of 16 plots, representing four different treatments of water (brackish well water and agricultural water) and compost inoculation. Half of the sensors are buried at 6 in. and half at 12 in. deep. The sensor network is datalogged using a Raspberry Pi in a setup similar to those described in Lab sessions 5 and 6. The SDI-12 sensors have addresses A, B, C,…, H for well water and a, b, c,…, h for agricultural water.

For analysis, we will use as an example the files in archive `teros12-datalog.zip` provided in the RTEM repository. Download this archive to your `labs/data` folder and extract to have a folder `teros-datalog` containing two files, the data file `20220909.csv` and a setup file `teros.csv`. We start by reading the data file and checking a few first records

```
X <- read.table("data/teros-datalog/20220909.csv" ,sep=",")
> X[1:5,]
                    V1 V2      V3   V4  V5
1 2022/09/09 06:07:23  A 2056.99 24.3 280
2 2022/09/09 06:07:28  B 2347.26 25.7 641
3 2022/09/09 06:07:33  a 2382.77 24.1 613
4 2022/09/09 06:07:39  b 2207.91 23.6 457
5 2022/09/09 06:07:44  C 2185.41 23.8 303
>
```

The columns are timestamp, SDI-12 sensor address, VWC as digital number (DN), soil temperature, and soil EC.

To convert digital number to VWC, we use a calibration equation (recall the concepts of first section of this guide) $VWC = 3.879 \times 10^{-4} \times DN - 0.6956$. We build a simple function

```
teros12.cal <- function(x){
 vwc <- 3.879*10^-4*x -0.6956
 return(round(vwc,3))
}
```

And apply it to the third column of X,

```
X[,3] <- teros12.cal(X[,3])
```

Assign names

```
names(X) <- c("Timestamp","Sensor","VWC","Temp","EC")
```

Verify that we have VWC in the third column

```
> X[1:5,]
            Timestamp Sensor  VWC Temp  EC
1 2022/09/09 06:07:23      A 0.102 24.3 280
```

```
2 2022/09/09 06:07:28      B 0.215 25.7 641
3 2022/09/09 06:07:33      a 0.229 24.1 613
4 2022/09/09 06:07:39      b 0.161 23.6 457
5 2022/09/09 06:07:44      C 0.152 23.8 303
>
```

We now use the setup file

```
setup <- read.table("data/teros-datalog/teros.csv",sep=',',header=TRUE)
```

which yields a data frame with three columns: treatment, sensor address, and depth

```
> setup[1:5,]
  Treat Teros Depth
1   WN    A    6
2   WC    B   12
3   AN    a   12
4   AC    b    6
5   WC    C    6
>
```

At this point, we want to convert the timestamp to be able to use it in a time series

```
X[,1] <- as.POSIXct(X[,1], format="%Y/%m/%d %H:%M")
```

Since we have a network of 16 sensors in the same file, we program a loop to separate each sensor using which(), store it in a list element, then make each element of the list into a time series. First use require() to ensure we load xts

```
require(xts)

X.t <- list()
for(i in 1:length(setup$Teros)){
    xx <- X[which(X[,2]==setup$Teros[i]),]
    tt <- xx[,1]
    X.t[[i]] <- xts(xx[,3:5],tt)
    names(X.t[[i]]) <- c("VWC","Temp","EC")
}
```

For example, verify the contents of the first element

```
> X.t[[1]][1:5,]
                      VWC Temp  EC
2022-09-09 06:07:00 0.102 24.3 280
2022-09-09 06:12:00 0.102 24.3 279
2022-09-09 06:17:00 0.102 24.3 278
2022-09-09 06:22:00 0.102 24.2 282
2022-09-09 06:27:00 0.102 24.2 278
>
```

A practical pproach to visualizing the results will be to plot all 16 sensors to a pdf, one page per sensor. Make sure you have a folder `labs/output` or edit the name of the output path below to conform with your own folder name. Note that to plot to a page of the pdf within a loop you wrap the `plot` function within a `print` function.

```
pdf('output/teros12.pdf',7,7)
  for (j in 1:length(X.t)) {
    title = paste('Sens ',setup$Teros[j],' @',setup$Depth[j],'in',' Treat',
setup$Treat[j],sep='')

print(plot(X.t[[j]],multi.panel=TRUE,yaxis.same=FALSE,ylab='',main=title,
col=1))
  }
dev.off()
```

Verify that you have a file `labs/output/teros12.pdf` of 16 pages, one page per sensor. For illustration, we show here sensor A buried at 6 in. treated with well water and no compost (Figure 13.5).

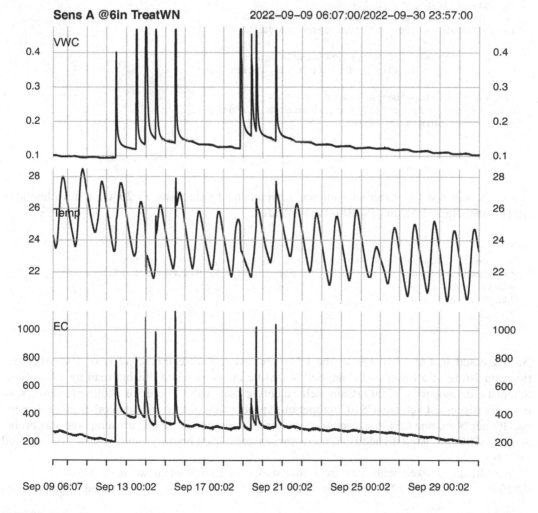

FIGURE 13.5 Example of VWC, temperature (in °C), and EC (in μS/cm) measurements using a TEROS 12 node part of a network. Sensor A at 6 in. in plot treated with well water and no compost.

EVAPOTRANSPIRATION

Write the following R function et0.fao in a file ET0-FAO-function.R. Store it in a folder lab13 within your working directory. It is also available in the eto.zip archive that you can download from the RTEM repository.

```
et0.fao <- function(albedo=0.23,soil=0.0, Temp.C, RH, Rad.MJ, BarP.kPa,
Wind2m.mps){
# Penman-Monteith
# Temp in degC, RH in %, Rad in MJ/m2, BarP in kPa, Wind2m in m/s
# convert to deg Kelvin
Temp.K <- Temp.C + 273

# water vapor pressure at saturation in kPa
satura <- 0.6108* exp(17.27*Temp.C/(Temp.C+237.3))
# slope in kPa/C
slope <- 4096*satura/(Temp.C+237.3)^2
# psychrometer constant a function of pressure Kpa
psychro <- 0.7*10^-3*BarP.kPa
# vapor pressure using RH
vap.press <- satura*RH/100
deficit <- satura - vap.press
latent.heat <- 2.50 - 0.0022*Temp.C # MJ/kg

denom <- slope+psychro*(1+0.34*Wind2m.mps)
netrad <- (Rad.MJ*(1-albedo))*(1-soil)/latent.heat

rad.term <- slope*netrad/denom
aero.term <- psychro*(37/Temp.K)*Wind2m.mps*deficit/denom
et0 <- rad.term+aero.term
return(et0)
}
```

Source it, and test this function with some hypothetical values say $T=20\,°C$, relative humidity (RH)=70%, Rad=1 MJ/m^2, $P=100\,kPa$, $u=2\,m/s$

```
source("R/lab13/ET0-FAO-function.R")
et0.fao(albedo=0.23, soil=0.0, Temp.C=20, RH=70, Rad.MJ=1, BarP.kPa=100,
Wind2m.mps=2)
[1] 0.2202244
>
```

As discussed in Labs 7 and 9, at the UNT Discovery Park campus, we have a weather station consisting of a tower and Campbell Scientific meteorological sensors read by a Campbell Scientific datalogger CR3000 (Campbell Scientific Inc. 2018). There are additional sensors we have not discussed yet: two CS107 temperature probes (one at the top and another just above the datalogger box), one CS106 barometric pressure sensor, one HMP45C air temp and RH probe, and a CS700 rain gage. In addition, there are sensors buried below ground, which we discussed in Lab 7.

We already worked with data from this station in Lab 7 when we used a csv file named CR3000 _ March28-2019.csv from October 2018 to March 2019. We studied solar radiation and its measurement with a pyranometer. In Lab 9, we worked with wind speed at 2 and 10 m and wind direction extracted from this file. In this lab session, we will look at other variables.

For this purpose, in this lab session, we will use a more extensive dataset DP Weather data CR3000 _ 13Feb2019 _ 13Jan2021.dat, which is available from the RTEM GitHhub repository as DP Weather data CR3000 _ 13Feb2019 _ 13Jan2021.zip. Store it in labs/data and open it in Notepad++ or Geany (Figure 13.6)

Browse through the file and identify the columns 9, 10, 16, 17, 20 corresponding to the sensors needed to calculate evapotranspiration: AirTC_Avg, RH, SlrkJ_Tot, BP_mmHg_Avg, and WS_mph_2m_Avg.

Now let us write R code to read the file and select these variables

```
folder <- "data/"
file <- "DP Weather data CR3000_13Feb2019_13Jan2021.dat"
folder.file <- paste(folder,file,sep="")

# read file headers
header.meta <- scan(folder.file, sep=",", what="", nlines=1)
header.vars <- scan(folder.file, sep=",", what="", nlines=1,skip=1)
header.units <- scan(folder.file, sep=",", what="", nlines=1,skip=2)
header.calc <- scan(folder.file, sep=",", what="", nlines=1,skip=3)

# read all after header
tt.xx <- read.table(folder.file, sep=",", skip=4)
names(tt.xx) <- paste(header.vars,"(",header.units,")",sep="")
```

FIGURE 13.6 File CR3000 _ 13Feb2019 _ 13Jan2021.dat

and verify

```
# PET weather variables
w <- tt.xx[,c(1,2,9,10,16,17,20)] # time,RecID,TC,RH,SlE,BarP,WS2m
names(w)
> names(w)
[1] "TIMESTAMP(TS)"         "RECORD(RN)"            "AirTC_Avg(Deg C)"
[4] "RH(%)"                 "SlrkJ_Tot(kJ/m²)"      "BP_mmHg_Avg(mmHg)"
[7] "WS_mph_2m_Avg(miles/hour)"
>
```

Now, we exclude no data values NA,

```
# read time sequence from file
tt.raw <- strptime(w[,1], format="%Y-%m-%d %H:%M:%S",tz="")
tt <- tt.raw[-which(is.na(tt.raw))]
# all data but time
ww <- w[-which(is.na(tt.raw)),-c(1,2)] # remove those with time = NA
names(ww)
```

create a time series using `xts` and select one year of data

```
require(xts)
# create time series using array with the data and time base tt
ww.tt <- xts(ww,tt)
# select Feb 2019 to Feb 2020
w.t <- ww.tt["2019-2/2020-2"]
```

Using `xts`, we calculate hourly values and convert units as needed. Note that the energy is total for the hour but the other variables are averages for the hour.

```
# hourly values
ends.h <- endpoints(w.t,"hours")
# average for TC,RH,BP,WS
wt.h <- period.apply(w.t[,c(1,2,4,5)], ends.h , mean)
# total for solar energy
slr.h <- period.apply(w.t[,c(3)], ends.h , sum)
# select and convert units as needed
Temp.C <- wt.h[,1]
RH <- wt.h[,2]
# rad in Mj from kJ
Rad.MJ <- slr.h/1000
#barometric pressure in kPa from mmHg
BarP.kPa <- wt.h[,3]*0.133322
# wind in m/s from mph
Wind2m.mps <- wt.h[,4]*0.4
```

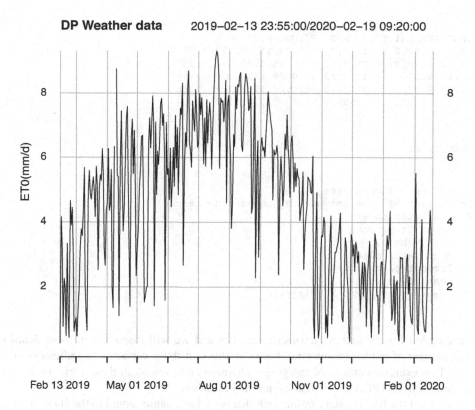

FIGURE 13.7   ET0 for DP weather station 2019–2020.

We are now ready to source the `ET0-FAO-function.R` file containing the `eto.fao` function and apply it to calculate hourly ET0, total to obtain daily, and plot to obtain Figure 13.7.

```
source("R/lab13/ET0-FAO-function.R")
# calculate hourly ET0
et0.h <- et0.fao(albedo=0.23,soil=0.0, Temp.C, RH, Rad.MJ, BarP.kPa, Wind2m.mps)
# total daily
et0.d <- apply.daily(et0.h,sum)
# plot
plot(et0.d,ylab="ET0(mm/d)",main="DP Weather data")
mtext("Daily ET0 (mm/d)",3,-1)
```

CANOPY MONITORING: DATA ANALYSIS

Let us work with data in the archive `canopy.zip` in the RTEM GitHub repository. Download and extract the file `canopy.csv` to the directory data in your working folder. The data set is from Acevedo et al. (2001) and the analysis from Acevedo (2013). It refers to fragmentation metrics of canopy images of a tropical cloud forest taken at different locations along two transects. Each image comes from a photograph of the canopy taken from the ground. A canopy opening is defined as the intensity of color in a pixel of the picture. Contiguous gaps make up patches of different sizes. The first rows the variable labels, and the first column has the sampling points `ID` label. This is segment illustrating the first few records.

```
ID,POC,PRL,LAI,NP,PD,MPS,LPI,MSI,MPFD
1n,2.98,8.02,2.29,1915,7.37,0.14,15.73,1.3,1.55
2n,1.72,8.76,2.94,2207,12.29,0.08,4.23,1.3,1.55
3n,1.21,5.33,2.78,1602,13.08,0.08,8.35,1.29,1.56
4n,1.71,7.27,2.53,2342,13.45,0.07,15.3,1.28,1.56
5n,5.19,16.57,3.02,2953,6.51,0.15,5.08,1.34,1.54
```

The variables are

```
POC= percent of open area
PRL= percent of area occupied by reflecting leaves
LAI= Leaf area index
NP= number of patches
PD= patch density
MPS= mean patch size
LPI= largest patch index
MSI= mean shape index
MPFD = mean patch fractal dimension
```

These are called fragmentation metrics or statistics and we will discuss with more detail in Lab 14 in the context of habitat monitoring. For now, think of these variables as indicators of canopy structure. This exercise consists of applying multivariate analysis to all these variables by principal components analysis (PCA), a method we learned in Lab 8.

First, we read the file as a data frame such that rows have names equal to the ID of the point

```
frag <- read.table("lab14/canopy.csv", header=T,row.names=1,sep=",")
```

By examining the data, we see that the variables have disparate ranges. It is important to standardize the variables before applying PCA or to perform the PCA using the correlation matrix instead of the covariance matrix. For simplicity, instead of standardizing, we will perform PCA using function princomp and setting argument cor=T.

```
> frag.pca <- princomp(frag,cor=T)
```

Check the summary

```
> summary(frag.pca)
Importance of components:
                          Comp.1     Comp.2     Comp.3     Comp.4     Comp.5
Standard deviation     2.0772715  1.7037539 0.84735546 0.81178307 0.52856526
Proportion of Variance 0.4794508  0.3225308 0.07977903 0.07322131 0.03104236
Cumulative Proportion  0.4794508  0.8019816 0.88176061 0.95498192 0.98602428
                          Comp.6       Comp.7      Comp.8       Comp.9
Standard deviation     0.25556321 0.169020175 0.155242159 0.0883232141
Proportion of Variance 0.00725695 0.003174202 0.002677792 0.0008667767
Cumulative Proportion  0.99328123 0.996455431 0.999133223 1.0000000000
>
```

We can see that it takes three components to explain almost 90% of the variance (88%) and that the first four components explain 95% of variance. Now, we plot loadings, variances, and biplots. We draw several biplots to include different pairwise combinations of components; for example, Comp2 vs. Comp 1 and Comp3 vs. Comp2.

```
barplot(loadings(frag.pca), beside=T)
plot(frag.pca)
biplot(frag.pca, choices=c(1,2))
biplot(frag.pca, choices=c(1,3))
```

which would yield plots shown in Figures 13.8–13.10.

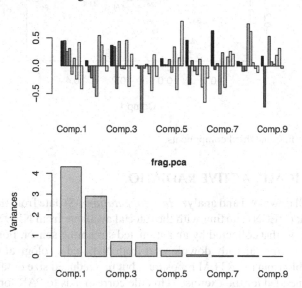

FIGURE 13.8 Plots for PCA of frag data using correlation.

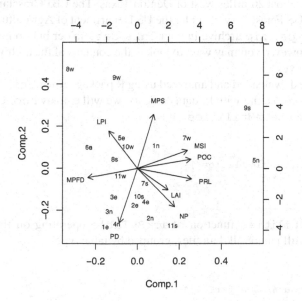

FIGURE 13.9 Biplot first and second components.

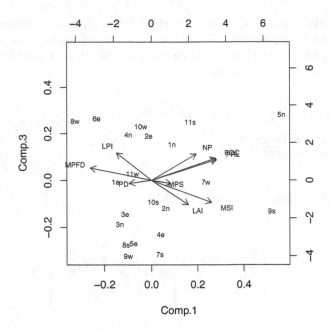

FIGURE 13.10 Biplot first and third components.

PHOTOSYNTHETICALLY ACTIVE RADIATION

In this section, we will download and analyze *Instrumentation* (IS) data from the National Ecological Observatory Network (NEON) starting with the tutorial available from NEON (2023a). The IS data category corresponds to that collected by an automated streaming sensor. For practice, we will use a NEON _ par.zip archive already downloaded from National Ecological Observatory Network (2023b) and is available from the RTEM repository, but in a reduced size obtained by deleting some of the large files not needed for the exercise. This file corresponds to PAR for the last three months of 2022 at the NEON LBJ National Grassland (CLBJ) site of Domain 11. This site is located near Decatur, Texas, USA, about 26 miles west of Denton Texas. The LBJ Grasslands is 16,800 acres of land managed by the US Forest Service under the US Department of Agriculture (AmeriFlux 2023).

Store this NEON _ par.zip archive in your labs/data folder but do not unzip yet, since we will use R to do so. However, you may want to look at the contents of the archive to become familiar with NEON data archives.

NEON data can be downloaded and analyzed using R package neonUtilities, neonOS, and raster. In this section, dealing with IS data category, we will employ neonUtilities (NEON 2023b). Install and load neonUtilities

```
install.packages("neonUtilities")
library(neonUtilities)
```

Now, we use neonUtilities function stackByTable operating on the NEON _ par.zip archive to perform a full join of all data files, grouped by table type.

```
> stackByTable("data/NEON_par.zip")
```

On the console, you would see the unzip process unfold and resulting in the following message

```
Merged the most recent publication of sensor position files for each site and
saved to /stackedFiles
Copied the most recent publication of variable definition file to /stackedFiles
Finished: Stacked 2 data tables and 3 metadata tables!
Stacking took 3.739974 secs
All unzipped monthly data folders have been removed.
```

In your `labs/data` folder, you should now have a folder `NEON_par` containing a folder `stackedFiles`, which contains several files including `PARPAR_30min.csv`, `PARPAR_1min.csv`, `sensor_positions.csv`, `variables.csv`, and `readme.txt`.

The plan is to read the 30-minute file using the function `readTableNEON()`. This function uses the `variables.csv` file to assign data types to each column of the dataset. It is useful to look at these variable specifications first; but instead of opening the file in an editor, we will read it into R as a data set using `read.csv`

```
parvar <- read.csv("data/NEON_par/stackedFiles/variables_00024.csv")
```

and then use `View` or `tibble` to inspect it.

```
> as_tibble(parvar)
# A tibble: 144 × 9
    table          fieldName    descr…¹ dataT…² units downl…³ pubFo…⁴ prima…⁵
categ…⁶
    <chr>          <chr>        <chr>   <chr>   <chr> <chr>   <chr>   <chr>   <lgl>
 1 PARPAR_1min domainID        Unique… string  <NA>  append… <NA>    N       NA
 2 PARPAR_1min siteID          NEON s… string  <NA>  append… <NA>    N       NA
 3 PARPAR_1min horizontal…     Index … string  <NA>  append… <NA>    N       NA
 4 PARPAR_1min verticalPo…     Index … string  <NA>  append… <NA>    N       NA
 5 PARPAR_1min startDateT…     Date a… dateTi… <NA>  basic   yyyy-M… <NA>    NA
 6 PARPAR_1min endDateTime     Date a… dateTi… <NA>  basic   yyyy-M… <NA>    NA
 7 PARPAR_1min PARMean         Arithm… real    micr… basic   *.##(r… <NA>    NA
 8 PARPAR_1min PARMinimum      Minimu… real    micr… basic   *.##(r… <NA>    NA
 9 PARPAR_1min PARMaximum      Maximu… real    micr… basic   *.##(r… <NA>    NA
10 PARPAR_1min PARVariance     Varian… real    micr… basic   *.##(r… <NA>    NA
# … with 134 more rows, and abbreviated variable names ¹description, ²
dataType,
#   ³downloadPkg, ⁴pubFormat, ⁵primaryKey, ⁶categoricalCodeName
# i Use 'print(n = ...)' to see more rows
>
```

You can browse through this `tibble` or use `View` to learn about the dataset we are about to work with. Importantly, we learn that there are several metrics of the 30-minute PAR, such as min, max, and mean, and that the PAR units are $\mu mol/(m^2 s)$.

Once we have done that, we are ready to use the function `readTableNEON` to read the 30-minute file as `dataFile` using the `varFile` which we just inspected.

```
par30 <- readTableNEON(
  dataFile=""data/NEON_par/stackedFiles/PARPAR_30min.csv"",
  varFile=""data/NEON_par/stackedFiles/variables_00024.csv""
)
```

The resulting `par30` object is a large dataset, so you can examine it with `View` or `as _ tibble`

```
> as_tibble(par30)
# A tibble: 29,280 × 24
   domainID siteID horizontalP…¹ verti…² startDateTime       endDateTime
   <chr>    <chr>  <chr>         <chr>   <dttm>              <dttm>
 1 D11      CLBJ   000           010     2022-09-01 00:00:00 2022-09-01 00:30:00
 2 D11      CLBJ   000           010     2022-09-01 00:30:00 2022-09-01 01:00:00
 3 D11      CLBJ   000           010     2022-09-01 01:00:00 2022-09-01 01:30:00
 4 D11      CLBJ   000           010     2022-09-01 01:30:00 2022-09-01 02:00:00
 5 D11      CLBJ   000           010     2022-09-01 02:00:00 2022-09-01 02:30:00
 6 D11      CLBJ   000           010     2022-09-01 02:30:00 2022-09-01 03:00:00
 7 D11      CLBJ   000           010     2022-09-01 03:00:00 2022-09-01 03:30:00
 8 D11      CLBJ   000           010     2022-09-01 03:30:00 2022-09-01 04:00:00
 9 D11      CLBJ   000           010     2022-09-01 04:00:00 2022-09-01 04:30:00
10 D11      CLBJ   000           010     2022-09-01 04:30:00 2022-09-01 05:00:00
# … with 29,270 more rows, 18 more variables: PARMean <dbl>, PARMinimum <dbl>,
#   PARMaximum <dbl>, PARVariance <dbl>, PARNumPts <dbl>, PARExpUncert <dbl>,
#   PARStdErMean <dbl>, PARFinalQF <dbl>, outPARMean <dbl>,
#   outPARMinimum <dbl>, outPARMaximum <dbl>, outPARVariance <dbl>,
#   outPARNumPts <dbl>, outPARExpUncert <dbl>, outPARStdErMean <dbl>,
#   outPARFinalQF <dbl>, publicationDate <chr>, release <chr>, and abbreviated
#   variable names ¹horizontalPosition, ²verticalPosition
# i Use 'print(n = ...)' to see more rows, and 'colnames()' to see all
variable names
>
```

We will focus on `verticalPosition` on the tower, which would correspond to attenuation by various canopy layers. Examine the unique values

```
pos <- unique(par30$verticalPosition)

pos
[1] "010" "020" "030" "040" "050"
>
```

concluding there are five unique vertical position levels. Note that there are multiple rows associated with each position; thus, we need to arrange as a data frame with columns given by each position level. One way of accomplishing this is to use `cbind` in a loop selecting the level by applying the `which` function

```
par30.h <- par30[which(par30$verticalPosition==pos[1]),]$PARMean
for (i in 2:5)
 par30.h <- cbind(par30.h,
        par30[which(par30$verticalPosition==pos[i]),]$PARMean)
```

and making the matrix into a data frame with names given by the vertical position labels

```
xx <- data.frame(par30.h)
names(xx) <- pos
```

double check the result by printing a few records

```
> xx[1:5,]
    010    020    030    040    050
1 3.25   5.01  27.12  27.40  29.87
2 0.63   0.91   4.91   4.79   5.38
3 0.04   0.01   0.20  -0.01   0.16
4 0.02  -0.02   0.06  -0.12   0.01
5 0.02  -0.03   0.05  -0.15   0.01
>
```

Next, we focus on the time stamp and note that it is given using GMT for the time zone; thus, we convert using the correct format and `tz="UTC"` which corresponds to GMT.

```
tt <- strptime(par30[which(par30$verticalPosition==pos[1]),
]$startDateTime, format=""%Y-%m-%d %H:%M:%S",tz="UTC")
```

We are interested in making a time series of all the vertical positions using package `xts`, which we make sure is loaded with `require`

```
require(xts)
x.t <- xts(xx,tt)
```

it is good practice to verify

```
> x.t[1:5,]
                      010    020    030    040    050
2022-09-01 00:00:00  3.25   5.01  27.12  27.40  29.87
2022-09-01 00:30:00  0.63   0.91   4.91   4.79   5.38
2022-09-01 01:00:00  0.04   0.01   0.20  -0.01   0.16
2022-09-01 01:30:00  0.02  -0.02   0.06  -0.12   0.01
2022-09-01 02:00:00  0.02  -0.03   0.05  -0.15   0.01
```

Create a title and label to use in plots

```
title ="CLBJ 30-min Avg PAR"
y.label = "PAR [umol/(m2 s)]"
pos.leg <- paste("Vpos=",pos,sep='')
```

We can start with a plot of the entire time series

```
plot(x.t,main=title,type="l",ylab=y.label,lwd=1,lty=1:5,col=1, grid.col=NA)
```

the result is not shown here, and after observing it you would realize that is difficult to visualize, but that there seems to be patterns for the levels. For better visualization, let us split the series by weeks and plot the first week (Figure 13.11)

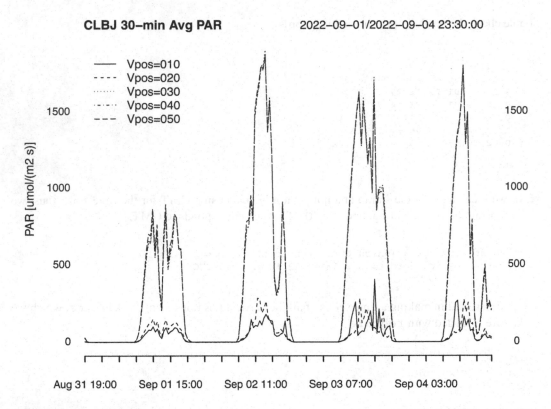

FIGURE 13.11 PAR 30-minute average at five vertical positions for the first week of September 2022. (Provisional Data from National Ecological Observatory Network (2023b).)

```
x.wk <- split(x.t, f="weeks")
plot(x.wk[[1]],main=title,ylab=y.label,lwd=1,lty=1:5,col=1, grid.col=NA)
addLegend("top",pos.leg,lty=1:5,col=1)
```

It is now easy to observe the dynamics of PAR for each day corresponding to daylight and nighttime as well as how the first two lower levels have much attenuated PAR with respect to the three top canopy levels, which is indicative of the canopy structure.

For more detailed views, we can split the series by days and plot one day using the same process (Figure 13.12)

```
x.d <- split(x.t, f="days")
plot(x.d[[1]],main=title,ylab=y.label,lwd=1,lty=1:5,col=1, grid.col=NA)
addLegend("top",pos.leg,lty=1:5,col=1)
```

As you have learned throughout the lab manual, there are many other interesting things that we can do once we have the data as an `xts` time series, such as applying daily totals, as well as studying light attenuation through the canopy.

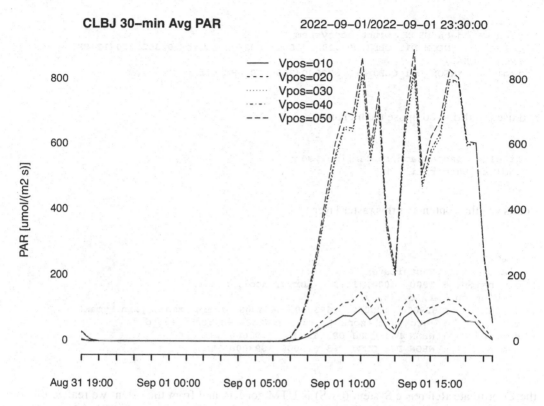

FIGURE 13.12 PAR 30-minute average at various vertical positions on the tower for one day in September. (Provisional Data from National Ecological Observatory Network (2023b).)

CANOPY HEIGHT BY LIDAR DATA

In this section, we will analyze remote sensing data collected by the *airborne observation platform* (AOP), e.g. LiDAR, from the NEON starting with the tutorial available from NEON (2023a). For practice, we will use a NEON _ struct-ecosystem.zip archive that we already downloaded from National Ecological Observatory Network (2023a) and is available from the RTEM repository. Download this archive to your labs/data folder. This file corresponds to the canopy height obtained by LiDAR in May 2022 at the NEON LBJ National Grassland (CLBJ) site of Domain 11 described in the previous section.

In this section, dealing with AOP data category we will use R package raster. We already used this package in Lab session 8 and should already be installed in your system. If you need to reinstall, you can use install.packages("raster"). Load this package

```
library(raster)
```

Extract the contents of the archive NEON _ struct-ecosystem.zip in your labs/data folder. You should now have a folder NEON _ struct-ecosystem containing a folder NEON. D11.CLBJ.DP3.30015.001.2022-05.basic.20230203T081218Z.PROVISIONAL/, which contains several files including NEON _ D11 _ CLBJ _ DP3 _ 629000 _ 3693000 _ CHM. tif. This is the file we will work with using raster. To facilitate addressing the file, we will define the following

```
path <- "data/NEON_struct-ecosystem/"
folder <- "NEON.D11.CLBJ.DP3.30015.001.2022-05.basic.20230203T081218Z.
PROVISIONAL/"
img1  <- "NEON_D11_CLBJ_DP3_629000_3693000_CHM.tif"
```

and use `paste` to compose the filename

```
file1 <- paste(path,folder,img1,sep='')
chm1 <- raster(file1)
```

Let us see the contents of this raster layer

```
> chm1
class       : RasterLayer
dimensions  : 1000, 1000, 1e+06   (nrow, ncol, ncell)
resolution  : 1, 1  (x, y)
extent      : 629000, 630000, 3693000, 3694000  (xmin, xmax, ymin, ymax)
crs         : +proj=utm +zone=14 +datum=WGS84 +units=m +no_defs
source      : NEON_D11_CLBJ_DP3_629000_3693000_CHM.tif
names       : NEON_D11_CLBJ_DP3_629000_3693000_CHM
```

the Coordinate Reference System (CRS) is UTM zone 14 and from the extent we realize this is a single layer of size $1000 \times 1000 \, \mathrm{m}^2$ with a spatial resolution of $1 \times 1 \, \mathrm{m}^2$ (refer to Lab 8 for a refresher of these concepts). We will visualize this layer using a plot to obtain Figure 13.13

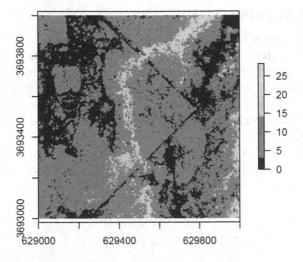

FIGURE 13.13   Canopy height (m) raster obtained from LiDAR data. (Data downloaded from National Ecological Observatory Network (2023a).)

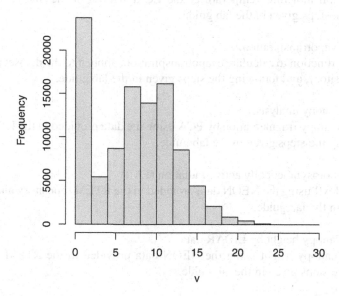

FIGURE 13.14 Histogram of canopy height from LiDAR. (Data downloaded from National Ecological Observatory Network (2023a).)

```
plot(chm1,col=topo.colors(6))
```

Taller canopy (more than 10 m) is present only in a strip pattern. Let us investigate the statistics of canopy height using a histogram (Figure 13.14)

```
hist(chm1,main="Canopy Height (m) NEON CLBJ")
```

Pixels with short canopy are very abundant and most of the taller canopy pixels are around 10 m in height.

EXERCISES

Exercise 13.1 Soil moisture sensor calibration.
   Perform calibration of an EC-5 probe using the data provided in the RTEM repository and following the steps given in the lab guide.

Exercise 13.2 Soil moisture analysis.
   Analyze the soil moisture dynamics of the four soil moisture sensors using the data provided in the RTEM repository and following the steps given in the lab guide.

Exercise 13.3 Soil conductivity sensor network.
　　Analyze soil moisture, temperature, and EC dynamics of the SDI-12 sensor network following the steps given in the lab guide.

Exercise 13.4 Evapotranspiration.
　　Build the function to calculate evapotranspiration, apply it to the dataset provided in the RTEM repository, and following the steps given in the lab guide.

Exercise 13.5 Canopy analysis.
　　Analyze canopy fragmentation by PCA using the data provided in the RTEM repository and following the steps given in the lab guide.

Exercise 13.6 Photosynthetically active radiation (PAR)
　　Analyze PAR using the NEON data provided in the RTEM repository and following the steps given in the lab guide.

Exercise 13.7 Canopy height by LiDAR data.
　　Analyze canopy height using the NEON data provided in the RTEM repository and following the steps given in the lab guide.

REFERENCES

Acevedo, M.F. 2013. *Data Analysis and Statistics for Geography, Environmental Science & Engineering. Applications to Sustainability*. Boca Raton, FL: CRC Press, Taylor & Francis Group, 535 pp.

Acevedo, M.F. 2024. *Real-Time Environmental Monitoring: Sensors and Systems - Textbook, Second Edition*. Boca Raton, FL: CRC Press, Taylor & Francis Group, 392 pp.

Acevedo, M.F., S. Monteleone, M. Ataroff, and C.A. Estrada. 2001. Aberturas del dosel y espectro de la luz en el sotobosque de una selva nublada andina de Venezuela. *Ciencia* 9(2):165–183.

AmeriFlux. 2023. *US-xCL: NEON LBJ National Grassland (CLBJ)*. Accessed January 2023. https://ameriflux.lbl.gov/sites/siteinfo/US-xCL/.

Campbell Scientific Inc. 2018. *CR300 Micrologger: Operators Manual*. Accessed March 2021. https://s.campbellsci.com/documents/us/manuals/cr3000.pdf.

Chen, L. 2008. Soil characteristics estimation and its application in water balance dynamics. Master of Science (Applied Geography), University of North Texas.

Meter. 2022a. *ECH20 EC-5 Soil Moisture Sensor*. Accessed April 2022. https://www.metergroup.com/environment/products/ec-5-soil-moisture-sensor/.

METER. 2022b. *TEROS12 Advanced Soil Moisture Sensor + Temperature and EC*. Accessed September 2022. https://www.metergroup.com/en/meter-environment/products/teros-12-soil-moisture-sensor.

National Ecological Observatory Network (NEON). 2023b. *Photosynthetically Active Radiation (PAR) (DP1.00024.001)*. Accessed January 2023. https://data.neonscience.org.

National Ecological Observatory Network (NEON). 2023a. *Ecosystem Structure (DP3.30015.001)*. https://data.neonscience.org/data-products/explore?site=CLBJ.

NEON. 2023a. *TUTORIAL Download and Explore NEON Data*. Accessed January 2023. https://www.neonscience.org/resources/learning-hub/tutorials/download-explore-neon-data.

NEON. 2023b. *TUTORIAL Use the NEON Utilities Package to Access NEON Data*. Accessed January 2023. https://www.neonscience.org/resources/learning-hub/tutorials/neondatastackr.

R Project. 2023. *The Comprehensive R Archive Network*. Accessed January 2023. http://cran.us.r-project.org/.

14 Wildlife Monitoring

INTRODUCTION

This lab session covers wildlife monitoring, and more generally topics of animal ecology that are useful for environmental monitoring. For this purpose, the guide uses an R program to look at details of the impact of the water–air interface on radio waves, which is important for radio telemetry in the aquatic environment. Subsequently, landscape metrics analysis using an R package is covered as a pathway to implement habitat monitoring. Spatial analysis including point patterns and kriging are covered, not because they relate exclusively to wildlife, but because they are general methods applicable to topics covered in the previous three labs (11, 12, and 13). The guide finishes analyzing breeding birds monitoring data, since this is very relevant to environmental indicators; data from the National Ecological Observatory Network (NEON) are employed as an example.

MATERIALS

READINGS

For theoretical background, you can use Chapter 14 of Acevedo, M.F. 2024. *Real-Time Environmental Monitoring: Sensors and Systems - Textbook, Second Edition* which is a companion to these guides (Acevedo 2024). Other bibliographical references are cited throughout the guide.

SOFTWARE (LINKS PROVIDED IN THE REFERENCES)

- R system (R Project 2023)
- R package `landscapemetrics` (Hesselbarth et al. 2019, 2022)
- R package for spatial analysis `spatstat`
- R package for geostatistics `sgeostat`
- R package `neonOS` for NEON OS data
- R package `neonUtilities` for NEON data
- R packages for other functions `renpow` and `tibble`

DATA AND PROGRAM FILES (AVAILABLE FROM THE GITHUB RTEM)

- Archive `water-air-interface.zip`
- Archive `spatial-functions.zip`
- Archive `spatial-data.zip`
- Archive `kriging-functions.zip`
- Archive `NEON _ count-landbird.zip`
- Archive `sppdiversity.zip`

SUPPLEMENTARY SUPPORT MATERIAL

Supplementary support material including additional screenshots, images, and procedures are available from the publisher's eResources web page provided for this book.

WATER–AIR INTERFACE

We can write a script to calculate transmission angle θ_t for a variety of incidence angle θ_i between 0° and 90° for two scenarios. The code is contained in `water-air-interface.zip` available from the RTEM repository.

```
# Snell's law water and air interface
perm.r <- 80
mat <- matrix(1:2, 2, 1, byrow = T)
layout(mat, widths = rep(30,2), heights = rep(30,2), respect=F)
par(mar = c(5, 8, 1, 8),cex=0.8)

theta.i <- seq(0,90,01); rad.i <- theta.i*pi/180
sin.rad.t <- sin(rad.i)/sqrt(perm.r)
rad.t <- asin(sin.rad.t)
theta.t <- rad.t*180/pi
plot(theta.i,theta.t,type="l",
     xlab="Incidence angle (degrees)",
     ylab= "Transmission angle (degrees)")
abline(h=6.42,lty=2)
text(50,3,"Air to Water",cex=0.8)
theta.i <- seq(0,8,0.01); rad.i <- theta.i*pi/180
sin.rad.t <- sqrt(perm.r)*sin(rad.i)
rad.t <- asin(sin.rad.t)
theta.t <- rad.t*180/pi
plot(theta.i,theta.t,type="l",
     xlab="Incidence angle (degrees)",
     ylab= "Transmission angle (degrees)")
abline(v=6.42,lty=2)
text(1,60,"Water to Air",cex=0.8)
```

We see in the resulting graphs (Figure 14.1) that due to water's high relative permittivity the transmission angle from air to water is low, below 6.41°, meaning that waves penetrate almost normal to the surface. Likewise, the transmission from water to air is such that waves with incidence angles higher than 6.41° emerge almost parallel to the water surface.

LANDSCAPE METRICS

In Chapter 14 of the companion textbook, we described the importance of tracking landscape metrics to assess changes in wildlife habitat. We will use R package `landscapemetrics` (Hesselbarth et al. 2019, 2022) to exercise these calculations. First, install and load the package.

```
install.packages("landscapemetrics")
library(landscapemetrics)
```

We will use the sample data `landscape` in the package as an example and will rename with a short label `xl` for convenience

```
# sample dataset in package
data(landscape)
# short name for convenience
xl <- landscape
```

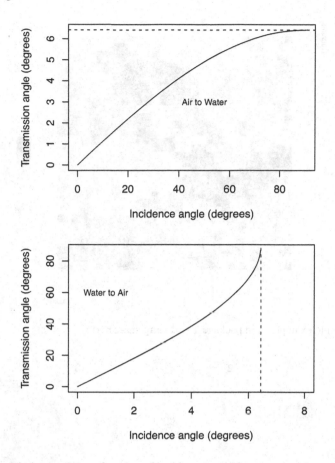

FIGURE 14.1 Transmission angle as a function of incidence angle for two scenarios.

inspect the dataset

```
> xl
class        : RasterLayer
dimensions   : 30, 30, 900  (nrow, ncol, ncell)
resolution   : 1, 1  (x, y)
extent       : 0, 30, 0, 30  (xmin, xmax, ymin, ymax)
crs          : NA
source       : memory
names        : clumps
values       : 1, 3  (min, max)

>
```

It is a small raster $30 \times 30 = 900$ cells. As discussed in Chapter 14 of the companion textbook, this simple landscape helps us understand the metrics. Now, we draw the image (Figure 14.2)

```
show_landscape(xl, discrete = TRUE )
```

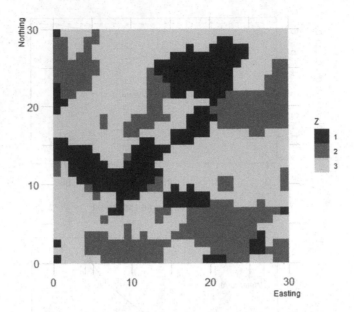

FIGURE 14.2   Sample example from package `landscapemetrics`.

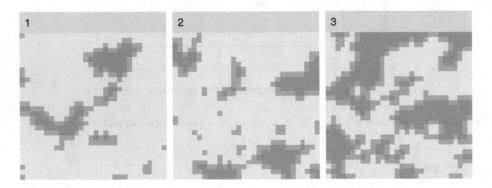

FIGURE 14.3   Classes by patch type.

We see three classes, labeled simply 1, 2, and 3. As explained in Chapter 14 of the textbook, observe how you can follow all contiguous cells of one class and distinguish them on the landscape into patches, which we can separate using cores into the three maps of Figure 14.3,

```
show_cores(x1)
```

one for each class showing all the patches you can trace for each in Figure 14.2. As explained in Chapter 14, each one of the three maps of Figure 14.3 is a *class* meaning is the layer representation for that patch type displaying the patches in that class.

There is a *boundary* around each patch, and the cells inside the boundary are defined as *core area*; this has to do with the rule that assigns a cell to a patch, which entails several decisions. First, a cell may look in all eight directions from it to its neighbors, like a queen move in chess, or only in four directions from it, as a rook moves in a chess board.

The functions of `landscapemetrics` have prefix `lsm _ c _` for a class level metric, `lsm _ p _` for a patch level metric, and `lsm _ l _` for a landscape level metric. We can find out the number of patches np for each class using the chess queen move or eight-directions and `lsm _ c _`

```
>lsm_c_np(x1, directions=8)
```

and looking at the result

```
# A tibble: 3 × 6
  layer level class      id metric value
  <int> <chr> <int> <int> <chr>   <dbl>
1     1 class     1    NA np          9
2     1 class     2    NA np         14
3     1 class     3    NA np          4
>
```

We note from this 3×6 `tibble`, each row is a class, and the last column has the number of patches for that class. Class 2 has more patches than the other two, and class 3 has only a few patches. To include a cell as part of a patch, the program calculates the number of sides of a cell that belong to the boundary and how far away from the patch that cell is.

Examine the core area for each class

```
> lsm_c_ca(x1, directions = 8)
# A tibble: 3 × 6
  layer level class      id metric  value
  <int> <chr> <int> <int> <chr>    <dbl>
1     1 class     1    NA ca      0.0179
2     1 class     2    NA ca      0.0242
3     1 class     3    NA ca      0.0479
>
```

That shows a larger core area for class 3. A cell is considered part of the core area if all the neighbors are of the same class. Calculations are done in meters with area in hectares (ha). Recall that $1\,\text{ha} = 10{,}000\,\text{m}^2$. This is convenient when the image is in UTM coordinates. A 30×30 pixel is

$$\frac{30 \times 30\,\text{m}^2}{10{,}000\,\text{m}^2/\text{ha}} = 0.09\,\text{ha or nearly a tenth of a hectare.}$$

Consider calculation of a patch metric based on area,

```
> pa <- lsm_p_area(x1, directions = 8)
> pa
# A tibble: 27 × 6
   layer level class      id metric  value
   <int> <chr> <int> <int> <chr>    <dbl>
 1     1 patch     1     1 area    0.0001
 2     1 patch     1     2 area    0.0005
 3     1 patch     1     3 area    0.0148
 4     1 patch     1     4 area    0.0001
 5     1 patch     1     5 area    0.0001
 6     1 patch     1     6 area    0.0014
 7     1 patch     1     7 area    0.0003
 8     1 patch     1     8 area    0.0005
 9     1 patch     1     9 area    0.0001
10     1 patch     2    10 area    0.0035
# … with 17 more rows
# i Use 'print(n = ...)' to see more rows
>
```

This is a `tibble` and we can use `print(n=27, pa)` to know more, not shown here for the sake of space, but you would see that column `id` is the patch, that column `class` is the class that the patch belongs to, and that column values are the area in ha. Then, the *core area index* (CAI) which is defined as the proportion of core area of the patch in relation to the total patch area expressed in percent can be calculated with

```
pcai <- lsm_p_cai(xl, directions = 8)
```

and you can examine the results with `print(n=27, pcai)`.

Now at the class level, we can use the statistics (mean, standard deviation, coefficient of variation) of CAI values for each class. For instance, the mean

```
> lsm_c_cai_mn(xl, directions = 8)
# A tibble: 3 × 6
  layer level class    id metric value
  <int> <chr> <int> <int> <chr>  <dbl>
1     1 class     1    NA cai_mn  6.92
2     1 class     2    NA cai_mn 11.2
3     1 class     3    NA cai_mn 14.8
>
```

In this example, the mean *CAI* for class 1 is 6.92%, which tells us that on average these patches have little core area compared to the patch size, whereas it is 14.8% for class 3, confirming our intuition that this class would be a better habitat. The coefficient of variation of the patches *CAI* for the class, which is the variability scaled by the mean, allows a better comparison. In this case, we get 233 for class 1 vs. 158 for class 3.

```
> lsm_c_cai_cv(xl, directions = 8)
# A tibble: 3 × 6
  layer level class    id metric value
  <int> <chr> <int> <int> <chr>  <dbl>
1     1 class     1    NA cai_cv  233.
2     1 class     2    NA cai_cv  160.
3     1 class     3    NA cai_cv  158.
```

One *shape* metric of the patches is their *fractal dimension*, which expresses a ratio of perimeter to area, thus measuring the patch complexity. At the class level, metrics may consist of the mean and coefficient of variation of the fractal dimension of all patches in the class. Computing the mean fractal dimension of the example

```
> lsm_c_frac_mn(xl, directions = 8)
# A tibble: 3 × 6
  layer level class    id metric  value
  <int> <chr> <int> <int> <chr>   <dbl>
1     1 class     1    NA frac_mn  1.15
2     1 class     2    NA frac_mn  1.23
3     1 class     3    NA frac_mn  1.30
>
```

we get 1.15, 1.23, and 1.30 for classes 1, 2, and 3, indicating an increase in complexity of the patches from classes 1 to 3.

In terms of *edge* of patches, one metric is *edge density* calculated by the sum of all edges of a class divided by the landscape area and given in m/ha.

```
> lsm_c_ed(x1, directions = 8)
# A tibble: 3 × 6
  layer level class      id metric value
  <int> <chr> <int> <int> <chr>   <dbl>
1     1 class     1    NA ed       2000
2     1 class     2    NA ed       2522.
3     1 class     3    NA ed       3567.
>
```

In the example, class 1 has an edge density of 2 km/ha whereas class 3 has 3.6 km/ha.

Lastly, we mention an *aggregation* metric clumpy, defined by the proportional deviation of the proportion of like adjacencies (of the class) from that expected under a spatially random distribution.

```
> lsm_c_clumpy(x1)
# A tibble: 3 × 6
  layer level class      id metric value
  <int> <chr> <int> <int> <chr>   <dbl>
1     1 class     1    NA clumpy 0.732
2     1 class     2    NA clumpy 0.697
3     1 class     3    NA clumpy 0.649
>
```

In the example, this metric for class 1 would have a value of 0.732 while for class 3 the value is lower than 0.649, indicating more aggregation in class 1.

SPATIAL ANALYSIS

This section is a brief summary of spatial analysis based on the material covered in detail in Acevedo (2013) plus some examples related to animal ecology. We will use the spatial analysis spatstat and geostatistics sgeostat R packages. First, install these packages

```
install.packages('spatstat')
install.packages('sgeostat')
```

and load

```
library(spatstat)
library(sgeostat)
```

Also, from the RTEM repository download archive `spatial-data.zip` and extract to `labs/data` to obtain folder `labs/data/spatial-data` containing several files. In addition, we will use `spatial-functions.zip` also available from the RTEM repository; extract its contents `spatial-functions.R` to the folder `labs/R/lab14` that you would have created for this session.

As a first example, we will look at file `spatial-data/unif100.csv`. It has header coordinate labels x, y, then the records

```
x,y
0.301,0.376
0.439,0.845
0.48,0.491
0.525,0.262
0.319,0.679
```

We can read it with `read.table`

```
path <- "data/spatial-data/"
file <- "unif100.csv"
filename <- paste(path,file,sep='')
unif100 <- read.table(filename,sep=',',header=TRUE)
```

and check the result

```
> unif100[1:5,]
      x     y
1 0.301 0.376
2 0.439 0.845
3 0.480 0.491
4 0.525 0.262
5 0.319 0.679
>
```

convert into a ppp object using function `ppp` which specifies the coordinates

```
unif100.p <- ppp(unif100$x,unif100$y)
```

We can check the contents

```
> unif100.p
Planar point pattern: 100 points
window: rectangle = [0, 1] x [0, 1] units
>
```

We can use summary on a ppp object

```
> summary(unif100.p)
Planar point pattern:  100 points
Average intensity 100 points per square unit

Coordinates are given to 3 decimal places
i.e. rounded to the nearest multiple of 0.001 units

Window: rectangle = [0, 1] x [0, 1] units
Window area = 1 square unit
```

To plot the ppp object we will use function panels of package renpow

```
require(renpow)
panels(6,6,1,1,pty='m')
plot(unif100.p$x,unif100.p$y, xlab="x",ylab="y")
title("unif100.p",cex.main=0.8)
```

which yields Figure 14.4.

QUADRAT ANALYSIS

From the file spatial-functions.R, we will use function quad.chisq.ppp, which requires a point pattern and a target density. Therefore, source this file

```
source("R/lab14/spatial-functions.R")
```

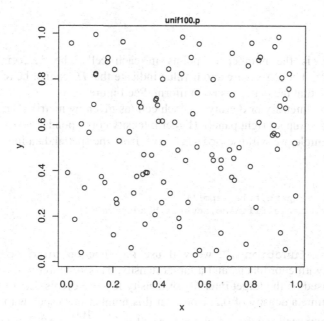

FIGURE 14.4 Point pattern plot from file.

We will apply to `unif100` obtained earlier. The function will define the number of cells in the grid based on the target intensity or density. Recall that chi-square requires five points per cell. Note that Xp complies with this minimum for a 4×4 grid because the expected number of points per cell is $\frac{m}{T} = \frac{100}{4 \times 4} = \frac{100}{16} = 6.25$, where m is the number of points and T is the number of cells. So, apply function `quad.chisq.ppp` with arguments Xp and 5

```
quad100 <- quad.chisq.ppp(unif100,5)
```

And check the results

```
> quad100
$pppset
 planar point pattern: 100 points
window: rectangle = [ 0 , 1 ] x [ 0 , 1 ]

$Xint
     [,1] [,2] [,3] [,4]
[1,]    8    4    4    9
[2,]    4   10    5    9
[3,]    6   11   10    3
[4,]    3    5    5    4

$intensity
[1]  6.25

$chisq
[1]  18.4

$p.value
[1]  0.1891652
>
```

The matrix Xint is the number of points in each cell. The `intensity` is the grand mean $= 100/16 = 6.25$. The chi-square and p-value indicate that H_0 cannot be rejected and therefore there is no evidence that the pattern is not uniform. See Figure 14.5

In this figure, the intensity or density for each cell (as given by matrix Xint) is visualized as a lattice or raster image (upper right panel). Higher intensity corresponds to darker gray.

As a second example, we will use `pois100.csv` from the spatial data files

```
file <- "pois100.csv"
filename <- paste(path,file,sep='')
rand100 <- read.table(filename,sep=',',header=TRUE)
```

from the `spatial-functions.R`, we will use the function `quad.poisson.ppp`, which requires a dataset with coordinates and a target density. This function will define the number of cells in the grid based on the target intensity or density for rare events. Let us use the sample data `rand100` and assume a density of 0.2. Note that this number corresponds to a grid of about 500 cells because the expected number of points per cell is $\frac{m}{T} = \frac{100}{500} = 0.2$. Let us apply the function with arguments `rand100` and 0.2

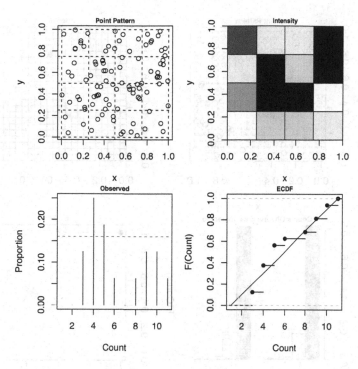

FIGURE 14.5 Quad analysis for uniformity.

```
pois100 <- quad.poisson.ppp(rand100,0.2)
```

And check the results

```
> pois100
$pppset
Planar point pattern: 100 points
window: rectangle = [0, 1] x [0, 1] units

$num.cells
[1] 484
```

We do not show $Xint for the sake of space

```
$chisq
[1] 0.02195254

$df
[1] 2

$p.value
[1] 0.9890837

$intensity
[1] 0.2066116

>
```

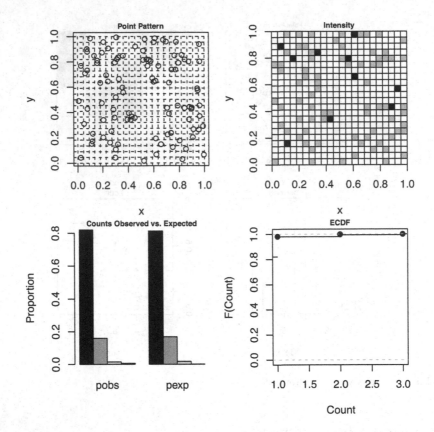

FIGURE 14.6   Poisson test for randomness.

The chi-square and p-value indicate that H_0 cannot be rejected and there is no evidence that the pattern is not random (Figure 14.6). As before, we are visualizing the intensity or density for each cell (i.e., lattice data) in the upper right panel using a gray scale for intensity.

NEAREST-NEIGHBOR ANALYSIS: G AND K FUNCTIONS

For random patterns, the number of points should follow a Poisson with mean λA, where λ is the rate or intensity (number of points per unit area) and A is the area. We will use the nearest-neighbor statistics G and K. First, convert the dataset `rand100` of the previous section to `ppp` object and apply `Gest` function with

```
file <- "pois100.csv"
filename <- paste(path,file,sep='')
rand100 <- read.table(filename,sep=',',header=TRUE)
pois100 <- quad.poisson.ppp(rand100,0.2)
"none" and "km" for edge correction

rand100.p <- ppp(rand100$x,rand100$y)
G.u <- Gest(rand100.p,correction=c("none","rs","km"))
```

Check the result

```
> G.u
Function value object (class 'fv')
for the function r -> G(r)
.............................................................
           Math.label       Description
r          r                distance argument r
theo       G[pois](r)       theoretical Poisson G(r)
raw        hat(G)[raw](r)   uncorrected estimate of G(r)
rs         hat(G)[bord](r)  border corrected estimate of G(r)
km         hat(G)[km](r)    Kaplan-Meier estimate of G(r)
hazard     hat(h)[km](r)    Kaplan-Meier estimate of hazard function h(r)
theohaz    h[pois](r)       theoretical Poisson hazard function h(r)
.............................................................
Default plot formula:  .~r
where "." stands for 'km', 'rs', 'raw', 'theo'
Recommended range of argument r: [0, 0.079639]
Available range of argument r: [0, 0.19143]
>
```

As indicated r is the distance, raw is the uncorrected G, km is the Kaplan–Meier estimate of G, rs is the reduced sample correction or border corrected G, and theo is the theoretical G that would be obtained if the data followed a Poisson.

Kest provides estimates of Ripley's K-function. There are several corrections: e.g., isotropic and border. Apply Kest function to rand100.p with "none" and "iso" for edge correction

```
> K.u <- Kest(rand100.p,correction=c("none","iso"))
> K.u
Function value object (class 'fv')
for the function r -> K(r)
.............................................................
           Math.label       Description
r          r                distance argument r
theo       K[pois](r)       theoretical Poisson K(r)
un         hat(K)[un](r)    uncorrected estimate of K(r)
iso        hat(K)[iso](r)   isotropic-corrected estimate of K(r)
.............................................................
Default plot formula:  .~r
where "." stands for 'iso', 'un', 'theo'
Recommended range of argument r: [0, 0.25]
Available range of argument r: [0, 0.25]
>
```

Note that r is the distance, un is the uncorrected K, iso is the isotropic corrected K, and theo is the theoretical K that would be obtained if the data followed a Poisson.

We can produce graphics for comparison of empirical to theoretical (Figure 14.7).

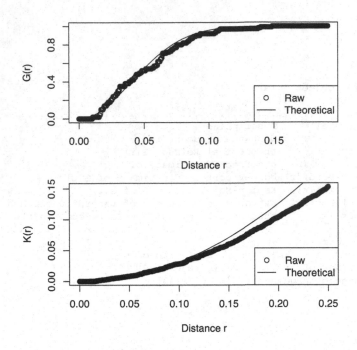

FIGURE 14.7 Ghat and Khat function: empirical compared to theoretical.

```
require(renpow)
panels(6,6,2,1,pty='m')

plot(G.u$r,G.u$raw,xlab="Distance r",ylab="G(r)")
lines(G.u$r,G.u$theo)
legend("bottomright",leg=c("Raw","Theoretical"),pch=c(1,-1),lty=c(-1,1),
merge=TRUE)

plot(K.u$r,K.u$un,xlab="Distance r",ylab="K(r)")
lines(K.u$r,K.u$theo)
legend("bottomright",leg=c("Raw","Theoretical"),pch=c(1,-1),lty=c(-1,1),
merge=TRUE)
```

Overall, the empirical plot is very close to the theoretical and therefore the pattern seems to be random.

VARIOGRAMS

Now we will work with point patterns that have continuous values at each point. As a first example, we will use the spatial-data/xyz.csv data file, which has a mark z at each point of coordinates x,y. This is how the first few records of the file should look like.

```
x,y,z
0.510 0.751 1.830
0.273 0.713 0.732
0.051 0.734 0.747
0.719 0.659 3.761
0.619 0.832 3.444
0.052 0.296 2.058
0.357 0.317 1.825
```

We will read this file

```
xyz <- read.table("data/spatial-data/xyz.csv",sep=",",header=TRUE)
```

use function `point`, of package `sgeostat` to generate a point object, plot it to visualize the point pattern (Figure 14.8)

```
xyz.p <- point(xyz)
plot.point(xyz.p,v='z',legend.pos=2,pch=c(21:24),cex=0.7)
```

Now use `pair` to generate a pair object based on the number of lags, max distance, and type. The pair object contains all pairs separated at each lag up to max distance, then estimate the variogram

```
xyz.pr <- pair(xyz.p,num.lags=10, type='isotropic', theta, dtheta, maxdist=0.45)
xyz.v <- est.variogram(xyz.p,xyz.pr,'z')
```

the variogram result has the static at various lags; we will use the classic variogram described in Chapter 14 of the companion textbook.

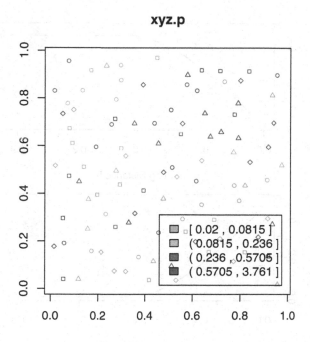

FIGURE 14.8 Marked point pattern. Symbols represent various quantiles.

```
> xyz.v
   lags  bins   classic      robust      med      n
1      1 0.0225 0.09990675 0.04330418 0.03840832   12
2      2 0.0675 0.70693031 0.23132770 0.14496928   68
3      3 0.1125 0.78272027 0.33766519 0.14451244  138
4      4 0.1575 0.74364678 0.29546471 0.13872255  192
5      5 0.2025 0.83539287 0.37358955 0.20924215  217
6      6 0.2475 0.96216997 0.34356786 0.13403939  267
7      7 0.2925 0.86799281 0.35400730 0.19816677  288
8      8 0.3375 0.67231396 0.29973627 0.14735033  279
9      9 0.3825 0.85489899 0.34439950 0.19108823  327
10    10 0.4275 0.92901103 0.33775452 0.15892916  292
>
```

Initial estimates of a semivariogram model parameters would be for the sill to be approximately 0.5 of the last value in classic $0.92/2 = 0.46$ and using the variance of the entire field

```
> var(xyz$z)
[1] 0.4144493
```

A reasonable first estimate is in between these two, i.e., $c_0 = 0.44$. The range would be that lag where the values in the classic column reaches nearly the corresponding value of the sill $0.44 \times 2 = 0.88$, which is in between 0.20 and 0.24, i.e., $\sim a = 0.22$. We will assume the nugget is zero. With these estimates, now we use function model.semivar.cov of the spatial-functions.R to plot the variogram together with an estimated spherical model, as well as the covariance (Figure 14.9).

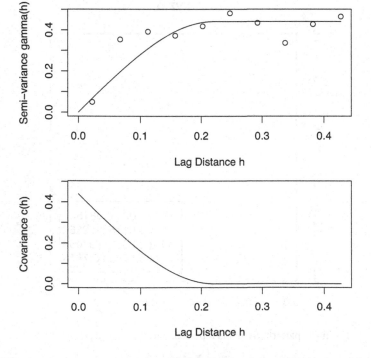

FIGURE 14.9   Calculated semivariogram along with spherical model estimate and covariance.

```
m.xyz.v <- model.semivar.cov(var=xyz.v, p=list(n0=0,c0=0.44,a=0.22))
```

variograms models can also be fit by nonlinear optimization but we do not discuss here for the sake of space; detailed descriptions are given in Acevedo (2013).

KRIGING

Extract the contents of `kriging-functions.zip` to your `lab14` folder and source it

```
source("R/lab14/kriging-functions.R")
```

Use `make.variogram` function from `kriging-functions.R` to make a variogram model with the above parameter values in the following manner

```
xyz.vsph <- make.variogram(nugget=0, sill=0.44, range=0.22)
```

Now, we use this model to perform ordinary kriging on a 100×100 grid using function `Okriging`. From `kriging-functions.R`. First, we must select a step for the grid for the prediction. Use minimum and maximum values in each axis to select a distance step. In this case, we will use step=0.01.

```
xyz.ok <- Okriging(xyz, xyz.vsph, step=0.01, maxdist=0.25)
```

We obtain a dataset of the kriged values of the variable z over the prediction grid together with the variance of the kriging error. Examine `xyz.ok` with `tibble` and note that it has the following contents

```
> as_tibble(xyz.ok)
# A tibble: 9,216 × 4
       x      y  zhat varhat
   <dbl> <dbl> <dbl>  <dbl>
 1 0.015 0.017    NA     NA
 2 0.025 0.017    NA     NA
 3 0.035 0.017    NA     NA
 4 0.045 0.017    NA     NA
 5 0.055 0.017    NA     NA
 6 0.065 0.017    NA     NA
 7 0.075 0.017    NA     NA
 8 0.085 0.017    NA     NA
 9 0.095 0.017    NA     NA
10 0.105 0.017    NA     NA
# … with 9,206 more rows
# i Use 'print(n = ...)' to see more rows
>
```

columns x and y contain the x and y coordinates of the predictions, zhat is the predicted value, and varhat is the variance estimate. Also, note that some of the prediction values are NA, which occur near the boundaries where there are no neighbors to interpolate the values.

Now, to obtain images just apply the function plot.kriged also from kriging-functions.R

```
plotkriged(xyz, xyz.ok,outpdf="output/xyz-kriged.pdf")
```

and obtain two pages in the output pdf file just declared. The first page (Figure 14.10) is the raster image of the kriged values. For additional visualization, we superimpose a contour map and a plot of the original point pattern (measured points). The second page (Figure 14.11) is the variance of the kriging error and provides a visual idea of how the error varies over the domain. The function produces also output that we can use for other purposes.

EXAMPLES RELATED TO ANIMAL ECOLOGY AND WILDLIFE MONITORING

Package spatstat includes some sample data that we can use to see the application of the techniques just learned to animal ecology and wildlife monitoring. Consider the dataset anemones that includes locations and diameters of sea anemones on a boulder (beadlet anemone *Actinia equina*) (Upton and Fingleton 1985). First look at the data and note it is a marked point pattern with marks equal to the diameter

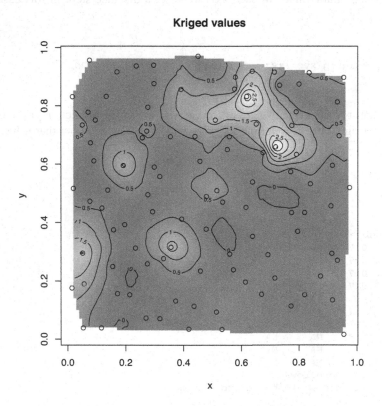

FIGURE 14.10 Ordinary kriging results.

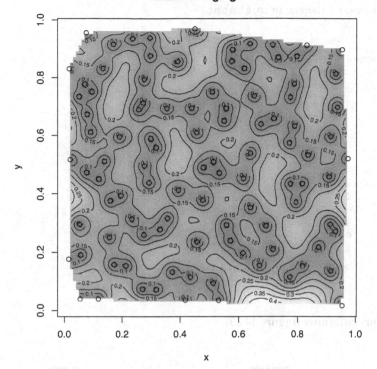

FIGURE 14.11 Variance of the kriging error.

```
> anemones
Marked planar point pattern: 231 points
marks are numeric, of storage type  'integer'
window: rectangle = [0, 280] x [0, 180] units
>
```

Let us plot it (Figure 14.12)

```
plot(anemones, markscale=1)
```

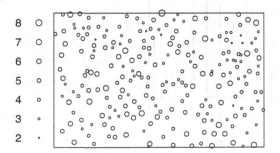

FIGURE 14.12 Anemones dataset from spatstat.

We will change the name for brevity and convert to a data frame so that we can manipulate it using other functions besides the one in spatstat.

```
an <- anemones
an.df <- data.frame(an)
```

Examine it

```
> an.df[1:5,]
     x  y marks
1  27  7     6
2 197  5     4
3  74 15     4
4 214 18     6
5 121 22     3
>
```

And analyze for uniformity (Figure 14.13)

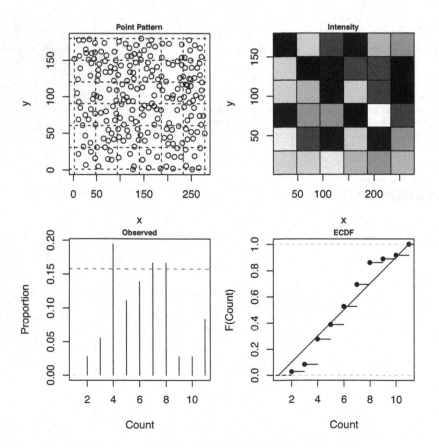

FIGURE 14.13 Quadrat analysis of anemones dataset.

```
an.q <- quad.chisq.ppp(an.df,5)

> an.q
$pppset
Planar point pattern: 231 points
window: rectangle = [2, 279] x [1, 180] units

$Xint
     [,1] [,2] [,3] [,4] [,5] [,6]
[1,]   11    4    7    8    5    6
[2,]    4   11    8    7    7   10
[3,]    4    5    8    4    7   11
[4,]    8    6    6    9    2    7
[5,]    3    7    8    4    8    6
[6,]    4    4    3    5    6    5

$intensity
[1] 6.333333

$chisq
[1] 30.31579

$p.value
[1] 0.6488844

>
```

Concluding that we cannot reject the H_0 of being uniform. If we test for randomness, we will see that the pattern departs from a Poisson distribution. More interesting, when calculating the variogram using geostat functions

```
an.p <- point(an.df)
an.pr <- pair(an.p,num.lags=20, type='isotropic', theta, dtheta, maxdist=100)
an.v <- est.variogram(an.p,an.pr,'marks')
```

results in a classic variogram that does not follow an increasing trend to a sill but rather relatively constant values for all lags.

```
> an.v
   lags bins  classic   robust      med   n
1     1  2.5 4.500000 1.722653 1.722653   4
2     2  7.5 2.794872 1.841863 2.158273  78
3     3 12.5 2.511013 1.588790 2.177813 227
4     4 17.5 2.883534 1.760099 2.178725 249
5     5 22.5 3.104225 2.181865 2.181541 355
6     6 27.5 2.837264 2.221193 2.182619 424
7     7 32.5 3.058190 1.990911 2.183098 464
8     8 37.5 3.049213 1.980225 2.183538 508
9     9 42.5 3.090253 2.064895 2.183923 554
10   10 47.5 2.815182 1.993024 2.184288 606
```

This means that the covariance is zero for lags other than $h=0$, which is the variance of the field. Indeed, we can check that the variance

```
> var(an$marks)
[1] 1.556446
```

is approximately the same as the mean of the variogram or ~1.5

```
> mean(an.v$classic/2)
[1] 1.493454
```

this is an example of a spatial pattern with no spatial dependence; in other words, the marks are spatially uncorrelated. In this case, we cannot interpolate by kriging since there is no relationship between neighboring point. An analogy for time series is one where the autocorrelation is zero for all lags or an AR(0) model which is basically white noise.

Another example in spatstat is the gorillas dataset, which is a marked point pattern with locations of 647 nesting sites of gorilla groups observed over time (Funwi-Gabga and Mateu 2012). This dataset has duplicated observations; we will study one way of dealing with this issue. Convert to a data frame, find the duplicates, and remove them

```
gordf <- data.frame(gorillas)
dup <- which(duplicated(gordf[,1:2])==TRUE)
X <- gordf[-dup,]
```

Select only the dry season points

```
X.dry <- X[which(X$season =="dry"),]
```

Determine extent, convert to ppp and plot (Figure 14.14)

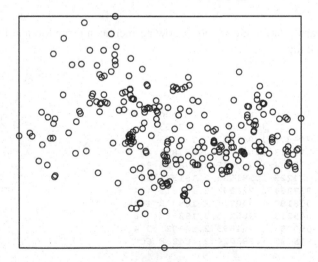

FIGURE 14.14 Plot of sample gorillas dataset. Dry season only.

```
xr <- c( floor(min(X.dry[,1])), ceiling(max(X.dry[,1])) )
yr <- c( floor(min(X.dry[,2])), ceiling(max(X.dry[,2])) )
Xp.dry <- ppp(X.dry$x,X.dry$y,xrange=xr,yrange=yr)
plot(Xp.dry)
```

Perform the nearest-neighbor analysis and plot (Figure 14.15)

```
require(renpow)
panels(6,6,2,1,pty='m')

G.u <- Gest(Xp.dry,correction=c("none","rs","km"))
plot(G.u$r,G.u$raw,xlab="Distance r",ylab="G(r)")
lines(G.u$r,G.u$theo)
legend("bottomright",leg=c("Raw","Theoretical"),pch=c(1,-1),
lty=c(-1,1), merge=TRUE)

K.u <- Kest(Xp.dry,correction=c("none","iso"))
plot(K.u$r,K.u$un,xlab="Distance r",ylab="K(r)")
lines(K.u$r,K.u$theo)
legend("bottomright",leg=c("Raw","Theoretical"),pch=c(1,-1),
lty=c(-1,1), merge=TRUE)
```

We can conclude that spatially, the pattern departs from a Poisson distribution and therefore it is not random. A better analysis would include using a polygonal boundary corresponding to the boundary of the sanctuary and analyzing the pattern only within that polygon.

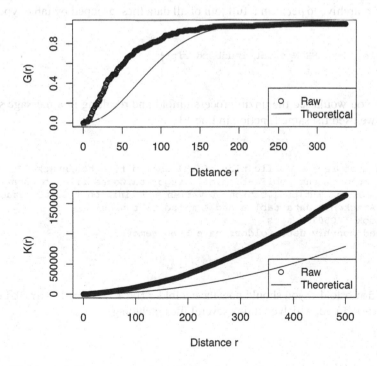

FIGURE 14.15   Nearest-neighbor analysis of gorillas dataset. Dry season only.

BREEDING BIRDS

Monitoring breeding birds has relevance because of several factors, among them being indicators of land use and climate change (NEON 2023). In this section, we will work with data on breeding birds from NEON (National Ecological Observatory Network 2023), which is part of the *Observational* (OS) data category. The OS data category corresponds to that collected by surveys, field work, and laboratory analysis. For practice, we will use a NEON _ count-landbird.zip archive that we already downloaded from National Ecological Observatory Network (2023) and placed in the RTEM repository. For simplicity, you can directly download from the repository and store in your labs/data folder but do not unzip yet since we will use R to do so. This file corresponds to breeding birds count data for May 2022 at the NEON LBJ National Grassland (CLBJ) site of Domain 11, which we already described in Chapter 13.

Package neonOS must be installed and loaded along with neonUtilities.

```
install.packages("neonUtilities")
install.packages("neonOS")

library(neonUtilities)
library(neonOS)
```

Also, set global option to not convert all character variables to factors

```
options(stringsAsFactors=F)
```

Now, we use neonUtilities function stackByTable operating on the NEON _ count-landbird.zip archive to perform a full join of all data files, grouped by table type.

```
stackByTable("data/NEON_count-landbird.zip")
```

On the console you would see the unzip process unfold and resulting in a message similar to that received when we used the same function in Lab 13.

```
Copied the most recent publication of validation file to /stackedFiles
Copied the most recent publication of categoricalCodes file to /stackedFiles
Copied the most recent publication of variable definition file to /stackedFiles
Finished: Stacked 2 data tables and 4 metadata tables!
Stacking took 1.050369 secs
All unzipped monthly data folders have been removed.
>
```

In your labs/data folder, you should now have a folder NEON _ count-landbird containing a folder stackedFiles, which contains several files including

```
"data/NEON_count-landbird/stackedFiles/brd_countdata.csv",
"data/NEON_count-landbird/stackedFiles/brd_perpoint.csv",
"data/NEON_count-landbird/stackedFiles/variables_10003.csv"
```

We will read the first two into data frames using the function `readTableNEON()`. As discussed in Chapter 13, this function uses the `variables.csv` file to assign data types to each column of the dataset. It is useful to look at these variable specifications first; but instead of opening the file in an editor, we will read it into R as a dataset using `read.csv`

```
parvar <- read.csv("data/NEON_par/stackedFiles/variables_00024.csv")
```

and then use `View` or `tibble` to inspect it. Once you browse through `parvar`, run the following read tables

```
brd_countdata <- readTableNEON(
  dataFile="data/NEON_count-landbird/stackedFiles/brd_countdata.csv",
  varFile="data/NEON_count-landbird/stackedFiles/variables_10003.csv"
)

brd_perpoint <- readTableNEON(
  dataFile="data/NEON_count-landbird/stackedFiles/brd_perpoint.csv",
  varFile="data/NEON_count-landbird/stackedFiles/variables_10003.csv"
)
```

Again, you can use `View` to inspect these data frames

```
View(brd_countdata)
View(brd_perpoint)
```

You would note that both datasets have fields in common, and that we need information from the `perpoint` set, such as coordinates, to analyze the information in `countdata`. Thus, we need to join these two tables, which is a function provided by `neonOS`

```
brdall <- joinTableNEON(brd_countdata, brd_perpoint)
```

and you can view

```
View(brdall)
```

```
bird.tb <- as_tibble(brdall)
colnames(bird.tb)
```

At this point, let us understand what the dataset consists of. There are nine plots coded by a plot ID, each with coordinates given by latitude and longitude. For each plot, there is a grid of nine points coded by point ID, and at each one of these points using six-minute timestamped observation intervals, a count is recorded with minutes into the interval, species, and distance to the observer.

Upon inspection, out of the many possibilities, we could pursue for analysis, would like to work with columns point ID, plot ID, latitude, longitude, point count minute, and species code. We will make a new data frame with these variables selected by the number of the column and change the names for simplicity.

```
X <- brdall[,c(11,10,5,2,35,37)]
names(X) <- c("lon","lat","pl","pt","ct","spp")
```

we can use View or make a tibble for inspection.

Look at the spatial structure of the plots using the unique coordinates since they are replicated many times

```
coord <- as.matrix(unique(X[,1:2]))
#determine range
xr <- c( min(X[,1]), max(X[,1]) )
yr <- c( min(X[,2]), max(X[,2]) )
```

just in case require(spatstat), and make a ppp object to plot

```
Xp <- ppp(X$lon,X$lat, xrange=xr,yrange=yr)
plot(Xp)
```

which we do not show here for the sake of space, but it could be brought into other tools for visualizing the location.

Our next step is to separate the observations by plot, using the coord or alternatively using the plot ID. Here, the list XX will contain the data by plot and u.o will have the number of observations or counts per plot

```
u.c <- list();u.o <- array();XX <- list()
lc <- dim(coord)[1]
```

the entries are calculated by looping through the plots and using which to select the records

```
for(i in 1:lc){
 u.c[[i]] <- which( X[,1] == coord[i,1] & X[,2] == coord[i,2] )
 u.o[i] <- length(u.c[[i]])
 XX[[i]] <- X[u.c[[i]],]
}
```

Just to make sure we can verify the total

```
> sum(u.o)
[1] 1462
>
```

which matches the number of records.

For each plot, we can breakdown the counts per point using the unique point ID, and store in a matrix ct.p

```
u.p <- unique(X$pt)
lp <- length(u.p)
ct.p <- matrix(nrow=lc,ncol=lp)
i=1; j=1
for (i in 1:lc){
 for(j in 1:lp){
  ct.p[i,j] <- length(which(XX[[i]]$pt==u.p[j]))
 }
}
```

The resulting matrix (plots are rows and points are columns) shows that there does not seem to be much difference in the counts per point across the plots.

```
> ct.p
      [,1] [,2] [,3] [,4] [,5] [,6] [,7] [,8] [,9]
[1,]   12   17   15   18   19   20   17   22   14
[2,]   17   17   16   18   12   22   13   19   16
[3,]   22   19   17   20   18   20   21   18   18
[4,]   21   19   21   22   22   20   17   22   17
[5,]   20   17   18   22   22   20   22   19   21
[6,]   15   17   17   17   17   17   12   21   20
[7,]   21   15   15   15   20   14   18   14   21
[8,]   19   16   21   14   15   23   13   14   16
[9,]   16   19   21   17   18   18   20   20   17
>
```

We are ready to look at species, using the unique codes and store in a matrix ct.s

```
u.s <- unique(X$spp)
ls <- length(u.s)
ct.s <- matrix(nrow=ls,ncol=lc)
for (j in 1:lc){
```

```
for(i in 1:ls){
 ct.s[i,j] <- length(which(XX[[j]]$spp==u.s[i]))
 }
}
```

The resulting matrix is 88 rows (species) by 9 columns (plots). Just to illustrate print a few rows

```
> ct.s[1:10,]
       [,1] [,2] [,3] [,4] [,5] [,6] [,7] [,8] [,9]
 [1,]     4   11    6   11    7    6   11    7   10
 [2,]    13   10   11   13   12   11   15   12   14
 [3,]     9    3    7    5    6    9    4    6    4
 [4,]    19   20   16   18   32   19   25   23   19
 [5,]     7    0    3    4    9   12    9    7    5
 [6,]    12   13   10   11   15   11    7    3    7
 [7,]     2    1    4    5    7    3    2    5    5
 [8,]     4   11    6    4    7   13   13   13    9
 [9,]     0    0    0    0    0    0    0    0    0
[10,]     9    0    6    5    9    4    1    5    1
>
```

We may be interested in the species richness and diversity by plot. For a first look, graph the data for each plot use `require (renpow)` as needed to be able to use function `panels` (Figure 14.16)

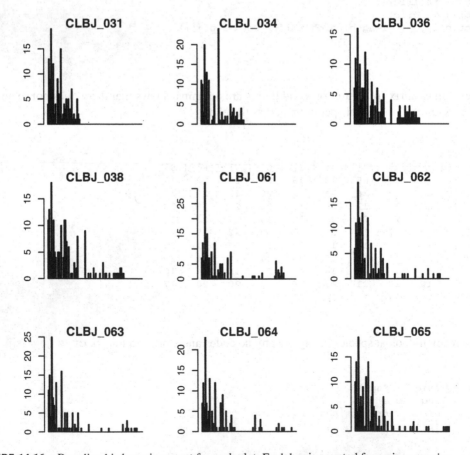

FIGURE 14.16 Breeding bird species count for each plot. Each bar is counted for a given species.

```
panels(6,6,3,3,pty='m')
for(i in 1:9) barplot(ct.s[,i])
```

You would notice that some plots have many more species than others and that, in general, there is a high count for a few species whereas others are infrequent. To quantify, build a function to calculate the number of species and diversity that we could apply to each plot. This function, based on Shannon's entropy, is also available from the RTEM repository in sppdiversity.zip if you wish to download it

```
spp.div <- function(x){
 nx <- length(x); sx <- sum(x)
 ns <-0; xx <- array()
 for (i in 1:nx){
  if (x[i]>0){
    ns<- ns+1
    xx[ns]<- x[i]
   }
  }
 # pmf
 p <- xx/sx
 # entropy
 H <- round(sum(-p*log2(p)),2)
 # ns is number of species or spp richness
 # H is species diversity measured by Shannon's entropy
 return(list(ns=ns,H=H))
}
```

Apply this function to each plot

```
ns <- array()
Hs <- array()

for(i in 1:lc){
 ns[i] <- spp.div(ct.s[,i])$ns
 Hs[i] <- spp.div(ct.s[,i])$H
 }
```

and combine results in a data frame

```
> data.frame(plotid,ns,Hs)
    plotid ns   Hs
1 CLBJ_031 29 4.36
2 CLBJ_034 30 4.12
3 CLBJ_036 45 5.02
4 CLBJ_038 43 4.80
5 CLBJ_061 33 4.38
6 CLBJ_062 33 4.38
7 CLBJ_063 34 4.23
8 CLBJ_064 32 4.34
9 CLBJ_065 38 4.59
```

We confirm that plot CLLBJ_036 has high species richness and diversity.

A dual concept is whether the species are ubiquitous across the plots, and for that we can use the same function

```
npl <- array()
Hpl <- array()
for(i in 1:ls){
 npl[i] <- spp.div(ct.s[i,])$ns
 Hpl[i] <- spp.div(ct.s[i,])$H
}
```

And build a data frame; we only show here the first 8 out of 88 rows

```
> spp <- unique(X$spp)
> data.frame(spp,npl,Hpl)
    spp npl  Hpl
1  BGGN   9 3.10
2  TUTI   9 3.16
3  INBU   9 3.08
4  NOCA   9 3.14
5  CARW   8 2.88
6  AMCR   9 3.07
7  CACH   9 2.99
8  WEVI   9 3.05
```

And these are the most ubiquitous species in the plots. For illustration, a couple of these are BGGN=Blue-gray Gnatcatcher (*Polioptila caerulea*) and TUTI=Tufted Titmouse (*Baeolophus bicolor*). You can find more from the countdata table.

EXERCISES

Exercise 14.1 Water-air interface.
Practice the water–air interface program as given in the guide. Produce plots and discuss.

Exercise 14.2 Landscape metrics.
Practice the landscape metrics analysis as given in the guide. Interpret results in terms of patches and classes.

Exercise 14.3 Quadrat and nearest neighbor analysis.
Practice spatial analysis using quadrat and nearest neighbor as given in the guide.

Exercise 14.4 Variograms and kriging.
Practice variogram analysis and perform kriging as given in the guide.

Exercise 14.5 Breeding birds.
Analyze the NEON breeding bird dataset as given in the guide.

REFERENCES

Acevedo, M.F. 2013. *Data Analysis and Statistics for Geography, Environmental Science & Engineering. Applications to Sustainability*. Boca Raton, FL: CRC Press, Taylor & Francis Group, 535 pp.

Acevedo, M.F. 2024. *Real-Time Environmental Monitoring: Sensors and Systems - Textbook, Second Edition*. Boca Raton, FL: CRC Press, Taylor & Francis Group, 392 pp.

Funwi-Gabga, N., and J. Mateu. 2012. Understanding the nesting spatial behaviour of gorillas in the Kagwene Sanctuary, Cameroon. *Stochastic Environmental Research and Risk Assessment* 26(6):793–811.

Hesselbarth, M.H.K., M. Sciaini, K. With, K.A. Wiegand, and J. Nowosad. 2019. Landscapemetrics: An open-source R tool to calculate landscape metrics (ver.0). *Ecography* 42:1648–1657.

Hesselbarth, M.H.K., M. Sciaini, K. With, K.A. Wiegand, and J. Nowosad. 2022. *Landscapemetrics*. Accessed April 2022. https://r-spatialecology.github.io/landscapemetrics/.

National Ecological Observatory Network (NEON). 2023. *Breeding Landbird Point Counts (DP1.10003.001)*. https://data.neonscience.org/data-products/explore.

NEON. 2023. *Birds*. Accessed January 2023. https://www.neonscience.org/data-collection/birds.

R Project. 2023. *The Comprehensive R Archive Network*. Accessed January 2023. http://cran.us.r-project.org/.

Upton, G.J.G., and B. Fingleton. 1985. *Spatial Data Analysis by Example, Volume 1: Point Pattern and Quantitative Data*. Chichester: John Wiley & Sons.

Index

915 MHz 171, 175–176, 178, 191, 204; *see also* Moteino

abline 33, 79–80, 83, 115–116, 233, 255, 353, 356–359, 428; *see also* regression
absorption 363, 377
action 393–394
active 14–15, 77, 369, 403, 418, 426; *see also* active sensor
ADC 49, 87–89, 125, 127, 182
Advanced IP Scanner 42, 59, 75, 78, 172
agglomerative 261, 264; *see also* cluster analysis
  hierarchical agglomerative clustering 264
aggregation 433
Ah 232; *see also* batteries; depth of discharge
air mass 211; *see also* atmospheric effects
airquality 30–31, 284–285, 287, 309
amplifier 113, 121, 124, 127–128, 133–134, 136, 142
  differential amplifier 121
  in-amp 113, 133–134
  instrumentation amplifier 113, 133–134, 136, 142
  op-amp 121, 123–124, 141
  operational amplifier 113, 124
analog channel 215; *see also* datalogger
  differential 121, 363, 377
  SE 39, 215, 318, 338, 395
anemometers 234, 237, 311; *see also* wind speed
antenna 173, 175–178, 182, 191, 204; *see also* dBi
Arduino 41–42, 48–51, 68–73, 75–78, 80, 87–88, 90–92, 95–98, 105, 111–114, 125–126, 129–131, 133–135, 140–146, 148–150, 152–154, 156–158, 160–166, 168–170, 172, 180, 183, 189–193, 196–198, 200, 204–205, 363, 404; *see also* Serial monitor; String literal; Shields
  Arduino-CLI 42, 68–72, 75–76, 78, 95, 97, 112, 114, 129–130, 134, 142, 144, 170, 172, 196–198, 200, 205
  UNO 41, 48–49, 70–72, 76–77, 88, 95, 97, 113, 130, 134, 143, 148, 157, 192, 198, 200
arrays 2, 12, 307
aspect 208, 338, 340–341, 347; *see also* GIS; QGIS; terrain
atmospheric effects 207–208; *see also* Air mass
attach 30, 35, 267, 285, 301, 306, 319, 368; *see also* detach
attenuated 246, 422
attributes 266–267, 299–300, 314, 321–323, 325, 327
autocorrelation 22, 39, 448
autoregressive 379; *see also* YW
  AR 59, 379, 393–396, 401, 448
azimuth 208–210, 212–213; *see also* sun path

B model 79, 82, 115–116, 128; *see also* B parameter; thermistor
B parameter 78–79, 82, 89, 111, 114; *see also* B model; Thermistor
balanced source divider 119–120; *see also* bridge circuit; half-bridge
bands 171, 239–249, 259–260, 269, 302, 304; *see also* green; near-infrared; red; remote sensing; SWIR
bandwidth 16, 21, 28

base rate 272, 275–276, 278; *see also* Bayes
batteries 207; *see also* Ah; depth of discharge
baud 51, 73, 166, 185, 195–196
Bayes 271–274, 276–279, 309; *see also* base rate; classification; confusion matrix; Naïve Bayes; probability; sensitivity
BCD 144
binomial 280–281, 309
biplot 257, 259–260, 417–418; *see also* principal components analysis
bipolar 113, 124–128, 133; *see also* unipolar
birds 427, 450, 456–457; *see also* NEON
boolean 149, 159
boost converter 113, 123–124, 142; *see also* DC-DC converter; step-down; step-up
boundary 430–431, 449; *see also* fragmentation; fragmentation metrics; landscapemetrics
boxplot 14–15, 18, 20, 23, 27; *see also* exploratory data analysis
bridge circuit 113, 120–121, 127
  half-bridge 122
  linearized bridge 113, 121–122, 125–126, 128, 132–133, 135, 141–142
  quarter-bridge 113, 120–121, 126
  Wheatstone bridge 119–120
broker 171, 178, 180, 182, 184–189, 204; *see also* MQ telemetry transport; MQTT
bulk 404
bulk density 404; *see also* gravimetric water content; volumetric water content
  BD 404

canopy 403, 415–416, 420, 422–426
capacity factor 228; *see also* electric power; energy
  CF 228, 230–231, 237
carbon dioxide 377
  CO_2 26, 349–357, 376
categories 403
chi-square 283, 309, 436, 438
  chi-square test 309
class 42, 92, 171, 240–241, 243, 265–267, 283–284, 297–308, 310, 336–337, 424, 429–433, 439; *see also* fragmentation metrics
class C 42, 171; *see also* dynamic IP; IP address; IPv4
classification 239, 261, 264–267, 270–271, 274, 283–284, 293, 295, 297, 303, 305–306, 310, 331
classification and regression trees 271, 295; *see also* complexity parameter
  CART 271, 301, 303–307, 310, 336 (*see also* complexity parameter)
  decision tree 295
  terminal nodes 304–305
client 57, 157, 159–162, 171, 180, 184, 186–190, 203–205; *see also* server
cloud 415
cluster 239, 261–265, 270
cluster analysis 239, 261, 270; *see also* dendrogram; agglomerative

Printed in the United States
by Baker & Taylor Publisher Services

Printed in the United States
by Baker & Taylor Publisher Services